T0328939

MEDICAL TECHNOLOGY AND ENVIRONMENTAL HEALTH

Medical Technology and Environmental Health

Edited by

Ade Gafar Abdullah, Isma Widiaty & Cep Ubad Abdullah

CRC Press
Taylor & Francis Group
Boca Raton London New York

CRC Press is an imprint of the
Taylor & Francis Group, an **informa** business

A BALKEMA BOOK

CRC Press/Balkema is an imprint of the Taylor & Francis Group, an informa business

© 2020 Taylor & Francis Group, London, UK

Typeset by Integra Software Services Pvt. Ltd., Pondicherry, India

Library of Congress Cataloging-in-Publication Data

Applied for

Published by: CRC Press/Balkema
       Schipholweg 107C, 2316XC Leiden, The Netherlands
       e-mail: Pub.NL@taylorandfrancis.com
       www.routledge.com – www.taylorandfrancis.com

ISBN: 978-0-367-86053-0 (Hbk)

ISBN: 978-0-367-54587-1 (pbk)

ISBN: 978-1-003-01670-0 (eBook)
DOI: 10.1201/9781003016700
https://doi.org/10.1201/9781003016700

Table of contents

Public health and occupational health

Hospital and nursing management

Preface

The 1st Medicine and Global Health Research Symposium (MoRes) 2019 is an international conference that took place in Bandung, Indonesia, on 23 October 2019. It was hosted by the Institute of Research and Community Service (LPPM Universitas Islam Bandung (UNISBA), Indonesia) in collaboration with the Asia Metropolitan University, Malaysia, Universiti Sains Islam Malaysia, Malaysia, and Cyberjaya University College of Medical Sciences, Malaysia. The theme of this year's conference was "Medicine and Global Health within Society 5.0", and the conference aimed (1) to bring together scientists, practitioners, students, researchers, and civil society organization representatives at the medical and global health symposium; and (2) to share and discuss theoretical and practical knowledge on innovation in medicine and global health. Participants from several countries such as Australia, Malaysia, Singapore, and many cities in Indonesia attented this year's conference.

Discussing its theme "Innovative collaboration on advanced medical technology and environmental health in the industrial revolution 4.0 era", the conference invited Prof. Dr. Ieva B. Akbar, a professor of physiology from Universitas Indonesia, Indonesia, Dr. Peter Davey from the School of Environment and Science, Griffith University, Australia, Prof. Dr. Amaluddin Ahmad, professor of healthcare and teaching industry from Asia Metropolitan University, Malaysia, Mej. Gen. Prof. Dato' Dr. Mohd. Zin Bidin, professor of Faculty of Medicine from Cyberjaya University College of Medical Sciences, Malaysia, Dr. Rafidah Bahari, assistant professor in psychiatry from Cyberjaya University College of Medical Sciences, Malaysia, and Dr. Roy Rillera Marzo, public health and medical education, Asia Metropolitan University, Malaysia, to share their knowledge as well as experiences with the audience.

There were approximately 200 submissions to the conference by authors from various countries. The committee went through a rigorous review procedure and selected 66 papers to be published in the proceedings of MoRes 2019, published by CRC Press/Balkema, Taylor & Francis Group.

Ade Gafar Abdullah, *Universitas Pendidikan Indonesia*
Isma Widiaty, *Universitas Pendidikan Indonesia*
Cep Ubad Abdullah, *Universitas Pendidikan Indonesia*

Scientific committee

1. Dr. Peter Davey *(Griffith University, Australia)*
2. Prof. Dr. Amaluddin bin Ahmad *(Asia Metropolitan University, Malaysia)*
3. Dr. Rafidah Bahari *(Cyberjaya University College of Medical Sciences, Malaysia)*
4. Dr. Roy Rillera Marzo *(Asia Metropolitan University, Malaysia)*
5. Jerico Fransiscus Pardosi, Ph.D *(School of Public Health and Social Work, QUT Brisbane, Australia)*
6. Prof. Dr. Ieva B. Akbar, dr., AIF. *(Universitas Islam Bandung, Indonesia)*
7. Dr. Titik Respati, drg., M.Sc.PH. *(Universitas Islam Bandung, Indonesia)*
8. Dr. Maya Tejasari, dr., M.Kes. *(Universitas Islam Bandung, Indonesia)*
9. Zulmansyah, dr., Sp.A., M.Kes. *(Universitas Islam Bandung, Indonesia)*

Medical Technology and Environmental Health – Abdullah, Widiaty & Abdullah (eds)
© 2020 Taylor & Francis Group, London, ISBN 978-0-367-86053-0

Organizing committee

1. Dr. Ike Junita Triwardhani, S.Sos., MSi (Chairman Committee)
2. Dadi Ahmadi, S.Sos., M.Ikom (Chairman Activities Committee)
3. Ahmad Arif Nurrahman, ST., MT (Vice Chairman Committee)
4. Rabiatul Adwiyah, SE., MSi (Secretary Committee)
5. Andalusia Neneng Permatasari, S.S., M.Hum (Vice Secretary Committee)
6. Sriyanti, ST., MT. (Treasurer Committee)
7. Dwi Agustin Nuriani,S, S.Si., M.Stat (Vice Treasurer Committee)
8. Siti Sunendiari, Dra., MS. (Secretariat Committee)
9. Dr. Alhamuddin M.M.Pd. (Sponsorship Committee)
10. Lilim Halimah, BHSc, M.HSPY (Call for Paper Committee)
11. Arba'iyah Satriani, S.Pi, M. A(Hons) (Call for Paper Committee)
12. Ninuk Pertamasari, S.Ked., M.H.Kes. (Call for Paper Committee)
13. Abdul Kudus, S.Si., M.Si., Ph.D. (Call for Paper Committee)
14. Iyan Bachtiar, ST., MT. (Call for Paper Committee)
15. Dr. Tresna Wiwitan, M.Si (Call for Paper Committee)
16. Riza Hernawati, S.Sos., M.Si. (Call for Paper Committee)

Acknowledgments

The organizing committee of MoRes 2019 would like to properly acknowledge the Institute of Research and Community Services (LPPM UNISBA) as well as all the co-hosting universities including Asia Metropolitan University, Malaysia, Universiti Sains Islam Malaysia, Malaysia, and Cyberjaya University College of Medical Sciences, Malaysia. The committee would also like to thank PT Rumah Publikasi Indonesia who has helped with the publication of the conference papers.

Basic and clinical medical sciences

Aedes-aegypti organophosphate resistance detection in the Rawasari subdistrict of Central Jakarta, Indonesia, as an effort for dengue hemorrhagic fever vector control

A. Hardjanti, I. Indrawati & E. Donanti
YARSI University, Jakarta, Indonesia

H. Wibowo & Z. Zulhasril
Indonesia University, Jakarta, Indonesia

ABSTRACT: Dengue hemorrhagic fever (DHF) is a public health problem in Indonesia. Jakarta has the most cases compared to other provinces, and Rawasari District has been declared an endemic area for DHF. Efforts to eradicate DHF have centered on vector (*Ae. aegypti*) control. Organophosphate has been the insecticide of choice for more than 25 years of vector control efforts. Insect resistance to organophosphate is marked by an increase of nonspecific esterase enzyme activity that can be detected using microplate assay. This study collected *Ae. aegypti* egg and larvae randomly from houses in the research area and conducted an alfa esterase activity assay using a microplate reader. Briefly, larvae homogenate was placed inside ice-cooled microplate wells. The prepared reagent was used to assess esterase activity from each homogenate and evaluated using a spectrophotometer at 450 nm. Organophosphate resistance patterns from Rawasari show a 22.4% rate of high resistance to organophosphate among the larvae, 41.1% moderate resistance, and 36.4% sensitivity. Resistance patterns shown in this study were similar to those found in research conducted in other areas in Jakarta. However, a large proportion of the vector has moderate resistance, which warrants a vector resistance surveillance program. Vector control programs conducted in Rawasari can still use organophosphate to eliminate the DHF vector.

1 INTRODUCTION

Dengue hemorrhagic fever (DHF) is a viral disease with an incidence rate that has increased 30 times in 50 years and has been spreading from urban to rural areas in the past decade (World Health Organization 2009). The first extraordinary DHF breakout was identified in Surabaya and Jakarta in 1986. As many as 58 children were infected, and 24 of them died. In Indonesia, DHF had become a public health problem in 32 (97%) districts and 382 (77%) cities by 2009 (Kementerian Kesehatan 2010). In December 2011, records showed a national decline in DHF incidence, and the nation's capital, Jakarta, still ranked first in incidence rate (2004–2009). This situation may have been caused by an increase in population density due to high mobility and developed transportation systems in the city.

DHF is caused by dengue virus infection using *Ae. aegypti* as an actual vector and *Ae. albopictus* as a potential vector. To date, no specific medication or vaccine exists to treat or prevent dengue infection. Vector control is the only possible measure to reduce DHF prevalence (Sukowati 2010). Insecticide is still the preferred agent for vector control aided by eradication of possible mosquito nests. Currently, DHF breaks out in a five-year cycle.

World Health Organization (WHO) guidelines suggests that organophosphate and pyrethroid are the insecticides of choice for adult mosquito control and as a larvacide. This type of insecticide is widely used in households and is easily obtained as a spray to prevent mosquito bites regardless of DHF incidence in the area. This practice helps increase organophosphate resistance.

Pimsamarn and colleagues (2009) report an organophosphate and parathyroid resistance in *Ae. aegypti* in northern Thailand. Shinta and colleagues also report malathion and pyrethroid resistance in Denpasar (Sukowati 2010). Organophosphate exposure will increase the level of nonspecific esterase enzymes (NSE) in mosquitos and reduce their susceptibility to insecticides. The level of these enzymes can be detected by microplate assay, and mosquito susceptibility can be assessed by measuring absorbance value from the ELISA test.

Reports also show DHF vector resistance to organophosphate in two subdistricts in Jakarta – namely Tanjung Priok and Mampang Prapatan in north and south Jakarta, respectively (Zulhasril & Suri 2007). This research completes the picture by detecting DHF vector organophosphate resistance in central Jakarta.

2 METHOD

2.1 *Egg and larva collection*

This research used a cross-sectional design. Mosquito eggs and larvae of the DHF vector were obtained from 13 houses from eight community areas randomly selected within the subdistrict. Each house coordinate was recorded in UTM format. The households of each house designated as a mosquito collection station were interviewed with reference to a prepared questionnaire. Mosquito eggs were collected using an egg trap comprising a container filled with clean water and lined with filter paper to facilitate the mosquitos laying their eggs. Filter paper containing eggs was removed and brought to the laboratory, where the egg were hatched in a 400 sq. cm enamel pan. Resulting larvae were given fish food, and the water inside the pan was changed every other day until the larvae reached the fourth stadium and were ready to undergo microplate assay. Some of the larvae were allowed to become adult mosquitos and put into a 50 sq. cm mosquito cage and fed using the blood of guinea pigs. These mosquitos will again lay their eggs, which will be collected for further assay.

2.2 *Microplate assay*

A prepared reagent comprised alpha-naphthyl acetate in acetone (6 g/L), phosphate buffer saline (0.02 M, pH 7.0) and fast blue B salt, aqudest, and a sodium dodecyl sulphate 35 mL (5%) as a coupling reagent. The NSE activity assay was conducted according to Lee's method (Lee 1990). The *Ae. aegypti* larvae were put inside the microplate assay, one larva per well. The associated liquid inside the well was dried using filter paper, and the larvae were finely ground using a mortar and added phosphate buffer saline in order to produce a homogenate. An aliquot of the resulting homogenate was transferred into the microwell plate at 4°C. Four replicas were used for each larva and added to the prepared reagent for 1 minute, followed by the coupling reagent for another minute in order to allow color change. Acetic acid (10%) was used to stop the reaction.

Results were assessed visually by observing color change and scored (0) no color, (1) light blue, (2) greenish, and (3) deep blue interpreted as very sensitive (0–2), moderate resistance (2.1–2.5), and high resistance (2.6–3.0). Assay results were also confirmed using a spectrophotometer at 450 nm with the resulting absorbance value interpreted as very sensitive (0–0.7), moderate resistance (0.7–0.9), and high resistance (> 0.9). Resulting data were analyzed with ANOVA using SPPS version 22. Resistance levels in the community area were presented in a geographical information system.

3 RESULTS AND DISCUSSION

The research found that 58 houses were positive for larvae. Table 1 shows the results within each community area (RT). Figure 1 shows the spatial distribution of houses with larvae.

This result shows that each community area (RT) had about 31–92% of houses with larvae. This is a rather high percentage attributed to the high-density settlement. Most of the houses

Table 1. Community area (RT) with larvae.

Community Area (RT)	Number of Houses	Houses with Larvae	Mean Larvae/RT
2	13	7	16
4	13	6	22
5	13	6	89
8	13	10	62
9	13	12	94
10	13	4	33
11	13	6	46
13	13	7	48

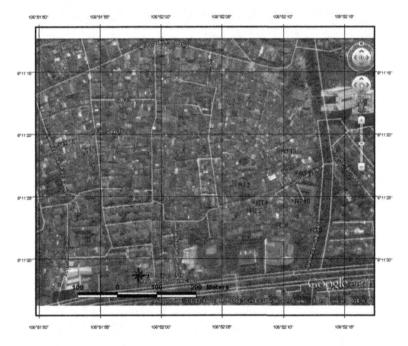

Figure 1. Houses with larvae.

have a shared wall, narrow alleys, and minimal ventilation and windows. These conditions cause unhealthy air circulation and minimal sunlight penetration into each house, which is a preferable condition for *Ae. aegypti* breeding in the house water reservoir. Figures 2 and 3 show the spatial distribution of meal larvae/RT and a comparison between a house with larvae and the mean larvae count/RT, respectively.

The high number of larvae found in this study may have been the cause of the high prevalence of DHF in Rawasari. Although a different number of houses have larvae, the ANOVA test shows no statistical difference in larvae number between RT (Table 2).

An epidemiology study in Jakarta between January 2012 and May 2013 revealed that Cempaka Putih District was one of the areas with a high DHF incidence rate of 86.54 per 100,000 residents in 2012 and a 0% case fatality rate. In 2013, the incidence rate dropped to 36.37 with a 3.23% case fatality rate. Data show Rawasari was the area with a consistent DHF case every month in 2014. A previous study conducted in the Pulogadung subdistrict of east Jakarta within the same time frame revealed a higher incidence rate (66.89%) compared to Cempaka

5

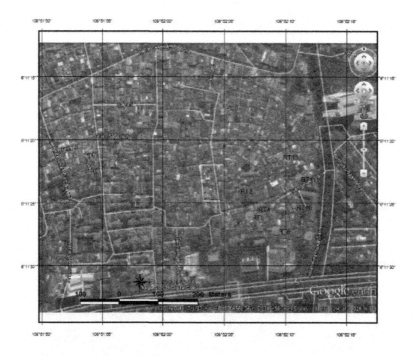

Figure 2. Mean larvae count/RT.

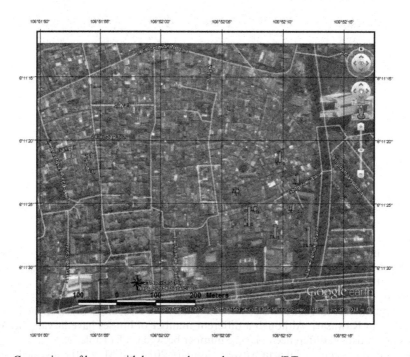

Figure 3. Comparison of houses with larvae and mean larvae count/RT.

Putih (36.37%). However, Cempaka Putih has a higher fatality rate (3.23%) compared to Pulogadung (1.14%).

An NSE assay shows that 22.5% of *Ae. aegypti* larvae have high resistance, 41.1% have moderate resistance, and 36.4% are sensitive to organophosphate (Table 3 and Figure 4).

Table 2.   ANOVA statistical test.

ANOVA

Number of Larva

	Sum of Squares	Df	Mean Square	F
Between Groups	45,005.044	7	6,429.292	1.714
Within Groups	183,811.938	49	3,751.264	
Total	228,816.982	56		

Table 3.   Larvae resistance.

| | Resistance | | |
RT	% Sensitive	% Moderate Resistance	% High Resistance
2	0	50	50
5	83	0	17
11	40	60	0
8	50	40	10
9	8	75	17
13	56	22	22
4	29	57	14
10	25	25	50

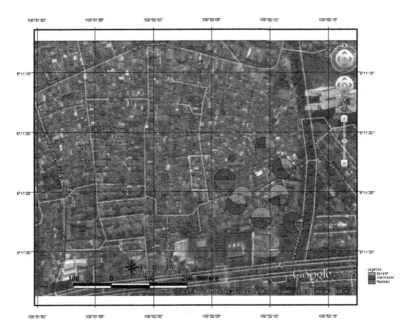

Figure 4.   Vector resistance mapping.

A vector control program has been conducted, including fogging, in order to reduce the DHF infection cycle and to eradicate adult *Ae. aegypti*. The Jakarta health office records that malathion is the typical insecticide used in this program in the Jakarta region. The national government has used malathion since 1969. The larvacide used in this program is Temefos, and it has been used since 1980. Malathion and Temefos are organophosphates that if used in the long term, could develop mosquito resistance. Insecticide resistance mechanisms depend on the genetic factor. The resistance genes coded in the assembly of particular enzymes such as acetyl-cholinesterase can reduce organophosphate to an inactive compound (Melo-Santos et al. 2010).

Organophosphate is an acetylcholinesterase inhibitor. Acetylcholinesterase is an enzyme that hydrolyzes acetylcholine to become acetic and choline. Organophosphate reacts with the active part of this enzyme and blocks its functions. The level of acetylcholine in nerve synapses will increase, causing a persistent postsynaptic stimulus. Acetylcholine is present in all nervous systems, including the autonomic nervous system, and acts as a neurotransmitter in sympathetic and parasympathetic nerve ganglion.

These results show a low percentage of highly resistant larvae. However, a large, moderate resistance percentage requires continuous insecticide resistance surveillance and insecticide usage control in order to prevent the development of high-resistance larvae. Recent reports from Brazil (Melo-Santos et al. 2010) and Colombia (Ocampo et al. 2011) show the surveillance program required in this situation. Aside from north and south Jakarta (Zulhasril & Suri 2010), reports from Saudi Arabia (Dieng et al. 2011), Trinidad and Tobago (Polson et al. 2010), and Malaysia (Shafie et al. 2012) also demonstrate *Ae. aegypti* resistance to organophosphate. Shetty and colleagues (2013) in Karnataka, India, and Karunaratne and colleagues (2013) in Sri Lanka show a gradient of resistance from high to moderate, which is similar to the data from this study.

Resistance data in this study compared with the previous study in Pulogadung show a similarly low percentage of high resistance. This finding suggests that organophosphate can still be used for a vector control program in Rawasari and Pulogadung. To conclude, the vector control program using insecticide can still be implemented along with an insecticide resistance surveillance in order to monitor vector resistance development.

REFERENCES

Dieng, H., Hassan, A. A., Satho, T., Miake, F., Salmah, M. R. C., & AbuBakar, S. 2011. Insecticide susceptibility of the dengue vector *Aedes aegypti* (Diptera: culicidae) in Makkah City, Saudi Arabia. *Asian Pacific journal of tropical disease* 1(2): 94–99.

Karunaratne, S. H. P. P., Weeraratne, T. C., Perera, M. D. B., & Surendran, S. N. 2013. Insecticide resistance and efficacy of space spraying and larviciding in the control of dengue vectors *Aedes aegypti* and *Aedes albopictus* in Sri Lanka. *Pesticide biochemistry and physiology* 107(1): 98–105.

Kementerian Kesehatan. 2010. Demam Berdarah Dengue di Indonesia Tahun 1968–2009. *Buletin Jendela Epidemiologi* 2.

Lee, H. L. 1990. A rapid and simple method for the detection of insecticide resistance due to elevated esterase activity in *Culex quinquefasciatus*. *Tropical biomedicine* 7(1): 21–28.

Melo-Santos, M. A. V., Varjal-Melo, J. J. M., Araújo, A. P., Gomes, T. C. S., Paiva, M. H. S., Regis, L. N., ... & Ayres, C. F. J. (2010). Resistance to the organophosphate temephos: Mechanisms, evolution and reversion in an *Aedes aegypti* laboratory strain from Brazil. *Acta tropica* 113(2): 180–189.

Ocampo, C. B., Salazar-Terreros, M. J., Mina, N. J., McAllister, J., & Brogdon, W. 2011. Insecticide resistance status of *Aedes aegypti* in 10 localities in Colombia. *Acta tropica* 118(1): 37–44.

Pimsamarn, S., Sornpeng, W., Akksilp, S., Paeporn, P., & Limpawitthayakul, M. 2009. Detection of insecticide resistance in *Aedes aegypti* to organophosphate and synthetic pyrethroid compounds in the north-east of Thailand.

Polson, K. A., Rawlins, S. C., Brogdon, W. G., & Chadee, D. D. 2010. Organophosphate resistance in Trinidad and Tobago strains of *Aedes aegypti*. *Journal of the American Mosquito Control Association* 26(4): 403–410.

Shafie, F. A., Tahir, M. P. M., & Sabri, N. M. 2012. *Aedes* mosquitoes resistance in urban community setting. *Procedia: Social and behavioral sciences* 36: 70–76.

Shetty, V., Sanil, D., & Shetty, N. J. 2013. Insecticide susceptibility status in three medically important species of mosquitoes, *Anopheles stephensi*, *Aedes aegypti* and *Culex quinquefasciatus*, from Bruhat Bengaluru Mahanagara Palike, Karnataka, India. *Pest management science* 69(2): 257–267.

Sukowati, S. 2010. Masalah vektor demam berdarah dengue (DBD) dan pengendaliannya di Indonesia. *Buletin Jendela Epidemiologi* 2(1): 26–30.

World Health Organization 2009. Dengue: Guidelines for di-agnosis, treatment, prevention and control: 57–87. France: World Health Organization.

Zulhasril & Suri, S. D., 2010. *Resistensi* Aedes aegypti *terhadap organofosfat di Tanjung Priok Jakarta Utara dan Mampang Prapatan Jakarta Selatan*. Jakarta: Majalah Kedokteran FK UKI. 27(3): 96–100.

Medical Technology and Environmental Health – Abdullah, Widiaty & Abdullah (eds)
© 2020 Taylor & Francis Group, London, ISBN 978-0-367-86053-0

Stimulatory effect of methanolic extract and N-hexane insoluble and soluble fraction of parijoto fruit (*Medinilla speciosa* Blume) on the spermatozoa quantity of male Sprague Dawley rats

R. Wijayanti
Universitas Islam Sultan Agung, Semarang, Indonesia
Unversitas Gadjah Mada, Yogyakarta, Indonesia

S. Wahyuono, I.P. Sari & D.M. Rizal
Unversitas Gadjah Mada, Yogyakarta, Indonesia

ABSTRACT: Infertility is a major health problem with an incidence of 15–20% in around 50 million couples, in which men account for about 50% of all cases. Parijoto fruit (*Medinilla speciosa* Blume) is empirically used to increase fertility. The purpose of this study was to determine the stimulatory effect of methanol extract and n-hexane insoluble and soluble fraction of parijoto on the quantity of spermatozoa. This study used 28 male Sprague Dawley (SD) rats divided into six groups, where group I was normal; groups II, III, and IV were methanol extract groups receiving doses of 100 mg/kgBW, 250 mg/kgBW, and 500 mg/kgBW; group V received n-hexane insoluble fraction 500 mg/kgBW; and group VI received n-hexane soluble fraction 500 mg/kgBW, each dosage given orally for 14 days. The parameter observed was the spermatozoa quantity. The obtained data were analyzed using Kruskal-Wallis and Mann-Whitney tests. The results showed the mean quantity of spermatozoa in groups I, II, III, IV, V, and VI was respectively 26.8 ± 1.6, 27.1 ± 1.3, 30.4 ± 1.3, 43.5 ± 1.3, 67.1 ± 2.8, and 50.1 ± 1.9 (million/ml ejaculate). Significant differences emerged between group I and groups IV, V, and VI. Methanol extract and n-hexane insoluble and soluble fraction of parijoto fruit provided a stimulating effect on the quantity of spermatozoa in SD male rats.

1   INTRODUCTION

In domestic life, children (offspring) are like a rope that can further strengthen the relationship of a husband and wife. But many married couples fail at getting offspring. The incidence of infertility is still a major health problem in the world, including in Indonesia. It is known that a total of 48.5 million couples of productive age cannot have children, including 19.2 million couples who are unable to have their first child, and 26.3 million couples who cannot have their second and subsequent children (Mascarenhas et al. 2012). About 15% of couples cannot reach pregnancy within 1 year and seek fertility treatment. In as many as 50% of couples who do not have children, male infertility factors are found together with abnormalities in semen (Jungwirth et al. 2014).

Male infertility arises from a number of factors including genetics, chronic disease, and lifestyle (Tunc 2011). Factors that influence male infertility include unhealthy lifestyles, cigarette smoking, disease (Murod et al. 2014), and oxidative stress (Agarwal & Prabakaran 2005). Exposure to free radicals over a period of time can also cause a decrease in fertility due to decreased testicular weight, increased lipid peroxidation, decreased antioxidant (vitamin C), and oxidative damage (Saryono & Santoso 2015). Free radicals are also one of the main causes of disruption of spermatogenesis in the testes (Anzila et al. 2017). In 30–40% of cases of male infertility, no cause is found (idiopathic male infertility). In such cases, men do not have abnormalities on physical examination or on examination of endocrine, genetic, and biochemical chemistry. Idiopathic male infertility occurs as a result of endocrine disruption due to environmental pollution, reactive oxygen species, or genetic and epigenetic disorders (Jungwirth et al. 2014). Many synthetic drugs have been used to overcome problems related to infertility. However, the use of synthetic drugs has side effects such as heart rhythm disorders, suicidal tendencies, mental disorders, tremors, and dilatation of blood vessels in other parts of the body resulting in headaches and fainting (Balamurugan et al. 2013). To minimize these side effects, traditional medicine is used.

Several types of plants in the Melastomataceae family have antioxidant activity (Wachidah, 2013). It has been reported that the Melastoma malabathricum plant (family Melastomataceae) at doses of 250 mg/kg body weight and 500 mg/kg body weight can affect the reproductive system of male rats (Balamurugan et al. 2013). Parijoto is a typical plant of the same family from Colo Village, Dawe District, Kudus, Central Java, Indonesia, which is empirically used by the community to increase fertility. Previous research proves that the ethyl acetate fraction of parijoto (FEABP) (*Medinilla speciosa* Blume) (Melastomataceae) contains polyphenol compounds and flavonoids as natural antioxidants (Wachidah 2013). Related research has established that parijoto fruit methanol extract at doses of 250 mg/kgBW and 500 mg/kgBW can reduce blood sugar levels and improve sexual function in a male Wistar strain of chronic DM (*Diabetes Mellitus*)models (Wijayanti & Lestari 2018). The results of this study contribute to the understanding of health science, especially in the fields of reproduction and pharmacology, regarding the effect of giving parijoto (*Medinilla speciosa* Blume) on increasing the number of spermatozoa. This research is a first step in knowing the active compounds found in parijoto fruit by utilizing local wisdom.

2 METHOD

2.1 *Research materials*

The research materials used in this study included parijoto (*Medinilla speciosa* Blume), methanol, n-hexane, NaCl, and aqua bidest. Sprague Dawley (SD) strain male rats aged 2 months and weighing 200 grams were randomly selected and placed in six groups of six rats each.

2.2 *Research equipment*

The research equipment employed in this study consisted of lenders, measuring cups, glass bottles, water baths, rotary evaporators, aluminum foil, microscopes (Olympus IX71), ovens, desiccators, an analytical balance, crushers, pinchers, porcelain plates, petri dishes, incubators, autoclaves, minor surgical instruments, syringes, mouse *sonde*, a glass beaker, measuring tubes, minor sets, small tubes, micro pipettes, object and deck glasses, and a hemocytometer nebauer.

2.3 *The research path*

2.3.1 *Plant determination*
Determination of plants was carried out at FMIPA UNNES Semarang, Indonesia, to ensure that the samples used were truly parijoto (*Medinilla speciosa* Blume).

2.3.2 *Making parijoto fruit extract (*Medinilla speciosa *Blume)*
Parijoto fresh fruit (*Medinilla speciosa* Blume) was finely macerated with methanol (1:10) for 24 hours. Maserat was evaporated using a rotary evaporator at a temperature of 45°C. Maceration was done until the maserat obtained was clear. Crude extract was obtained (Wachidah 2013). The crude extract was tested for chemical compound analysis and activity as a fertility agent, with doses of 100 mg/kgBW, 250 mg/kgBW, and 500 mg/kgBW.

2.3.3 *Fractionation of parijoto fruit (*Medinilla speciosa *Blume)*

The crude extract was re-dissolved with 100 mL of 50% methanol, put into a separating funnel with 100 mL of n-hexane, shaken a few moments, then allowed to stand apart with the n-hexane phase at the top and the methanol phase at the bottom. Partitioning was done many times until the n-hexane phase was colorless. The n-hexane phase was collected and concentrated until a soluble fraction of n-hexane was obtained. The methanol phase was collected and concentrated until an n-hexane insoluble fraction was obtained. The soluble and insoluble fractions of n-hexane were tested for activity as a fertility agent with a dose of 500 mg/kgBW.

2.3.4 *Test for activity as a fertility agent*

The test for activity as a fertility agent was carried out on male elementary strain rats with the following group divisions: the K1 group (normal) was not given any treatment; Groups K2, K3, and K4 were given parijoto fruit extract (*Medinilla speciosa* Blume) respectively at doses of 100 mg/kgBW/day, 250 mg/kgBW/day, and 500 mg/kgBW/day for 14 days; Group K5 was given an insoluble fraction of n-hexane parijoto at a dose of 500 mg/kgBW for 14 days. Group K6 was given a soluble fraction of n-hexane parijoto at a dose of 500 mg/kgBW for 14 days.

2.3.5 *Calculation of spermatozoa*

Test animals were sacrificed by cervical dislocation, then dissected using a dissecting kit for the removal of the testicular organs and the right cauda epididymis. Next, the cauda epididymis was put into a petri dish containing 1 mL of 0.9% NaCl and then cut into pieces. The semen that came out of the cauda epididymis was stirred in 0.9% NaCl solution with a stirring glass until homogeneous, so that a suspension of spermatozoa was formed as a stock.

2.3.6 *Data analysis and statistics*

Data on the number of spermatozoa in this study were tested for normality and homogeneity using the Lavene test. The data showed that it was not normally distributed and/or not homogeneous, so the analysis was continued through a nonparametric test, the Kruskal-Wallis test, and continued with the Mann-Whitney test at the 95% confidence level.

3 RESULTS

The results of the study concerning the number of spermatozoa are presented in Table 1.

Table 1. Results of calculation of spermatozoa.

No	Number of Spermatozoa (Million/ml ejaculate)					
	K1	K2	K3	K4	K5	K6
1	25.6	25.6	30.4	44	76	52.8
2	27.2	29.6	32.8	48	56.8	50.4
3	28.2	32	31.2	41.6	65.6	52
4	24	24	32	46.4	63.2	52
5	33.6	24	25.6	39.2	68.8	52.8
6	22.4	27.2	30.4	41.6	72	40.8
Average	26.8	27.1	30.4	43.5[a]	67.1[a, b]	50.1[a, b, c]
SD	1.6	1.3	1.3	1.3	2.8	1.9

Information:
a = significantly different from the normal group
b = significantly different from the extract group that received a dose of 500 mg/kg
c = significantly different from the group that received an n-hexane non-soluble fraction dose of 500 mg/kgBW

4 DISCUSSION

This research was conducted in July–September 2019 in the laboratory of the Faculty of Medicine, UNISSULA, Indonesia. The purpose of this study was to determine the stimulatory effects of extract and n-hexane insoluble and soluble fraction of parijoto fruit on the number of spermatozoa in male Sprague Dawley (SD) strain rats. The initial stage of this study, determining the plants, was conducted at the biology laboratory of the Faculty of Mathematics and Natural Sciences, Semarang State University. The results of plant determination showed that the parijoto fruit used was from the Melastomataceae family of the species *Medinilla speciosa* Blume. This study used the fruit from purplish red parijoto 3–4 months old, obtained from Colo Village, Dawe District, Kudus, Central Java, precisely at Mount Muria Kudus, Indonesia.

Statistical analyses showed that the data were not normally distributed ($P < 0.05$) and not homogeneous ($P < 0.05$), then analysis was continued with the Kruskal Wallis nonparametric test ($P < 0.05$), followed by Mann Whitney. Statistical results indicated a significant difference between the 500 mg/kgBW dose extract group, the 500 mg/kgBW n-hexane insoluble fraction group, and the 500 mg/kgBW n-hexane soluble fraction group with the normal group ($P < 0.05$). This shows an increase in the number of spermatozoa in elementary male strain rats. The increase in spermatozoa concentration occurred because SD rats were treated with methanol extract and n-hexane insoluble and soluble fraction of parijoto fruit each with a dose of 500 mg/kgBW for 14 days.

Statistical analyses also showed significant differences between the 500 mg/KgBW n-hexane non-soluble fraction group, and the 500 mg/kgBW n-hexane soluble fraction group with the 500 mg/kgBW dose extract group. This shows the ability to increase the number of spermatozoa in that the fraction group was higher than the crude methanol extract group. Significant differences in the statistical results also appeared between the n-hexane non-soluble fraction group with the dose of 500 mg/kgBW against the n-hexane soluble fraction group with a dose of 500 mg/kgBW. These results indicate that the n-hexane insoluble fraction has a better ability to increase the number of spermatozoa compared to the n-hexane soluble fraction.

The results prove the ability of methanol extract and n-hexane insoluble and soluble fractions of parijoto fruit to increase the number of spermatozoa due to the content of secondary metabolites. Based on research conducted by Wachidah (2013), the results of phytochemical screening tests of positive crude extracts contain saponins, glycosides, flavonoids, and tannins. The n-hexane (methanol fraction) positive insoluble fraction contains saponins, glycosides, flavonoids, and tannins. The n-hexane soluble fraction showed negative results on all phytochemical screening tests. Wachidah (2013) also concluded that the antioxidant activity is best owned by the n-hexane insoluble fraction, followed by crude extracts and the n-hexane soluble fraction; the total phenolic contents of the crude extract and the n-hexane insoluble and soluble fraction were 408, 388, and 86 (mg GAE/g extract), respectively.

The ability of parijoto fruit to increase the number of spermatozoa is due to its flavonoid content. Flavonoids have free radical capture activity because they have hydroxyl groups on the B ring (Abdullah & Ifayanti 2018). The existence of these groups also makes flavonoids polar so that they can provide hydrogen with a lipid head on the cell membrane. Flavonoids can bind to the main components of cell membranes so that cells are protected from free radical attacks that have the potential to damage cells (Abdullah & Ifayanti 2018). Flavonoids can also increase the work of antioxidant enzymes in the body such as GSH, which can convert H_2O_2 molecules and lipid peroxidants into H_2O. GSH enzymes in the cytoplasm will work on phospholipid membranes that are oxidized to free radicals (Anggi & Herlina 2016).

5 CONCLUSION

The results of this study provide scientific evidence regarding the activity of extracts and n-hexane insoluble and soluble fractions of parijoto fruit on increasing the number of spermatozoa in male Sprague Dawley strain rats. This research is a first step in knowing the active compounds found in parijoto fruit by utilizing local wisdom. The conclusion of this study is that methanol extract and n-hexane insoluble and soluble fractions from parijoto (*Medinilla speciosa* Blume), each at a dose of 500 mg/KgBW, can provide a stimulating effect in increasing the number of spermatozoa in male Sprague Dawley rats.

REFERENCES

Abdullah, F. & Ifayanti. 2018. Pengaruh Pemberian Ekstrak Rosella (*Hibiscus Sabdariffa* Linn) Terhadap Jumlah dan Kecepatan Spermatozoa, Berat Testis Tikus Jantan Strain Wistar Yang Terpapar Karbon Tetraklorida (CCl4). *Jurnal Ilmu Kesehatan* 2(1): 124–129.

Agarwal, A. & Prabakaran, S. A. 2005. Mechanism, measurement, and prevention of oxidative stress in male reproductive physiology: *Indoan Journal of Experimental Biology* 43(11): 963–974.

Anggi, A. R. & Herlina, E. C. 2016. Pengaruh Pemberian Dark Chocolate Terhadap Jumlah Spermatozoa Mencit Balb/C Jantan yang Dipapar Asap Rokok. *Journal of medical and environmental sciences* 5 (4): 475–484.

Anzila, I., Soewondo, A., & Rahayu, S. 2017. Pengaruh Ekstrak Ethanol Kemangi (*Ocimum canum* Sims.) terhadap Struktur Histologi Testis Mencit (*Mus musculus*) Jantan. *Biotropika: Journal of tropical biology* 5(1): 22–26.

Balamurugan, K., Sakthidevi, G., & Mohan, V. R. 2013. Stimulatory effect of the ethanol extract of *Melastoma malabathricum* L. (Melastomataceae) leaf on the reproductive system of male albino rats. *Journal of applied pharmaceutical science* 3(2): 160–165.

Jungwirth, A., Diemer, T., Dohle, G. R., Giwercman, A., Kopa, Z., & Torunaye, H. 2014. Guidelines on male infer-tility. Male Infertility. European Association of Urology.

Mascarenhas, M. N., Flaxman, S. R., Boerma, T., Vanderpoel, S., & Stevens, G. A. 2012. National, regional, and global trends in infertility prevalence since 1990: A systematic analysis of 277 health surveys. *PLoS medicine* 9(12): 1–12.

Murod, A. M. 2014. *Uji Aktivitas Ekstrak Air Herba Kemangi (Ocimum americanum L.) terhadap Kualitas Sperma dan Densitas Sel Spermatogenesis Tikus Sprague-Dawley Jantan secara in vivo (Skripsi)* Jakarta: UIN Syarif Hi-dayatullah Jakarta.

Saryono, R. H. & Santoso, D. 2015. Seduhan biji kurma (*Phoenix dactlifera*) memperkuat membran plasma. *Jurnal Ners* 10(2): 355–359.

Tunc, O. 2011. Investigation of the role of oxidative stress in male infertility (Doctoral dissertation). Adelaide: University of Adelaide, Australia.

Wachidah, L. N. 2013. *Uji Aktivitas Antioksidan Serta Penen-tuan Kandungan Fenolat dan Flavonoid Total dari Buah Pa-rijoto (Medinilla speciosa Blume) (Skripsi)* Jakarta: UIN Syarif Hidayatullah Jakarta.

Wijayanti, R. & Lestari, A. P. (2018). Pengaruh Ekstrak Etanolik Buah Parijoto (*Medinilla Speciosa* Blume) Terhadap Kadar Gula Darah dan Fungsi Seksual Tikus Jantan Galur Wistar Model Diabetes Mellitus Kronik. *Jiffk: Jurnal Ilmu Farmasi dan Farmasi Klinik* 15(2): 1–7.

The impact of purple sweet potato water extract on excess weight gain in pregnant mice

U.A. Lantika, R. Damailia, T. Bhatara, R.R. Ekowati & A.B. Yulianti
Universitas Islam Bandung, Bandung, Indonesia

ABSTRACT: This research aimed to observe the effect of purple sweet potato extract supplementation during pregnancy on body weight of pregnant mice (*Mus musculus*). This was an experimental study with a completely randomized design. Thirty pregnant mice were divided into three groups: a control group, a group receiving an intervention of purple sweet potato water extract at a dose of 60 mg/g BW, and a group receiving an intervention of purple sweet potato water extract at a dose of 120 mg/g BB. The extract was given from the sixth day until the eighteenth. During the study, the mice were weighed and their weight gain was calculated. Statistical analysis was performed on data obtained through the ANOVA test in order to measure the significance of the treatment. The average body weight of mice was found to be the greatest in the group receiving the purple sweet potato water extract at a dose of 120 mg/g BW and the smallest in the control group. A statistical analysis revealed no significant difference in the mean body weight between groups ($p > 0.05$). Supplementation of purple sweet potato water extract does not cause weight gain in pregnant mice.

1 INTRODUCTION

The usage of natural ingredients as a standardized herbal medicine has been accepted around the world, including Indonesia. Indonesia as a tropical country with a variety of flora has used plants as medicine for generations (Jumiarni & Komalasari 2017). One of the plants proven to give many benefits to human health is purple sweet potato (*Ipomea batatas*). Its purple color represents the anthocyanin pigment. Anthocyanins are phenolic compounds that have antioxidant activity. Purple sweet potato has a higher anthocyanin content than other types of tubers. Total anthocyanin content of purple sweet potato is 519 mg/100 g wet weight (El Sheikha & Ray 2017).

Purple sweet potato is a source of calories because it contains saccharides such as sucrose, glucose, and fructose, and maltose, fat, and protein. It also contains vitamins and minerals, including vitamin A, vitamin C, thiamin (vitamin B1), riboflavin, iron (Fe), phosphorus (P), and calcium (Ca) (Mohanraj & Sivasankar 2014; Simonne et al. 1993). Purple sweet potato is rich in nutrients, making this plant one of the candidates for supplementation during pregnancy.

Pregnancy is a condition that occurs in women after fertilization. One good indicator of pregnancy is maternal weight gain (Kaiser & Campbell 2014). Decrease in weight gain during pregnancy will cause low birth weight in infants. Conversely, excessive weight gain during pregnancy can lead to gestational diabetes, which will have a negative impact on both the mother and the baby. Mothers can suffer metabolic and cardiovascular disease, whereas infants can experience macrosomia, which can have increase the risk of obesity and long-term metabolic abnormalities (Moll et al. 2017; Nahar et al. 2009; Rong et al. 2015). For this reason, the selection of supplements and proper nutrition is a factor in the prevention of these conditions. The purpose of this study was to examine the effect of purple sweet potato extract (*Ipomea batatas*) on the weight of pregnant mice (*Mus musculus*).

2 METHODS

2.1 *Animal handling*

This research was an experimental study using mice DDY strains as experimental animals from the laboratory of Biofarma, Cisarua, Bandung, West Java, Indonesia. This research was conducted May–September 2019 at Biomedic Laboratory, Faculty of Medicine, Islamic University of Bandung, Bandung, West Java. The animals were kept under controlled conditions as to their cages, light (12-h light and 12-h dark), and temperature. During the research, animals were fed a standard diet. The research procedures were approved by the Islamic University of Bandung Faculty of Medicine Ethics Committee. The ethical approval number of this research was 371/Komite Etik.FK/III/2018.

After the mice had adapted to the conditions, mating between male and female mice was carried out. Female mice in the estrous period were put together with male mice in the afternoon. Every morning, the female mice were examined, and female mice with a vaginal plug were moved into the maintenance cage and the time was noted as gestational day 0 (GD 0).

2.2 *The making of purple sweet potato water extract*

The Murizaki variant of the purple sweet potato was used in this experiment. The purple sweet potatoes were washed, then chopped into small pieces. Pieces of the sweet potatoes were dried using an oven at a temperature of 30–40°C. After being dried, the pieces were blended until smooth and then macerated with water for 72 hours. The solution was collected after 72 hours and evaporated using a vacuum rotary evaporator.

2.3 *Experimental animal intervention*

Thirty pregnant mice were divided into three groups: control (n = 10), dose 1 (n = 10), and dose 2 (n = 10). Mice in the dose 1 group were given 60 mg/g BW sweet potato extract and mice in the dose 2 group were given 120 mg/g BW sweet potato extract. Administration of extract began on the sixth day of pregnancy and ended on the eighteenth day of pregnancy.

2.4 *Body weight measurement*

Weight gain was scaled before mating and before and after the intervention using a digital weight scale. Weight gain was calculated with the formula:

$$\Delta BB = BB2 - BB1 \tag{1}$$

ΔBB: Weight gain
BB1: Body weight of pregnant mice before intervention
BB2: Body weight of pregnant mice after 12 days' intervention

2.5 *Statistical analysis*

Weight gain data presented mean ± standard error of mean (SEM). The statistical significance of differences between the mean of the sample of the groups was assessed using one-way ANOVA. Differences were considered significant if $p < 0.005$. This analysis was compared to the control group.

3 RESULTS

Female mice that had reached the estrous phase were weighed and mated with male mice. After the vaginal plug (GD0) was visible, female mice were separated. On the sixth day of pregnancy, the purple sweet potato water extract was given to the pregnant mice after they had been weighed. After 12 days of intervention (about the eighteenth day of pregnancy), the weight of the pregnant mice was once again measured. Weight measurement results can be seen in Figure 1. The mean weight gains in all groups show an increase during pregnancy and intervention.

Body weight of pregnant mice in all groups increased during the intervention. Data included mean ± standard error of mean (SEM), BB0 (body weight of female mice before mating), BB1 (body weight of pregnant mice before intervention), BB2 (body weight of pregnant mice after 12 days' intervention), control (control group, only received water), dose 1 (group given 60 mg/g BB of water extract of purple sweet potato), and dose 2 (group given 120 mg/g BB of water extract of purple sweet potato).

Weight gain during the intervention was obtained from the subtraction of the body weight before the intervention from the final body weight. The analysis found no difference in weight gain between groups ($p > 0.05$) (see Figure 2). This shows that giving purple sweet potato water extract during pregnancy does not cause weight gain.

No differences of weight gain were observed in pregnant mice in all groups after intervention with water extract of purple sweet potato for 12 days ($p > 0.005$) (one-way ANOVA test). Data were mean ± standard error of mean (SEM), control (control group, only received water), dose 1 (group given 60 mg/g BB of water extract of purple sweet potato), and dose 2 (group given 120 mg/g BB of water extract of purple sweet potato).

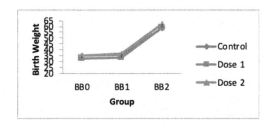

Figure 1.   Pregnant mice body weight during intervention.

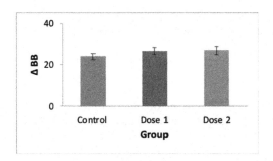

Figure 2.   Comparison of weight gain in pregnant mice after intervention.

4   DISCUSSION

Good nutrition is very necessary during pregnancy. Additional supplementation is often needed to meet nutritional requirements. Adequate nutrition will result in a normal weight gain during pregnancy. Weight gain during pregnancy is due to fetal growth and development, amniotic fluid, and fat collection in adipose tissue of pregnant women (Plante et al. 2020; Yang et al.

2017). This study revealed weight gain during pregnancy in all groups (see Figure 1). We can conclude that nutritional needs in pregnant mice in all groups were fulfilled.

The research showed that pregnant mice did not present differences in body weight and weight gain was associated with purple sweet potato water extract supplementation (see Figure 2). However, other studies observed reduction in body weight or delta weight when comparing consumption of green tea extract with a control diet and water with a control diet (Cunha et al. 2013; Hachul et al. 2018a; Hachul et al. 2018b). A study conducted by Iyare and Iyare (2007) also showed that supplementation with aqueous extract of *Hibiscus sabdariffa* during pregnancy caused a decrease in body weight.

Purple sweet potato has sufficient carbohydrate and fat content. However, it did not cause excess weight gain in this study. The presence of anthocyanins in purple sweet potato is suspected to be one of the causes of the condition. Anthocyanins have been shown to play a role in weight loss by inhibiting adipogenesis (Azzini et al. 2017). To prove the effect of supplementation with purple sweet potato water extract on the potential for weight gain inhibition during pregnancy, further research needs to be conducted by adding more parameters such as fat weight and cytokine levels measurement that influence those conditions.

5 CONCLUSION

Consumption of purple sweet potato (*Ipomea batatas*) water extract during pregnancy does not cause excess weight gain. This result could be a fundamental reason for making purple sweet potato water extract a candidate supplement during pregnancy.

CONFLICT OF INTEREST

There were no conflicts of interest during the implementation of the research.

ACKNOWLEDGMENT

This research was funded by the Medical Faculty of the Islamic University Bandung, Bandung, West Java. We appreciate and thank the various parties who helped carry out this research.

REFERENCES

Azzini, E., Giacometti, J., & Russo, G. L. 2017. Antiobesity effects of anthocyanins in preclinical and clinical studies. *Oxidative medicine and cellular longevity* 2017: 1–11.

Cunha, C. A., Lira, F. S., Rosa Neto, J. C., Pimentel, G. D., Souza, G. I., da Silva, C. M. G., & Rodrigues, B. 2013. Green tea extract supplementation induces the lipolytic pathway, attenuates obesity, and reduces low-grade inflammation in mice fed a high-fat diet. *Mediators of inflammation* 2013: 1–8.

El Sheikha, A. F. & Ray, R. C. 2017. Potential impacts of bioprocessing of sweet potato. *Critical reviews in food science and nutrition* 57(3): 455–471.

Hachul, A. C., Boldarine, V. T., Neto, N. I., Moreno, M. F., Carvalho, P. O., Sawaya, A. C., & Oyama, L. M. 2018a. Effect of the consumption of green tea extract during pregnancy and lactation on metabolism of mothers and 28d-old offspring. *Scientific reports* 8(1): 1–7.

Hachul, A. C., Boldarine, V. T., Neto, N. I., Moreno, M. F., Ribeiro, E. B., do Nascimento, C. M., & Oyama, L. M. 2018b. Maternal consumption of green tea extract during pregnancy and lactation alters offspring's metabolism in rats. *PloS One* 13(7).

Iyare, E. E. & Iyare, F. E. 2007. Maternal consumption of aqueous extract of *hibiscus sabdariffa* during pregnancy attenuates pregnancy weight gain and postpartum weight loss. *African journal of biomedical research* 10(3).

Jumiarni, W. O. & Komalasari, O. 2017. Eksplorasi Jenis dan Pemanfaatan Tumbuhan Obat Pada Masyarakat Suku Muna di Permukiman Kota Wuna. *Traditional medicine journal* 22(1): 45–56.

Kaiser, L. L. & Campbell, C. G. 2014. Practice paper of the Academy of Nutrition and Dietetics abstract: Nutrition and lifestyle for a healthy pregnancy outcome. *Journal of the Academy of Nutrition and Dietetics* 114(9): 1447.

Mohanraj, R. & Sivasankar, S. 2014. Sweet potato (*Ipomoea batatas* [L.] Lam): A valuable medicinal food: A review. *Journal of medicinal food* 17(7): 733–741.

Moll, U., Olsson, H., & Landin-Olsson, M. 2017. Impact of pregestational weight and weight gain during pregnancy on long-term risk for diseases. *PLoS One* 12(1).

Nahar, S., Mascie-Taylor, C. N., & Begum, H. A. 2009. Impact of targeted food supplementation on pregnancy weight gain and birth weight in rural Bangladesh: An assessment of the Bangladesh Integrated Nutrition Program (BINP). *Public health nutrition* 12(8): 1205–1212.

Plante, A. S., Lemieux, S., Drouin-Chartier, J. P., Weisnagel, S. J., Robitaille, J., Drapeau, V., & Morisset, A. S. 2020. Changes in eating behaviours throughout pregnancy: Associations with gestational weight gain and pre-pregnancy body mass index. *Journal of obstetrics and gynaecology Canada* 42(1): 54–60.

Rong, K., Yu, K., Han, X., Szeto, I. M., Qin, X., Wang, J., & Ma, D. 2015. Pre-pregnancy BMI, gestational weight gain and postpartum weight retention: A meta-analysis of observational studies. *Public health nutrition* 18(12): 2172–2182.

Simonne, A. H., Kays, S. J., Koehler, P. E., & Eitenmiller, R. R. 1993. Assessment of β-carotene content in sweetpotato breeding lines in relation to dietary requirements. *Journal of food composition and analysis* 6(4): 336–345.

Yang, W., Han, F., Gao, X., Chen, Y., Ji, L., & Cai, X. 2017. Relationship between gestational weight gain and pregnancy complications or delivery outcome. *Scientific reports* 7(1): 1–9.

Effects of aqueous extract of unripe papaya (*Carica papaya* L.) on mice milk production

Y. Kharisma, S.B. Rahimah & H.S. Sastramihardja
Universitas Islam Bandung, Bandung, Jawa Barat, Indonesia

ABSTRACT: Breast milk is the best food to meet the nutritional needs of infants. According to the Indonesia Health Survey in 2005, 4–12% of babies in urban areas were exclusively breastfed and 4–5% of babies in rural areas were exclusively breastfed. The objective of this study was to examine the effect of aqueous extract of unripe papaya on milk production by measuring the weight gain and growth of baby mice. An experimental study was conducted with 21 lactating mice divided randomly into three treatment groups with 10 babies each. The first group served as a negative control; it was started at lactation days 4–16. The result was analyzed using ANOVA and followed by a Tukey test with a 95% level of confidence. The baby mice's weight gains in groups I, II, and III were 1.25 ± 0.62 g/6 hours, 2.25 ± 0.29 g/6 hours, and 2.21 ± 0.28 g/6 hours, respectively. The research showed that aqueous extract of unripe papaya has a better effect than the negative control and has an equivalent effect in improving weight gain and growth of baby mice given luteotropin.

1   INTRODUCTION

The World Health Organization (WHO) and governments around the world recommend to give infants only breast milk for the first 6 months (exclusive breastfeeding) and to continue breastfeeding until they are at least 2 years old (Abdel-Hafeez et al. 2013). Breast milk is rich in nutrients, contains immunological factors, and has higher bioavailability than formula. Breastfeeding also provides many benefits for mothers, such as uterus involution and contraception (Sardjono et al. 1996; Smith 2017).

Breastfeeding in Indonesia remains low. Based on data from the Indonesia Demographic and Health Survey in 2007, the scope of exclusive breastfeeding was 38%, a decrease from the 39.5% reported in 2002–2003. The number of infants under 6 months given formula increased from 16.7% in 2002–2003 to 27.9% in 2007 (Directorate General of Health Department of Community Health 2007). Poor exclusive breastfeeding causes nutritional problems in infants. Nationally, in 2005, the problem of malnutrition occurred in 110 of the 440 districts/cities in Indonesia that have a prevalence above 30% (weight for age). This would threaten the quality of human resources in the future (Ministry of Health, Republic of Indonesia 2011). The government is making efforts to handle such problems by increasing the quantity and quality of breast milk of breastfeeding mothers. This can be done by improving mothers' diets and increasing exclusive breastfeeding (Ministry of Health, Republic of Indonesia 2011).

Food high in calories and rich in nutrients plays an important role in lactation. These types of foods should contain the chemical compounds involved in lactation such as phytoestrogens, sterols, saponins, alkaloids, and tryptophan (Krishna et al. 2008; Oloyede 2005), which have been used traditionally to increase milk production (Jacobson 2014; Siswoyo 2004; Tietze 2002; Yakubu et al. 2008). In Asia, unripe papaya has been used as a lactagogue, a substance that can increase mothers' milk production (Jacobson 2014). Papaya is a widely cultivated plant in Indonesia that contains pro-vitamin A, vitamin C, calcium, and minerals (Oloyede 2005). It also contains saponins, alkaloids, and enzymes that help the digestion of protein, carbohydrates, and fats (Tietze 2002). Provision of water extracts *Cnidoscolous aconitifolius*

(Miller), which contain alkaloids, and saponins to female mice can raise the levels of prolactin due to the dopamine antagonist effect of the extract. Increased levels of prolactin will stimulate lactation. Saponin can increase the activity of alkaline phosphatase in the mammary glands of rats. This enzyme is found in most mammals' plasma membranes and plays a role in regulating oxytocin-mediated milk ejection (Leung et al. 1989). Sap (latex) of unripe papaya has the same effect as oxytocin in the uteri of both pregnant and nonpregnant rats, which causes an increase in uterine contractions. It also plays a role in milk ejection by stimulating the smooth muscles around the alveoli of the mammary gland (Adebiyi & Adaikan 2004).

At the beginning of a child's life, breast milk is a source of nutrients that are essential for the child to survive. Increased milk secretion will cover the nutritional needs of growing children. In this period, there is a positive correlation between milk consumption and weight gain (Lompo-Ouedraogo et al. 2004). This has an indirect effect on growth acceleration. The purpose of this study was to assess the effect of aqueous extract of unripe papaya on milk production by measuring weight gain and growth of mice and compare it to luteotropin preparations.

2 METHODS

This was an experimental study conducted in the Clinical Pharmacology Research Laboratory and Health Research Unit of Hasan Sadikin Hospital in Bandung, Indonesia. The experiment was performed on 21 breeding strains of Swiss Webster breastfeeding mice. Research subjects were randomly selected and divided into three treatment groups; each group consisted of 7 breeding mice with 10 babies each. Group I acted as a negative control. Group II was given luteotropin (Moloco®) with a dose of 6 mg/30 g/day orally, and group III was given aqueous extract of unripe papaya with a dose of 20 mg/30 g/day orally. This treatment was given on days 4–16 of lactation. This study considered the ethical aspects of research that can be accounted for and have been studied during the research proposal seminar. Providing treatment began with separating the dams and babies for 6 hours. One hour before regathering, the breastfeeding mice were given distilled water 1 cc in group I, and luteotropin preparation of 6 mg/30 g/day in group II, and aqueous extract of unripe papaya of 20 mg/30 g/day in group III.

Milk production of the mice was indirectly calculated through the babies' weight gain in each group by measuring the babies' weight gain differentiation before and after breastfeeding. Weighing was carried out collectively on the baby mice of the same dams. Growth of baby mice was also measured to complement the effectiveness of the test preparation and to increase milk production. Measurement was done by reducing the weight of the babies of each dam on day treatment by their weight of two days earlier.

Data were first tested by the Kolmogorov Smirnov normality test. Research data were displayed in tables and statistically tested with ANOVA and Tukey tests.

3 RESULTS

A series of experiments were performed on 21 breastfeeding breeding mice. Subjects were randomly divided into three treatment groups, each group consisting of 7 breeding mice with 10 babies each. Group I acted as a negative control, without any special treatment. Group II was given luteotropin of 6 mg/30 g/day orally, and group III was given aqueous extract of unripe papaya of 20 mg/30 g/day orally. This treatment was given on days 4–16 of lactation.

Comparison of the babies' weight gain in all three treatment groups is shown in Table 1, which shows that the average of baby mice's weight gain given 20 mg extract (2.21 ± 0.28) g/6 hours is higher than that of the negative control group (1.25 ± 0.62) g/6 hours, but less than that of the luteotropin group (2.25 ± 0.29) g/6 hours. The results of statistical tests using ANOVA at a 95% degree of confidence indicate significant differences among the three treatment groups with $p < 0.001$ (p-value ≤ 0.05). Further analysis of the effect of aqueous extract

Table 1.   The comparison of milk production among the groups.

Variable	Group						
	Negative control		Extract		Luteotropin		
	Average (g/6 hours)	SD	Average (g/6 hours)	SD	Average (g/6 hours)	SD	Value
Milk production	1.25	0.28	2.21	0.62	2.25	0.29	< 0.001*

Description: p-value = based on ANOVA (analysis of variance)
* p < 0.005 = significantly different
SD = standard deviation

Table 2.   Effects of aqueous extract of unripe papaya on increasing milk production.

Variable	Group	
	Negative control	Luteotropin
Milk production of extract group	0.002	0.984*

Note: * p > 0.05: not significantly different

of unripe papaya on the average breastfed baby mice's weight gain compared to negative controls and the luteotropin group is explained in Table 2. The results of statistical tests using Tukey's post hoc test on the degree of confidence 95% indicate significant differences between the average weight gain of baby mice given aqueous extract of unripe papaya and the negative control group with a p-value of 0.002 (p-value ≤ 0.05). Aqueous extract of unripe papaya does not cause a significant difference in the average weight gain of baby mice when compared to the luteotropin group with a p-value of 0.984 (p-value > 0.05).

Table 3 shows that the average of baby mice's growth in the group given aqueous extract of unripe papaya of 20 mg is (5.52 ± 1.53 g/2 days), higher than the growth of the negative control group (2.95 ± 0.82 g/2 days), but lower compared to the positive controls (7.10 ± 0.89) g/2 days. The results of statistical tests using ANOVA at 95% degree confidence indicate significance differences between the three groups on the growth of baby mice with p-value < 0.001 (p-value ≤ 0.05). Further analysis of the effect of aqueous extract of unripe papaya on the baby mice's growth compared to that of the negative control and luteotropin groups is described in Table 4. The results of statistical tests using Tukey's post hoc test with a 95% degree of confidence indicate that the effect of aqueous extract of unripe papaya on the baby mice's growth was significantly different compared to the negative control group with p-value <0.001 (p-value ≤ 0.05), but no significant differences occurred compared to the luteotropin group with p-value of 0.943 (p-value> 0.05).

Table 3.   Growth of baby mice.

Variable	Group						
	Negative control		Extract				
	Average (g/2 days)	SD	Average (g/2 days)	SD	Average (g/2 days)	SD	p-value*
Growth	2.95	0.82	5.52	1.53	7.10	0.89	< 0.001

Description: p-value = based on ANOVA (analysis of variance)
* p < 0.005 = significantly different
SD = Standard Deviation

Table 4. Effects of aqueous extract of unripe papaya on baby mice's growth.

Variable	Group	
	Negative control	Luteotropin
Milk production of extract group	< 0.001	0.943

4 DISCUSSION

Indonesia is rich in medicinal plants such as lactagogue plants, for example, katuk, lampes, bitter black cumin, moringa, jackfruit, Pulai, ginger, Turi, and unripe papaya (Siswoyo 2004; Tietze 2002). The increase of mothers' milk production will cover the nutritional needs of growing children. In one study, the high production of milk increased the weight of breastfed infants and was used to measure lactagogues' effect. Mice milk production was measured by calculating the weight change of baby mice before and after breastfeeding.

In this study, based on statistical analysis, the average weight gain of breastfed baby mice increased significantly in the group given aqueous extract of unripe papaya (2.21 g/6 hours ± 0.28) when compared to the negative control group (1.25 g/6 hours ± 0.62), and an equivalent effect occurred in the luteotropin group (2.25 g/6 hours ± 0.29) in increasing milk production.

Increased milk production of mice through the measurement of baby mice's weight gain in groups given unripe papaya aqueous extract was due to the saponins and alkaloids contained in the extract. Both can increase prolactin by inhibiting dopamine mechanisms. Prolactin plays a role in milk synthesis in cells' secretorius alveoli. Saponins can increase the activity of the hormone oxytocin in cells containing mioepitel around the alveoli and ducts. Alkaloids have the same action as α-adrenergic receptor agonists presenting in mammary gland ducts that work synergistically with the oxytocin hormone in milk ejection. A study revealed a dual effect on the mechanism of action of oxytocin, which plays a role in the mechanism of milk ejection and accelerates the secretion of milk components through the acceleration of intracellular transport of casein (Lollivier et al. 2006). In this study, significant differences were seen between the groups receiving aqueous extract of unripe papaya compared to the negative control group. The mechanism of hormonal stimulation in the study mentioned earlier is also a scientific reason to conduct this study. In this study, the milk production effect generated by aqueous extract of unripe papaya was equal to the effect created from luteotropin. This may be due to the content of other substances in the aqueous extract of unripe papaya, which has antagonist effects on milk production.

Indirect observation of the increase of milk produced by the dams can be viewed through the growth rate of baby mice. Growth of baby mice prior to weaning is affected by the secretion of milk. At the age of 1–14 days, baby mice have not been able to open their eyes (Robbins 2007). It can be assumed that the growth achieved at this age comes from energy provided by milk secretion. The rate of growth from birth to weaning is largely influenced by the amount of milk produced and by the individual's health. The average growth of baby mice given aqueous extract of unripe papaya was significantly better than the negative control group and did not differ significantly from the luteotropin group.

The average baby mice's growth in the group given aqueous extract of unripe papaya of 20 mg was (5.52 ± 1.53) g/2 days ≈ (2.76 ± 0.76) g/day, higher than that of the negative control group (2.95 ± 0.82) g/2 days ≈ (1.48 ± 0.41) g/day, but lower than that of the positive controls (7.10 ± 0.89) g/day ≈ (3.55 ± 0.45) g/day. The results of statistical tests using ANOVA at a 95% degree of confidence indicate significant differences in growth between the three groups with p-values < 0.001 (p-value ≤ 0.05).

Research conducted by Lompo-Ouedraogo and colleagues (2004) on the effects of aqueous extract of *Acacia nilotica ssp adansonii* on milk production in rats mentioned that the growth of baby rats in the test preparation group was significantly better compared to the control group – namely 0.86 ± 0.08 g/day (control group), 1.42 ± 0.12 g/day (group of 280 mg extract), and 1.43 ± 0.11 g/day (group of 560 mg extract) (Lompo-Ouedraogo et al. 2004). In this study, the average growth rate of baby mice of 20 mg aqueous extract of unripe papaya group was 5.52 ± 1.53 g/2 days, equivalent to 2.76 ± 0.76 g/day. Thus it is two times higher in comparison with the administration of *Acacia nilotica ssp adansonii* to rats.

5 CONCLUSION

It is concluded that aqueous extract of unripe papaya increases weight gain and growth of mice better than in the negative control and has an effect comparable to luteotropin preparation.

ACKNOWLEDGMENTS

We acknowldge the Laboratory of Clinical Pharmacology and Health Research Unit, Hasan Sadikin Hospital, Bandung, Indonesia, for all facilities provided during this study, and all parties who helped with the research.

REFERENCES

Abdel-Hafeez, E. H., Belal, U. S., Abdellatif, M. Z. M., Naoi, K., & Norose, K. 2013. Breast-feeding protects infantile diarrhea caused by intestinal protozoan infections. *Korean journal of parasitology* 51(5): 519–524.

Adebiyi, A. & Adaikan, P. G. 2004. Mechanisms of the oxytocic activity of papaya proteinases. *Pharmaceutical Biology* 42(8): 646–655.

Directorate General of Health, Department of Community Health. 2007. *Guidelines for training administration and training of nursing counseling facilitators.* Jakarta: Ministry of Health, 1.

Jacobson, H. 2014. Lactogenic foods and herbs. Mobimotherhood.org. Retrieved February 2.

Krishna, K. L., Paridhavi, M., & Patel, J. A. 2008. Review on nutritional, medicinal and pharmacological properties of Papaya (*Carica papaya Linn.*). *Natural Product Radiance* 7(4): 364–373.

Leung, C. T., Maleeff, B. E., & Farrell, H. M. (1989). Subcellular and ultrastructural localization of alkaline phosphatase in lactating rat mammary glands. *Journal of dairy science* 72(10): 2495–2509.

Lollivier, V., Marnet, P. G., Delpal, S., Rainteau, D., Achard, C., Rabot, A., & Ollivier-Bousquet, M. 2006. Oxytocin stimulates secretory processes in lactating rabbit mammary epithelial cells. *Journal of physiology* 570(1): 125–140.

Lompo-Ouedraogo, Z., Van der Heide, D., Van der Beek, E. M., Swarts, H. J., Mattheij, J. A., & Sawadogo, L. 2004. Effect of aqueous extract of *Acacia nilotica ssp adansonii* on milk production and prolactin release in the rat. *Journal of endocrinology* 182(2): 257–266.

Ministry of Health, Republic of Indonesia. 2011. National action plan to prevent and control malnutrition, 2005–2009. Ministry of Health. Republic of Indonesia. [Online Journal] [downloaded March 9, 2011]. Available from: www.dinkespurworejo.go.id.Website.ResmiDinasKesehatanKab.Purworejo.

Oloyede, O. I. 2005. Chemical profile of unripe pulp of Carica papaya. *Pakistan journal of nutrition* 4(6): 379–381.

Robbins, K. 2007. *Baby rat development* [Online Journal] [downloaded April 3, 2011]. Available from: www.afrma.org/pdfposters/babyratdev.pdf.

Sardjono, O. S., Hasanah, M., Yuliani, S., & Setiawati, A. 1996. *Produksi Sediaan dari daun katuk (Sauropus androgynus Merr) sebagai obat untuk meningkatkan produksi dan kualitas ASI. RUT.* Jakarta: Kantor Menteri Negara Riset & Teknologi.

Siswoyo, P. 2004. *Alternatif Obat Dengan Tumbuhan Alami: Tumbuhan Berkhasiat Obat.* Yogyakarta: Penerbit Absolut, 11–12.

Smith, R. P. 2017. *Netter's obstetrics and gynecology e-book.* Philadelphia: Elsevier Health Sciences.

Tietze, H. W. 2002. *Papaya therapy.* Jakarta: PT. Prestasi Pustaka, Indonesia.

Yakubu, M. T., Akanji, M. A., Oladiji, A. T., Olatinwo, A. O., Adesokan, A. A., Yakubu, M. O., ... & Ajao, M. S. 2008. Effect of *Cnidoscolous aconitifolius* (Miller) IM Johnston leaf extract on reproductive hormones of female rats. *Summer* 6(3):149–155.

Histopathological and microbiological analyses of an extrapulmonary tuberculosis diagnostic scoring model design

W. Purbaningsih, S. Masria, Y. Triyani & M. Tejasari
Universitas Islam Bandung, Bandung, Jawa Barat, Indonesia

ABSTRACT: Diagnosing extrapulmonary tuberculosis remains difficult because not all medical care facilities can do it as the microbiological examination of tissue materials requires invasive action and special equipment. Extrapulmonary tuberculosis diagnosis should be made by observing clinical manifestation and through microbiological and/or histopathological examinations of biopsies taken from infected organs, but in practice, this is not entirely possible. The purpose of this study was to analyze the histopathological features and microbiological examination of tissue biopsy to establish the extrapulmonary tuberculosis diagnosis in designing parameters for an extrapulmonary tuberculosis diagnostic scoring model. Samples consisted of tissue biopsies from extrapulmonary tuberculosis patients. Histopathological preparations used Hematoxylin Eosin staining along with the microbiological examination of Acid Fast Bacilli stains via the Ziehl Neelsen method. The results showed that the Acid Fast Bacilli stains mostly have a degree of positivity of +1. The histopathological feature was almost entirely granuloma and a relationship emerged between the degree of Acid Fast Bacilli positivity and the histopathology feature. This study concluded that the histopathological feature is related to the results of the microbiological examination so that it can be used as a parameter for an extrapulmonary tuberculosis diagnostic scoring model design.

1 INTRODUCTION

Tuberculosis (TB) is one of the most frequent infectious diseases and ranks among the top 10 causes of death in the world. The World Health Organization (WHO) reported that the highest TB incidence in 2017 occurred in Southeast Asia (44.2%), followed by the western Pacific (20.5%), Africa (19.7%), the eastern Mediterranean (8%), Europe (3.9%), and America (3.6%). In that year, 60% of new cases occurred in six countries – namely India, China, Indonesia, Nigeria, Pakistan, and South Africa. The number of TB cases reported in Indonesia in 2017 was 446,732 in an estimated population of around 264 million (Kementrian Kesehatan Republik Indonesia 2018). In the same year, WHO reported an estimated 842,000 TB cases in Indonesia. Based on these calculations, 395,268 (47%) TB cases in Indonesia have not been reported, so efforts are needed to improve TB diagnoses. Unreported and untreated cases of TB in the community will be a source of transmission (Kementrian Kesehatan Republik Indonesia 2018; World Health Organization 2018).

Over the past decades, extrapulmonary cases of TB have grown more frequent, typically attributed to the expanded predominance of procured immune deficiency disorder and the expanded number of organ transplants (Sener & Erdem 2019).

Establishing an appropriate TB diagnosis early is one of the characteristics of quality TB services and is important for achieving a target of eliminating TB by 2035 (Kementrian Kesehatan Republik Indonesia 2016). Prompt TB diagnosis can provide timely treatment and reduce the transmission of sputum droplets containing *Mycobacterium tuberculosis* (*M. tuberculosis*) from untreated patients (Hegde et al. 2014; Ong et al. 2014).

The incidence of extrapulmonary TB in developing and developed countries including the United States has increased since the mid-1980s (Handa et al. 2012). Global prevalence in 2017 ranged from 8% to 24% of all TB cases, namely 24% in the eastern Mediterranean, 16% in Europe and Africa, 15% in America and Southeast Asia, and 8% in the western Pacific (Handa et al. 2012; Ma et al. 2010; World Health Organization 2018). The prevalence of extrapulmonary TB in Brazil in 2011 was reported at 13.37% of all TB cases; in India in 2012, that number was 20% (Gomes et al. 2014; Handa et al. 2012). Most extrapulmonary TB occurs in the lymph nodes, followed in the pleura, genitourinary, bones and joints, meninges, peritoneum, and pericardium. Tuberculosis in lymph nodes is referred to as TB lymphadenitis (Handa et al. 2012; Hegde et al. 2014; Mohapatra & Janmeja 2009; Popescu et al. 2014).

Diagnosis of extrapulmonary TB is still difficult as not all medical care facilities can do it because the microbiological examination of tissue materials requires invasive action and special equipment. Extrapulmonary TB diagnoses should be made by observing clinical manifestations and through microbiological and/or histopathological examinations of biopsies from infected organs, but in practice, this is not entirely possible.

2 METHOD

This research comprised a retrospective descriptive observational study. The method used was an exploratory observation of tissue results from a biopsy in patients with extrapulmonary TB by analyzing the histopathological picture. Researchers also performed an explorative description of the clinical symptoms of extrapulmonary TB. The subjects in this study were patients diagnosed with extrapulmonary TB by a surgeon at the Al-Islam Hospital in Bandung, Indonesia. The object of the study included biopsy results from extrapulmonary TB patients at Al-Islam Hospital.

The data examined in this study came from the medical records of patients diagnosed with extrapulmonary TB based on the reading of preparations in the Anatomy Pathology Section of Al-Islam Hospital from January 2016 to December 2018.

The study population consisted of patients diagnosed with extrapulmonary TB. All populations were subject to inclusion and exclusion criteria.

Data on patients examined in the Clinical Pathology and Anatomy Pathology Sections of Al-Islam Hospital in 2018 were collected. These data were then sorted to determine the number of patients suffering from extrapulmonary TB with positive smear findings. Then the medical records were collected and the data recorded.

Data taken in this study included clinical manifestations, microscopic images of Hematoxylin Eosin (HE) staining, and sputum smear examination. Data were processed manually and presented in the form of images.

3 RESULTS

3.1 *Clinical manifestations of patients*

Figure 1 shows the clinical manifestations detected in extrapulmonary TB patients. Manifestations were divided into general and special manifestations. Common manifestations consisted of fever, night sweats, fatigue, and weight loss. It appears that the number of patients who complain of fever is the same as that of patients who complain of fatigue. On the other hand, the number of patients who complain of night sweats is the same as that of patients who complain of weight loss. In addition to clinical manifestations, local manifestations also emerged. In these local manifestations, the symptoms depended on the origin of the organ where extrapulmonary TB was located. For example, patients diagnosed with TBC KGB (*Tuberkulosis Kelenjar Getah Bening*, lymph node tuberculosis) will complain of enlargement of KGB with pain, whereas patients diagnosed with TBM mammae complain of lumps and pain. Such is the case with patients suffering with extrapulmonary TB that has other organ origins. Overall, all extrapulmonary TB patients complain of symptoms related to the organ of origin is located, although the number of complaints differs for each patient.

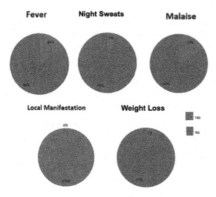

Figure 1.   Clinical manifestations.

3.2   *Histopathological and AFS features*

The data of this study came from biopsy tissue from patients diagnosed with extrapulmonary TB and non-extrapulmonary TB. After HE staining via the Ziehl Neelsen (ZN) method, which researchers observed under a light microscope, in some HE preparations, extrapulmonary TB patients have mature granulomas and varying Acid Fast Bacilli (AFB), ranging from negative to positive. The BTA in extrapulmonary TB tissue is less compared to the biopsy tissue of patients who are not diagnosed with extrapulmonary TB and does not contain granulomas, as shown in Figure 2.

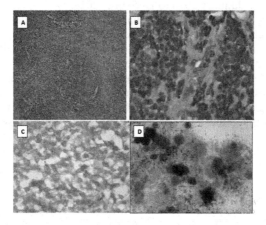

Figure 2.   Histopathological and AFS features.

4   DISCUSSION

This study showed that the amount of AFB in biopsy tissue that did not contain granulomas was less than that of tissue that did not contain granulomas. This is consistent with the results of Handa's research in 2017 (Mitra 2017).

The existence of granulomas comes with pros and cons, whether it is the success of the human body in isolating Mtb, or the ability of Mtb to escape from the immune system. Granuloma in the tissue resulting from the excisional biopsy is one of the histopathological parameters to establish a diagnosis of extrapulmonary TB in addition to necrosis of cancer and data Langhans. In the FNAB tissue, the presence of granulomas cannot be identified, because the specimen is only a smear from the aspirated tissue. The BTA examination can therefore be considered to help establish a diagnosis of extrapulmonary TB from the FNAB.

5 CONCLUSION

This study concluded that the histopathological features are related to the results of microbiological examination so that they can be used as a parameter for an extrapulmonary TB diagnostic scoring model design.

REFERENCES

Gomes, T., Reis-Santos, B., Bertolde, A., Johnson, J. L., Riley, L. W., & Maciel, E. L. 2014. Epidemiology of extrapulmonary tuberculosis in Brazil: A hierarchical model. *BMC Inf Dis* 14(9): 1471–2334.

Handa, U., Mundi, I., & Mohan, S. 2012. Nodal tuberculosis revisited: A Review. *J. Inf Dev Countries* 6(1): 6–12.

Hegde, S., Rithesh, K. B., Baroudi, K., & Umar, D. 2014. Tuberculous lymphadenitis: Early diagnosis and intervention. *Journal of international oral health* 6(6): 96–98.

Kementrian Kesehatan Republik Indonesia. 2016. *Peraturan Menteri Kesehatan Republik Indonesia No. 67 Tahun 2016 Tentang Penanggulangan Tuberkulosis.* Jakarta: Kemenkes RI.

Kementrian Kesehatan Republik Indonesia. 2018. *Infodatin Tuberkulosis.* Jakarta: Kemenkes RI.

Ma, Z., Lienhardt, C., McIlleron, H., Nunn, A. J. & Wang, X. 2010. Global tuberculosis drug development pipeline: The need and the reality. *Lancet* 375(9731): 2100–2109.

Mitra, S. K., Misra, R. K., & Rai, P. 2017. Cytomorphological patterns of tubercular lymphadenitis and its comparison with Ziehl-Neelsen staining and culture in Eastern UP. (Gorakhpur region): Cytological study of 400 cases. *J Cytol* 34(3): 139–143.

Mohapatra, P. R. & Janmeja, A. K. 2009. Tuberculous lymphadenitis. *J. Assoc Physicians India* 57: 585–590.

Ong, C. W., Elkington, P. T., & Friedland, J. S. 2014. Tuberculosis, pulmonary cavitation, and matrix metalloproteinases. *American journal of respiratory and critical care medicine* 190(1): 9–18.

Popescu, M. R., Călin, G. I., Strâmbu, I., Olaru, M., Bălășoiu, M., Huplea, V., . . . & Pleșea, I. E. 2014. Lymph node tuberculosis: An attempt of clinico-morphological study and review of the literature. *Romanian journal of morphology and embryology* 55(2): 553–567.

Sener, A. & Erdem, H., eds. 2019. *Extrapulmonary tuberculosis.* Basel: Springer Nature Switzerland.

World Health Organization. 2018. *Global tuberculosis teport.* Geneva: World Health Organization.

Case report of perioperative bronchospasm

I. Indrianto & S. Trisnadi
Universitas Islam Bandung, Bandung, Indonesia
Al Ihsan West Java Province Hospital Bandung, Bandung, Indonesia

ABSTRACT: Bronchospasm is a potentially serious perioperative problem for those under-going general anesthesia. Bronchospasm likely occurs when the respiratory tract is still sensitive to inhalation after an infection, or due to the release of histamine after the administration of atracurium. In developing countries all over the world, the incidence of trauma due to traffic accidents remains high, especially limb and head trauma in addition to multiple other traumas. A boy aged 13 years old and weighing 48 kg suffered a left thighbone fracture due to a traffic accident, and doctors planned to install plates and screws through surgery under general anesthesia. The preoperative examination was obtained: GCS 15, history of accidents 4 days prior, history of respiratory tract infections that had healed 2 days prior, history of asthma (−), history of drug allergy (−), family history of asthma (+). The physical examination found: BP 115/74 mmHg, pulse 84 x/min, respiration 18 x/min, and temperature afebrile. The laboratory blood tests were within normal limits. A pulmonary X-ray was normal. Induction was carried out by giving Fentanyl 50 mcg iv, Propofol 100 mg iv, and Atracurium 20 mg iv, and the patient was manually ventilated using O_2, N_2O, and sevoflurane gas. After 1 minute, the ventilation felt heavy, 20% SpO_2, BP 50/30 mmHg, pulse 130 x/min, acral appeared cyanotic, and the patient was immediately intubated with ETT no 6.5. Manual ventilation still felt severe, so 2 minutes later, the patient was given a bronchodilator Ventolin spray through ETT, Methylprednisolone 125 mg iv, Ephedrine 10 mg iv, and Dexamethasone 4 mg iv. Manual ventilation continued, gradually becoming lighter; after 15 minutes, BP 90/60 mmHg, pulse 120 x/min, SpO_2 95–97%, acral cyanosis (−). Conditions returned to normal and the operation continued with good results.

1 INTRODUCTION

Bronchospasm is a potentially serious perioperative problem for patients undergoing general anesthesia. Although volatile agents are bronchodilators, bronchospasm can still occur and may have grave consequences. Leaving the symptoms untreated can cause hypoxia or hypotension and can increase morbidity and mortality. Suspected bronchospasm should be assessed and treated promptly. Ongoing management should address the underlying cause (Looseley 2011).

2 CASE REPORT

A boy aged 13 years old and weighing 48 kg suffered a left thighbone fracture due to a traffic accident, and doctors planned to install plates and screws through surgery under general anesthesia. The preoperative examination was obtained: GCS 15, history of accidents 4 days prior, history of respiratory tract infections that had healed 2 days prior, history of asthma (−), history of drug allergy (−), family history of asthma (+). A physical examination found: BP 115/74 mmHg, pulse 84 x/min, respiration 18 x/min, temperature afebrile. Laboratory blood tests were within normal limits. A pulmonary X-ray was normal. Induction was carried out by giving Fentanyl 50 mcg iv, Propofol 100 mg iv, and Atracurium 20 mg iv, and the patient was manually ventilated using O_2, N_2O, and sevoflurane gas. After 1 minute, the ventilation felt heavy, SpO_2 20%, BP 50/30 mmHg, pulse 130 x/min, acral appeared cyanotic, and the patient was immediately intubated with ETT no 6.5. Manual ventilation still felt severe, so 2 minutes later, the patient was given a bronchodilator Ventolin puff through ETT, Methylprednisolone 125 mg iv, Ephedrine 10 mg iv, and Dexamethasone 4 mg iv. Manual ventilation continued, becoming gradually lighter; after 15 minutes, BP 90/60 mmHg, pulse 120 x/min, SpO_2 95–97%, acral cyanosis (−). Conditions returned to normal, and the operation continued with good results.

3 DISCUSSION

Bronchospasm is a common feature of reactive airway disease. Patients with bronchial asthma, chronic obstructive pulmonary disease, or a history of atopy, and immediately after viral upper respiratory infection, show hyperreactive airway responses to mechanical and chemical irritants, which all increase the risk of bronchospasm during general anesthesia. A combination occurs of constriction of bronchial smooth muscle, mucosal edema, and mucous hypersecretions with plugging. Perioperative bronchospasm in patients with reactive airway disease is relatively uncommon. The overall incidence of bronchospasm during general anesthesia is approximately 0.2%. Many patients with bronchospasm during general anesthesia have no history of reactive airway disease (Looseley 2011).

The management of the case revealed some essential concerns: the first concern was that the patient was intubated, which deepened the plane of anesthesia; a Dexamethasone injection a Salbutamol puff via an endotracheal tube could have been used (Marschall & Hines 2018).

Atracurium is part of a wide selection of nondepolarizing muscle relaxants, their chemical structure classified as benzylisoquinolinium, steroidal, and other compounds. Benzylisoquinoline tends to release histamine from mast cells and can result in bronchospasm, skin flushing, or hypotension from peripheral vascular dilation. Atracurium should be avoided in asthmatic patients. Severe bronchospasm is occasionally seen in patients without a history of asthma (Butterworth, Mackey, & Wasnick 2013).

4 SUMMARY

Perioperative bronchospasm is a serious problem during general anesthesia that can cause morbidity and mortality. Management begins with switching to 100% oxygen and calling for help early. Stop all potential precipitants and deepen anesthesia. Exclude mechanical obstruction or occlusion of the breathing circuit. Aim to prevent/correct hypoxemia and reverse bronchoconstriction. Consider a wide range of differential diagnoses including anaphylaxis, aspiration, or acute pulmonary edema.

REFERENCES

Butterworth, J. F., Mackey, D. C., & Wasnick, J. D. 2013. *Morgan & Mikhail's clinical anesthesiology*, 5th edition. New York: McGraw Hill Lange.

Looseley, A. 2011. Management of bronchospasm during general anaesthesia. *Update in anaesthesia* 27(1): 17–21.

Marschall, K. & Hines, R. 2018. *Stoelting's anesthesia and co-existing disease*, 7th edition. New York: Elsevier.

Medical Technology and Environmental Health – Abdullah, Widiaty & Abdullah (eds)
© 2020 Taylor & Francis Group, London, ISBN 978-0-367-86053-0

Correlation between ferritin levels and Peak Expiratory Flow (PEF) value in thalassemia patients at a private hospital in Indonesia in 2018

Y.D. Suryani, W. Risakti & R.G. Ibnusantosa
Universitas Islam Bandung, Bandung, Indonesia

ABSTRACT: Abnormality of hemoglobin synthesis such as thalassemia has recently been found in Indonesia. Thalassemia patients need blood transfusions regularly during their lifetime. This will increase ferritin levels and cause hemosiderosis in various organs, including the lungs. The aim of this study was to find a correlation between ferritin levels and pulmonary function in thalassemia patients as seen from peak expiratory flow (PEF) measurements. The methodology of this study employed observational analytics with a cross-sectional design. Ferritin levels were taken from medical records and PEF measurement using a peak flow meter was performed on 35 thalassemia patients. Data were analyzed using a Pearson test with statistical and data science analysis (STATA) version 11.0; the results obtained a mean ferritin level of 4,917.043 (SD ± 2,493.99) ng/dL. Meanwhile, the mean value of PEF was 126.57 (SD ± 51.86) L/min. This result indicates no significant correlation between these two variables. An increase in ferritin value because of multiple transfusions and hemolysis will induce hemosiderosis and pulmonary dysfunction. Besides pulmonary dysfunction, another condition like hepatosplenomegaly will affect the pulmonary function measurement, which is reflected in the PEF value.

1 INTRODUCTION

Thalassemia is a hereditary hemoglobin disease that is found in large numbers nowadays (Kohne 2011). This kind of hemoglobin abnormality was once limited to the tropics and to subtropical areas in the Mediterranean belt, but now it is found throughout the world (Modell & Darlison 2008). Thalassemia is now a major genetic problem in Asian countries, including Indonesia (Ariani et al. 2017). Based on data from the Ministry of Health of the Republic of Indonesia in 2017, the number of cases of thalassemia in Indonesia has increased continuously (Kementerian Kesehatan Republik Indonesia 2017).

Patients with thalassemia need repeated blood transfusions to maintain hemoglobin levels (Cossio et al. 2016). Patients have a risk of accumulating iron (Fe), which can be reflected in ferritin levels. This issue stems from ineffective physiological processes of removal of excess iron in the body due to the effects of long-term transfusions (Cusick et al. 2008; Muncie & Campbell 2009). The complication often caused by the deposition of iron in the organs is called hemosiderosis, and it can attack various target organs, including the lungs (Yaman et al. 2013).

Pulmonary hemosiderosis will cause progressively restrictive pulmonary disorders and decreased lung capacity. Restrictive pulmonary disorders are the main characteristics of lung function abnormalities that can occur in thalassemia major patients. Pulmonary dysfunction in patients with thalassemia major is caused mainly by lung fibrosis and interstitial lung edema associated with iron overload (Bourli et al. 2012; Uçar et al. 2014).

In addition to restrictive pulmonary abnormalities, pulmonary function examination in thalassemia patients can show obstructive abnormalities, although most patients are asymptomatic (Vij & Machado 2010). Examination of pulmonary function could be done using spirometry, but alternative tools can be used to screen for lung disorders, including a peak flow meter.

Based on this, research needs to be carried out in order to determine the relationship between ferritin levels and lung function as seen from the value of peak expiratory flow (PEF) in thalassemia patients in Al-Ihsan Regional Hospital (RSUD Al-Ihsan), West Java, in 2018.

2 METHOD

This was an observational analytic study with a cross-sectional design. The population in this study consisted of thalassemia patients in a private hospital – namely RSUD Al-Ihsan located in West Java, Indonesia – in 2018. The sampling technique used total sampling of as many as 35 thalassemia patients in RSUD Al-Ihsan. Ferritin levels examined in this research were obtained from the medical records of thalassemia patients and direct measurements of patients' lung function using a peak flow meter.

The results of the study were processed using a Pearson test if the results of the Shapiro Wilk Test for normality showed data were normally distributed with a P-value > 0.05. Data analysis was performed with the help of statistical and data science analysis (STATA) software version 11.0.

This study has obtained ethical approval from the Faculty of Medicine of the Islamic University of Bandung, number 46/Ethics Committee FK/III/2018.

3 RESULTS

Characteristics of respondents are presented in Table 1.

The age and height variables have a normal distribution, so they are presented with mean values, while weight variable data that are not normally distributed are presented with median values. Table 1 shows the number of respondents in the sample of 35 people with the male gender dominating. The average height indicates a shorter stature when viewed by age.

Table 2 shows that the mean value of ferritin levels of all patients had a higher value when compared to the normal reference value of 20–200 ng/dL. The results of lung function examinations were obtained by measuring the PEF on 35 respondents. The mean PEF result was lower than the normal values table based on height and age. The result of statistical calculation using the Pearson test in Table 2 show no significant correlation between ferritin levels with PEF values with $p = 0.86$ ($p \geq 0.05$).

Table 1. Characteristics of subjects based on age, sex, height, and weight.

Variables	Result
Age (year)	
Mean ± SD	8.57 ± 3.689
Median (min–max)	9 (2–19)
Height (cm)	
Mean ± SD	118.63 ± 16.81
Median (min–max)	121 (90–153.5)
Weight (kg)	
Mean ± SD	22.33
Median (min–max)	23 (10–53)
Gender, n (%)	
Male	22 (62.86)*
Female	13 (37.14)*

* Presented in proportions and percentages

31

Table 2. Relationship between ferritin levels with peak expiratory flow (PEF) values in thalassemia patients in Al-Ihsan Regional Hospital, 2018.

Variables	Mean ± SD	p*
PEF (L/min)	126.57 ± 51.86	
Ferritin (ng/dL)	4,917.043 ± 2,493.99	0.86

* Pearson test

4 DISCUSSION

Patients with hemoglobin abnormalities such as thalassemia need a blood transfusion routinely to keep their red blood cells and hemoglobin normal (National Heart, Lung, and Blood Institute 2014). One of the long-term effects of repeated transfusions is an increase of ferritin levels, which can trigger hemosiderosis in various tissues (Lanzkowsky et al. 2016). When hemosiderosis occurs in the lung parenchyma, it can cause functional abnormalities of the lung (Bourli et al. 2012).

The results of this study show no relationship between ferritin levels and lung function in thalassemia patients at Al-Ihsan Regional Hospital. The respondents' age factor and the use of chelating agents can influence the results of the study. The use of chelating agents is considered effective for maintaining iron levels in the body. Although it does not reduce iron to normal levels, it can reduce the likelihood of complications of hemosiderosis. Respondents in this study consumed chelating agents early and regularly, so they could minimize the possibility of hemosiderosis in the lungs that can cause fibrosis.

The results of this study are in accordance with those of research conducted by Guidotti and colleagues (2017) in thalassemia patients who undergo routine transfusions and receive adequate chelating agent therapy. This research on pulmonary dysfunction in thalassemia major patients shows the results of restrictive pulmonary abnormalities in relation to ferritin levels. It was found that high serum ferritin levels were more common in thalassemia major patients with restrictive lung disorders. Although the final results of the study did not find any relationship with the parameters of ferritin levels, the possibility of chronic effects of iron can play a role in the pathogenesis of lung disorders (Guidotti et al. 2017).

Research conducted by Boddu and colleagues (2015) a cross-sectional study at the hospital in beta major thalassemia patients with a total of 42 study subjects with a mean age of 12 years, got different results. The statistical analysis used was Spearman's correlation coefficient, an ANOVA test, and a T dependent test. The analysis found that the degree of pulmonary dysfunction was significantly related to high serum ferritin levels. In addition, studies of cardiorespiratory function tests in thalassemia major patients conducted by Nandurkar and colleagues (2018) showed similar results. Thalassemia patients suffer both obstructive and restrictive pulmonary abnormalities; the results of all lung function test parameters decreased when ferritin levels in patients above 12 years were increased (Nandurkar et al. 2018).

Transfusion frequency in thalassemia patients can be one of the factors that influence the results of the study. In studies with patients who were older than the patients in this study and whose ferritin levels were above normal, the results showed worse lung function. This situation arises because the older the patient, the greater the frequency of transfusions, so that the risk of hemosiderosis in the lungs is greater. This is consistent with research by Yilmazel and colleagues. Reports of cases of interstitial lung disease secondary to iron deposition in the lungs of thalassemia patients who have hemosiderosis show the results of research with respondents aged 29 years who have received repeated transfusions. Biopsy examination found secondary pulmonary hemosiderosis caused by repeated blood transfusions in patients; besides that, this study suggested that lung disease was very likely to occur as a result of repeated transfusions and accumulation of iron in the lungs (Uçar et al. 2014).

Thalassemia patients with a greater number of blood transfusions allow the accumulation of excess iron in the form of ferritin in the body for a longer period of time. The long-term effects of iron deposition on the lung parenchyma cause hemosiderosis, which can lead to fibrosis, which affects lung function. When pulmonary function is disrupted, it will certainly affect the PEF, which becomes lower as a result of lung restrictions. This is supported by the results of research by Bourli and colleagues (2012). Regarding the factors affecting restrictive pulmonary dysfunction in young beta major thalassemia patients, despite measurements of other parameters, serum ferritin remains a major component for predicting restrictive pulmonary dysfunction. Pulmonary dysfunction has been recognized as a complication that can occur in patients with beta major thalassemia, but the causes, pathogenesis, and natural mechanism of the condition are still debated. Long-term (chronic) effects of excessive iron levels are the most likely cause of restrictive pulmonary dysfunction, although this study cannot confirm the relationship between the two variables. It is well known that excessive iron levels have the potential to become free radicals that trigger tissue injury and fibrosis (Bourli et al. 2012).

5 CONCLUSIONS

Based on these data, we can conclude that there is no significant relationship between ferritin levels and PEF values. However, the long-term (chronic) effects of excessive iron levels can cause hemosiderosis in the lung tissue that triggers the pulmonary abnormalities that can interfere with its function so it could reduce the PEF.

REFERENCES

Ariani, Y., Soeharso, P., & Sjarif, D. R. 2017. Genetics and genomic medicine in Indonesia. *Molecular genetics & genomic medicine* 5(2): 103–109.

Boddu, A., Kumble, A., Mahalingam, S., Baliga, B. S., & Achappa, B. 2015. Pulmonary dysfunction in children with beta thalassemia major in relation with iron overload: A cross sectional hospital based study. *Asian journal of medical sciences* 6(5): 47–50.

Bourli, E., Dimitriadou, M., Economou, M., Vlachaki, E., Christoforidis, A., Maratou, E., ... & Aivazis, V. 2012. Restrictive pulmonary dysfunction and its predictors in young patients with thalassaemia major. *Pediatric pulmonology* 47(8): 801–807.

Cossio, M. L. T., Giesen, L. F., & Araya, G. 2016. *Hoffbrands essential haematology*, vol. 33. Ed. A. V. Hoffbrand & P. A. H. Moss. UK: John Wiley & Sons, Ltd.

Cusick, S. E., Looker, A. C., Cogswell, M. E., Pfeiffer, C. M., & Grummer-Strawn, L. 2008. Iron-status indicators. *Pediatrics* 121(3): 651–652.

Guidotti, F., Piatti, G., Marcon, A., Cassinerio, E., Giuditta, M., Roghi, A., ... & Cappellini, M. D. 2017. Pulmonary dysfunction in thalassaemia major: Is there any relationship with body iron stores? *British journal of haematology* 176(2): 309–314.

Kementerian Kesehatan Republik Indonesia. 2017. *Skrining penting untuk cegah thalassemia*. Retrieved from: www.kemkes.go.id.

Kohne, E. 2011. Hemoglobinopathies: Clinical manifestations, diagnosis, and treatment. *Deutsches Ärzteblatt International* 108(31–32): 532–540.

Lanzkowsky, P., Lipton, J. M., & Fish, J. D., eds. 2016. *Lanzkowsky's manual of pediatric hematology and oncology*. UK: Academic Press.

Modell, B. & Darlison, M. 2008. Global epidemiology of haemoglobin disorders and derived service indicators. *Bulletin of the World Health Organization* 86: 480–487.

Muncie Jr., H. L. & Campbell, J. S. 2009. Alpha and beta thalassemia. *American family physician* 80(4): 339–344.

Nandurkar, P., Goel, M., & Sharma, S. 2018. A study on cardiopulmonary function tests in thalassemia major patients (6–14 years) and its correlation to serum ferritin. *Journal of Pulmonary and Respiratory Medicine* 8(1):1–4.

National Heart, Lung, and Blood Institute (NHLBI). 2018. [Online]. [accessed February 10, 2018]. Retrieved from: www.nhlbi.nih.gov/health-topics/thalassemias.

Uçar, E. Y., Akgün, M., Araz, Ö., Barın, R. B., Demirci, E., & Alper, F. 2014. Interstitial lung disease secondary to iron deposition in the lungs in a patient with hemosiderosis. *Respiratory case reports* 3(3): 138–140.

Vij, R. & Machado, R. F. 2010. Pulmonary complications of hemoglobinopathies. *Chest* 138(4): 973–983.

Yaman, A., Pamir, I., Yarali, N., Karademir, S., Cetinkaya, S., Ali, B., ... & Bahattin, T. 2013. Common complications in beta-thalassemia patients. *International journal of hematology and oncology* 28(4): 193–199.

Yilmazel, U. E., Akgun, M., Araz, O., Bayraktar, B. R., Demirci, E., Alper, F. 2014. Interstitial lung disease secondary to iron deposition in the lungs in a patient with hemosiderosis. *Respir Case Reports.* 3(3): 138–40.

Effect of tomato juice on the sperm quality of mice exposed to tertiary cigarette smoke

A.R. Furqaani, A.K. Sari, R. Ekowati, L.H. Siswanti, A. Triamullah & T. Sugiartini
Universitas Islam Bandung, Bandung, Indonesia

ABSTRACT: Cigarette residues can accumulate in semen and sperm thereby reducing sperm quality and causing infertility in men. Meanwhile, lycopene in tomatoes (*Solanum lycopersicum*) is an antioxidant that serves as a protective mechanism against oxidative stress, acts against lipid peroxidation, and regulates gene function through non-oxidative mechanisms. Therefore, this study was conducted to determine the effect of tomato juice administration on sperm quality in mice exposed to thirdhand smoke. The research comprised an experimental study with a completely randomized design. Seventeen adult male mice (8–10 weeks in age, 35–40 g in weight) were randomly divided into three groups, the control group (K) and two treatment groups (T). Treatment group 1 (T1) was exposed to tertiary cigarette smoke for 14 days and treatment group 2 (T2) was exposed to tertiary cigarette smoke and received tomato juice for 14 days. After the treatment period, the animals were sacrificed by cervical dislocation. Spermatozoa were taken from the cauda epididymis; sperm quality examination was done immediately. Parameters measured in sperm quality examination included the number, morphology, and motility of the spermatozoa. The results showed that the number of the sperm in T1 tended to be lower than that in the control group and in T2, but did not differ significantly ($p > 0.05$). The sperm motility tended not to differ among the groups ($p > 0.05$). Abnormal sperm morphology in the test group was significantly higher than that in the control group ($p < 0.05$). These results indicate that tomato juice has potential in improving sperm quality in mice exposed to thirdhand smoke.

1 INTRODUCTION

Cigarettes and smoking behavior have to be a serious concern of all parties due to the increasing number of active smokers showing negative impacts on global public health. Data from the World Health Organization show that in 2012, there were around 967 million or nearly 1 billion active smokers with consumption of 6.25 trillion cigarettes. Meanwhile, basic health research from the Ministry of Health shows that active smokers' prevalence increased, from 34.7% in 2010 to 36.3% in 2013. Problems caused by smoking behavior become increasingly alarming because the adverse effects of smoking will be felt not only by active smokers but also by passive and tertiary smokers. Cigarette smoke exhaled by active smokers will leave residue attached to various surfaces of objects in a room or in the environment. Cigarette residue could furthermore react with other substances in the air, yielding toxic, mutagenic, and carcinogenic substances. Research shows that toxicant in cigarettes could accumulate in semen and sperm, decreasing sperm motility and concentration and causing male infertility (Day et al. 2016; Vallaster et al. 2017). Exposure to cigarette smoke also causes apoptosis and seminiferous tubule disruption so that spermatogenesis is impaired and sperm quality decreases.

Dai and colleagues (2015) have shown that toxins in cigarettes are compounds that play a role in activating the oxidative stress mechanism so that there is an increase in free radicals in the body. This free radical increase is positively correlated with DNA fragmentation and damage. Benzo(a)pyrene (B[a]P) found in cigarette smoke is mutagenic and carcinogenic. The compound will bind with DNA covalently to form a DNA adduct called benzo(a)pyrene diol epoxide-DNA (BPDE-DNA). It is known that BPDE-DNA increases in smokers and significantly decreases the percentage of halo acrosomes and is a contributing factor of DNA damage in smokers' sperm (Hammadeh et al. 2016). Increasing environmental pollutants, including cigarette residues, expose humans to various substances that threaten health. Therefore, people need to change their lifestyle and consume food and nutrients that can minimize these adverse effects.

Tomatoes are one of the most widely produced and consumed vegetable crops around the world, including in various regions in Indonesia. The tomato (*Solanum lycopersicum*) has major phytochemical content in the form of lycopene. Tomatoes and food products derived from tomatoes are the main source of lycopene. Lycopene is an antioxidant from a group of carotenoid compounds soluble in lipids; 100 g of fresh tomatoes contains 0.9–4.2 mg of lycopene (Chauhan et al. 2011). Tomatoes also contain coumaric and chlorogenic acids, which protect cells from the carcinogenic compounds derived from cigarette smoke (Bhowmik et al. 2012). Carotenoids, including lycopene, are antioxidants with the potential to decrease reactive oxygen species (ROS), especially in the form of oxygen singlets (Kong et al. 2010). Lycopene also plays a role in inhibiting DNA damage caused by oxidative stress activity, acts against lipid peroxidation, and regulates gene function through non-oxidative mechanisms (Lu et al. 2011; Palozza et al. 2011).

Due to data limitations and research related to the mechanism of tertiary cigarette smoke in causing pathological conditions in the reproductive system of nonsmokers, this study encourages researchers to study this topic further. In addition, the use of natural materials as a treatment or as a method to anticipate the adverse effects of smoking is the focus of researchers. Tomatoes and their derivative products are high in antioxidant contents and are easily available at affordable prices so these foods are chosen by researchers. Therefore, this research was conducted to examine the potential of tomato juice consumption in improving the sperm quality in mice exposed to tertiary cigarette smoke.

2 METHOD

This research consisted of an in vivo laboratory experimental study with a completely randomized design. Seventeen adult male mice were randomly divided into three groups, the control group and two treatment groups. Treatment group 1 (T1) was exposed to tertiary cigarette smoke for 14 days, and treatment group 2 (T2) was exposed to tertiary cigarette smoke and given 0.64 mL of tomato juice for 14 days. After completion of treatment, the animals were sacrificed by cervical dislocation. Spermatozoa were taken from the cauda epididymis. The parameters measured in sperm examination included number, morphology, and motility of spermatozoa. At the time of surgery, testes were also isolated in order to identify cells' density in testicular tissue. Data are presented in mean ± SD (standard deviation). Each parameter was analyzed using an ANOVA one-way test and the Tukey post hoc test at the 95% confidence interval ($\alpha = 0.05$). This study was reviewed and approved by the Ethical Committee of the Medical Faculty at the Islamic University of Bandung with approval number 058/Komite Etik.FK/IV/2019.

3 RESULTS

Data obtained in this study featured observations of sperm quality and testicular tissue. Subjects of the study included 18 male mice, but during treatment one mouse from the T1 group died, thus the total of experimental animals at the end of the study was 17 mice. Results of the sperm quality and testicular tissue examinations are listed in Table 1.

The data in Table 1 show that, after exposure to tertiary cigarette smoke, risks of decrease in sperm quality, sperm count, and motility in the T1 group tended to be lower than in the control group, but not significantly different (p > 0.05). However, significant differences

Table 1. Observation of sperm quality and testicular tissue.

Parameter	Groups			p-value
	Control (N = 6)	T1 (N = 5)	T2 (N = 6)	
Sperm count (10^6)	20.82 ±4.30	17.52 ±3.43	21.63 ±4.07	0.239
Sperm morphology (%)				
Normal	50.33 ±7.61[a]	34.40 ±4.51[b]	44.00 ±6.69[ab]	0,004*
Abnormal	49.67 ±7.61[a]	65.60 ±4.51[b]	56.00 ±6.69[ab]	0,004*
Sperm motility (%)				
Normal forward progression (a)	36.67 ±11.69	29.00 ±17.10	35.83 ±9.70	0,581
Nonprogressive motile (b)	28.33 ±6.83	36.00 ±7.42	35.83 ±5.85	0,118
Immotile (c)	35.00 ±11.83	35.00 ±18.03	28.33 ±9.31	0,619
Cell density (%)	0.20 ±0.03[a]	0.15± 0.01[b]	0.165 ±0.04[ab]	0,025*

* Shows significant differences. Letters a, b: a parameter that is not labeled with the same letters indicates significant differences. N: number of samples

appeared in the normal and abnormal morphological numbers between the control group and T1. The normal sperm morphology count in the T1 group was significantly smaller compared to the control group ($p < 0.05$). This has implications for significantly higher sperm counts with abnormal morphology in T1 compared to the control group. Testicular tissue examination showed that cell density in T1 was more diminutive than in the control group. These results indicate that the treatment in this study, exposure to tertiary cigarette smoke, causes a decrease in sperm quality in experimental animals.

Table 1 also shows that all the parameters of sperm quality and testicular tissue cell density of the T2 group did not differ significantly from those of the control group and T1. But there tends to be an increase in sperm count, normal morphological counts, sperm counts with normal forward progression, and cell density in testicular tissue. This indicates that the treatment given to the T2 group, provision of tomato juice, has potential to improve sperm quality.

4 DISCUSSION

Exposure to tertiary cigarette smoke in animals could affect sperm quality and cell density in testicular tissue. Decreased sperm quality and changes in testicular tissue microstructure are one implication of exposure to cigarette smoke, including tertiary cigarette smoke. Tertiary cigarette smoke or thirdhand smoke (THS) is residual cigarette smoke in the form of gases and particles attached to material surfaces such as carpets, walls, clothing, toys, etc., and it can be released back into the air so that it undergoes a series of chemical transformations and/ or accumulates to form hazardous by-products (Jacob et al. 2017). These residues include nicotine and tobacco-specific nitrosamines that can last up to 6 months after quitting smoking (Matt et al. 2017). Paternal exposure to residues can cause accumulation of toxic compounds in semen and sperm thereby reducing sperm quality (Matt et al. 2017; Vallaster et al. 2017). Lipophilic nicotine can easily penetrate the cell membrane barrier, activate the nicotinic acetylcholine receptor (nAChR) receptor, and potentially cause pathological effects (Albuquerque et al. 2009). Exposure to nicotine can induce stress on the endoplasmic reticulum, which causes disruption of protein folding and activates the NFκB pathway, resulting in changes in gene expression (Wong et al. 2015). One protein that may be affected is a protein that plays a role in spermatogenesis, Pebp1 (phosphatidylethanolamine-binding protein 1). In mice exposed to cigarette smoke, abnormal Pebp1 expression occurs involving the mitogen-activated protein kinase (MAPK) pathway through abnormal extracellular signal-regulated protein kinase (ERK), 1/2 expression causing spermatogenesis disorders including proliferation of spermatogonia, spermatocyte meiosis, capacitation, and acquisition of sperm motility (Xu et al. 2013). Although during spermatogenesis the sperm is protected by the cytoplasm of Sertoli cells, the blood–testes barrier is still permeable to many xenobiotics that can induce DNA damage, resulting in infertility (Beal et al. 2017). A series of these processes is thought to trigger morphological changes and decrease sperm density in the epididymis.

It was previously known that oxidative stress plays a role in the pathogenesis of infertility in men such as decreased sperm motility and viability, and increased sperm apoptosis (Durairajanayagam et al. 2014). Oxidative stress is a condition of imbalance between free radicals and antioxidants. Normally, free radicals (ROS) in the testicles are by-products produced during the sperm differentiation process and the formation of steroid hormones (Mathur & d'Cruz 2011). ROS is a molecule that has one or more unpaired electrons, thus it is very reactive, especially to the lipid component (Clément et al. 2012). Oxidative stress can affect spermatozoa through three pathways: lipid peroxidation in cell membranes, DNA damage, and apoptosis (Durairajanayagam et al. 2014; Mathur & d'Cruz 2011). In the Polβ and β-Globin mouse genes, TSH exposure causes DNA oxidation to increase malondialdehyde (MDA), a lipid peroxidation marker, and to increase the 8-oxo-dG (8OhdG) level, which is a DNA oxidation product (Jacob et al. 2017). Sperm apoptosis can cause sperm density/density in the epididymis to decrease. Cigarette smoke is also known to interfere with sperm maturation so that spermatogenesis is disturbed, thus sperm decreases (Shrivastava et al. 2014). Research by Esakky and colleagues (2016) showed that exposure to cigarette smoke condensate (CSC) is genotoxic to spermatocytes, causing apoptosis and seminiferous tubule disruption through oxidative stress mechanisms.

The results of this study are in line with those of previously published research, stating that sperm quality and testicular tissue cell density in the T1 group exposed to tertiary cigarette smoke decreased compared with the control group, although potential for improvement in sperm quality and testicular tissue cell density began to be seen in T2. Provision of tomato juice to the T2 group demonstrated the potential effect to inhibit or restore the adverse effects of THS. Tomatoes are a source of phenolic compounds (phenolic acids and flavonoids), carotenoids (lycopene, carotene-α, and -β), vitamins (ascorbic acid and vitamin C), and glycalkaloid. These compounds have anti-inflammatory, anti-genotoxicity, antimutagenic, antiproliferative, and chemopreventive effects (Chaudhary et al. 2018). Lycopene has been known to act as a potent antioxidant (Yamamoto et al., 2017). Reactive oxygen will be bound by lycopene and increase antioxidant potential so that it can reduce oxidative damage to lipids including lipoproteins and lipid membranes (Yusni et al. 2017). This occurs through two mechanisms: (a) neutralizing ROS thereby reducing lipid oxidation, protein, and DNA; and (b) increasing the antioxidant potential that will trigger a decrease in oxidative stress (Durairajanayagam et al. 2014; Yamamoto et al. 2017). In vitro, lycopene can protect lipids and DNA from oxidative damage (Zini et al. 2010). In addition, oxidative stress in mice spermatozoa caused by aflatoxin B1 can reduce sperm motility (Muzaffer et al. 2010). Lycopene supplementation in vitro is known to reduce the DNA fragmentation index of spermatozoa induced by H_2O_2 (Zini et al. 2010). Thus, lycopene can prevent oxidative stress that decreases sperm motility and viability, and increased apoptosis.

5 CONCLUSION

The results of this research showed that the exposure to cigarette smoke examined in this study impairs sperm morphology and viability so that the cell density of testicles is decreased. Furthermore, the administration of tomato juice has the potential to improve sperm quality and to increase testicular tissue cell density in mice exposed to thirdhand cigarette smoke. The number of sperm and sperm motility tended to increase in mice receiving tomato juice. Significant improvement in sperm morphology and testicular tissue cell density was also observed in those mice.

ACKNOWLEDGMENT

This study has been funded by LPPM Unisba. The experiment was held at the biomedical laboratory of the Faculty of Medicine Unisba.

REFERENCES

Albuquerque, E. X., Pereira, E. F., Alkondon, M., & Rogers, S. W. 2009. Mammalian nicotinic acetylcholine receptors: From structure to function. *Physiological reviews* 89(1): 73–120.

Beal, M. A., Yauk, C. L., & Marchetti, F. 2017. From sperm to offspring: Assessing the heritable genetic consequences of paternal smoking and potential public health impacts. *Mutation research/Reviews in mutation research* 773: 26–50.

Bhowmik, D., Kumar, K. S., Paswan, S., & Srivastava, S. 2012. Tomato: A natural medicine and its health benefits. *Journal of Pharmacognosy and phytochemistry* 1(1): 33–43.

Chaudhary, P., Sharma, A., Singh, B., & Nagpal, A. K. 2018. Bioactivities of phytochemicals present in tomato. *Journal of food science and technology* 55(8): 2833–2849.

Chauhan, K., Sharma, S., Agarwal, N., & Chauhan, B. 2011. Lycopene of tomato fame: Its role in health and disease. *International journal of pharmaceutical sciences review and Research* 10(1): 99–115.

Clément, C., Witschi, U., & Kreuzer, M. 2012. The potential influence of plant-based feed supplements on sperm quantity and quality in livestock: A review. *Animal reproduction science* 132(1–2): 1–10.

Dai, J. B., Wang, Z. X., & Qiao, Z. D. 2015. The hazardous effects of tobacco smoking on male fertility. *Asian journal of andrology* 17(6): 954–960.

Day, J., Savani, S., Krempley, B. D., Nguyen, M., & Kitlinska, J. B. 2016. Influence of paternal preconception exposures on their offspring: through epigenetics to phenotype. *American journal of stem cells* 5(1): 11.

Durairajanayagam, D., Agarwal, A., Ong, C., & Prashast, P. 2014. Lycopene and male infertility. *Asian journal of andrology* 16(3): 420.

Esakky, P., Hansen, D. A., Drury, A. M., Felder, P., Cusumano, A., & Moley, K. H. 2016. Paternal exposure to cigarette smoke condensate leads to reproductive sequelae and developmental abnormalities in the offspring of mice. *Reproductive toxicology* 65: 283–294.

Hammadeh, M.E., Amor, H., & Montenarh, M. 2016. Cigarette Smoking and Structural, Biochemical, Functional Alterations of Spermatozoa and their Consequences for ART. *Austin J Invitro Fertilization* 3(1): 1–16.

Jacob III, P., Benowitz, N. L., Destaillats, H., Gundel, L., Hang, B., Martins-Green, M., ... & Talbot, P. 2017. Thirdhand smoke: New evidence, challenges, and future directions. *Chemical research in toxicology* 30(1): 270–294.

Kong, K. W., Khoo, H. E., Prasad, K. N., Ismail, A., Tan, C. P., & Rajab, N. F. 2010. Revealing the power of the natural red pigment lycopene. *Molecules* 15(2): 959–987.

Lu, R., Dan, H., Wu, R., Meng, W., Liu, N., Jin, X., ... & Chen, Q. 2011. Lycopene: Features and potential significance in the oral cancer and precancerous lesions. *Journal of oral pathology & medicine* 40(5): 361–368.

Mathur, P. P. & d'Cruz, S. C. 2011. The effect of environmental contaminants on testicular function. *Asian journal of andrology* 13(4): 585.

Matt, G. E., Quintana, P. J., Zakarian, J. M., Hoh, E., Hovell, M. F., Mahabee-Gittens, M., ... & Chatfield, D. A. 2017. When smokers quit: Exposure to nicotine and carcinogens persists from thirdhand smoke pollution. *Tobacco control* 26(5): 548–556.

Mohamed, E. H., Amor, H., & Montenarh, M. 2016. Cigarette smoking and structural, biochemical, functional alterations of spermatozoa and their consequences for ART. *Austin journal of in vitro fertilization* 3(1): 1027.

Muzaffer, T., Berna, G.S., Doğan, K., Beran. Y., Muzaffer, D. 2010. Protective Role of Lycopene on Aflatoxin B1 Induced Changes Sperm Characteristics and Testicular Damages in Rats. *Kafkas Univ Vet Fak Derg* 16(4): 597–604.

Palozza, P., Simone, R. E., Catalano, A., & Mele, M. C. 2011. Tomato lycopene and lung cancer prevention: From experimental to human studies. *Cancers* 3(2): 2333–2357.

Shrivastava, V., Marmor, H., Chernyak, S., Goldstein, M., Feliciano, M., & Vigodner, M. 2014. Cigarette smoke affects posttranslational modifications and inhibits capacitation-induced changes in human sperm proteins. *Reproductive toxicology* 43: 125–129.

Taş, M., Saruhan, B. G., Kurt, D., Yokuş, B., & Denlİ, M. 2010. Protective role of lycopene on aflatoxin B1 induced changes sperm characteristics and testicular damages in rats. *Kafkas Üniversitesi Veteriner Fakültesi Dergisi* 16(4): 597–604.

Vallaster, M. P., Kukreja, S., Bing, X. Y., Ngolab, J., Zhao-Shea, R., Gardner, P. D., . . . & Rando, O. J. 2017. Paternal nicotine exposure alters hepatic xenobiotic metabolism in offspring. *Elife* 6: e24771.

Wong, M. K., Barra, N. G., Alfaidy, N., Hardy, D. B., & Holloway, A. C. 2015. Adverse effects of perinatal nicotine exposure on reproductive outcomes. *Reproduction* 150(6): R185–R193.

Xu, W., Fang, P., Zhu, Z., Dai, J., Nie, D., Chen, Z., . . . & Qiao, Z. 2013. Cigarette smoking exposure alters pebp1 DNA methylation and protein profile involved in MAPK signaling pathway in mice testis. *Biology of reproduction* 89(6): 1–11.

Yamamoto, Y., Aizawa, K., Mieno, M., Karamatsu, M., Hirano, Y., Furui, K., . . . & Suganuma, H. 2017. The effects of tomato juice on male infertility. *Asia Pacific journal of clinical nutrition* 26(1): 65.

Yusni, Y., Akbar, I. B., Rezania, R., & Fahlevi, R. 2017. Penurunan kadar gula darah akibat pemberian ekstrak manggis (*Garcinia mangostana*) dan tomat (*Lycopersicum esculentum* Mill) pada tikus diabetes. *Global medical & health communication* 5(1): 57–63.

Zini, A., San Gabriel, M., & Libman, J. 2010. Lycopene supplementation in vitro can protect human sperm deoxyribonucleic acid from oxidative damage. *Fertility and sterility* 94(3): 1033–1036.

Acute toxicity test for the ethanolic extract of the white oyster mushroom

S.B. Rahimah, Y. Kharisma, M.K. Dewi, J. Hartati & W. Maharani
Universitas Islam Bandung, Bandung, Jawa Barat, Indonesia

ABSTRACT: The white oyster mushroom (*Pleurotus ostreatus* Jacq: Fr Kumm) is a mushroom that contains high nutrition and secondary metabolites that have antioxidant and anti-inflammatory effects. The mushroom is widely used for the prevention of various diseases such as hypertension, hypercholesterolemia, and carcinoma. This study aimed to assess the acute toxicity of the ethanolic extract of the white oyster mushroom. The method used in this study was a new method recommended for the acute toxicity test of natural substances. Doses tested in this study ranged from 10 mg/kg body weight to a maximum dose of 5,000 mg/kg body weight. Observations of toxicity tests were conducted within the first 24 hours at each stage and the parameters observed were changes in behavior and the mortality of experimental animals. The results of the acute toxicity test showed that no experimental animal died during the range of dosing, since the smallest dose reached the maximum dose; researchers observed no other changes in animal behavior during ethanol extract administration. White oyster mushroom ethanol extract is relatively safe and does not show any acute toxicity, because it has a wide dose range.

1 INTRODUCTION

The oyster mushroom (*Pleurotus* sp.) is a commercial source of food that belongs to the Basidiomycetes class. It is white in color and usually has a 3–15 cm round cup. This mushroom is easily cultivated and contains high nutritional value with a large protein content containing nine essential amino acids. Its unsaturated fat and fiber are very good for digestion. The oyster mushroom also contains vitamins B and C, provitamin D2, and the minerals potassium, phosphorus, sodium, calcium, and magnesium (Iwalokun et al. 2007; Jayakumar et al. 2009; Jayakumar et al. 2011; Rahimah et al. 2019). Consumption of 10 mg of white oyster mushroom on a regular basis can reduce serum cholesterol and triglycerol levels, and its protein content can increase milk secretion in mice (Jayakumar et al. 2011). This oyster mushroom has antimicrobial potential against some bacteria, for example *Escherichia coli* (Jayakumar et al. 2009). The phenolic content, vitamins C and E, beta-carotene, and selenium in this fungus cause scavenging effects on primary and secondary radicals in certain concentrations. Ethanolic extract of *Pleurotus ostreatus* can prevent the negative effects of cigarette smoke that induce rat lungworm and it also has proven antioxidant effects that prevent hepatotoxicity because of CCL_4 (Jayakumar et al. 2009; Rahimah et al. 2019). Extract of *Pleurotus ostreatus* has a protective effect on the liver, kidneys, brain, and lungs (Jayakumar et al. 2009; Jayakumar et al. 2011; Kähkönen et al. 1999; Rahimah et al. 2019). The development of medicinal plants into herbal medicines must be supplemented with data from acute, subacute, and sub-chronic to chronic toxicity tests. This is to ensure the safety of this herb for use (Chinedu et al. 2013; Ifeoma & Oluwakanyinsola 2013).

2 METHOD

An acute toxicity test was carried out using the proposed method. This method was a new method recommended for testing the acute toxicity of drugs or natural substances. Acute toxicity is an undesirable effect due to the administration of certain substances, in single or repeated doses, in a short time in the first 24 hours. A dosage curve for undesirable effects can also be seen using this method, in addition to assessing mortality. The method used was the Proposed (New) Method (Chinedu et al. 2013). Experimental animals were divided into several stages; the next stage depended on the results of the previous stage. The first or initial stage used four groups containing one mouse for each group. All four were given different doses and were seen 1 hour after administration and in periodic examinations for 24 hours to assess their mortality. If no deaths occurred, the testing could be continued in the second stage. The second stage involved three animals given a higher dose than in the first stage. Observation was carried out the same as in the previous stage. If no mortality took place, then the testing continued to the third stage using three mice. The maximum dose given was 5,000 mg/kg body weight (Chinedu et al. 2013).

3 RESULTS

Table 1 shows that starting from the smallest dose at 10 mg/kg body weight to the largest dose at 5,000 mg/kg body weight, researcher observed no abnormalities of behavior or deaths in the mice.

Table 1. Dose on body weight.

Groups	Animal label	Doses (mg/KgBB)	Behavioral changes	Death of the animal
I	1	10	-	-
	2	100	-	-
	3	300	-	-
	4	600	-	-
II	1	50	-	-
	2	200	-	-
	3	400	-	-
	4	600	-	-
III	1	1,000	-	-
	2	1,500	-	-
	3	2,000	-	-
IV	1	3,000	-	-
	2	4,000	-	-
	3	5,000	-	-

4 DISCUSSION

The use of traditional medicines has been carried out in Indonesia for thousands of years, so it is only natural that the development of Indonesian herbal medicines must be increased. Indonesian traditional medicine is divided into three categories – namely herbal medicine, standardized herbs, and phytopharmaca. The development of traditional Indonesian medicine from herbal medicine to standardized herbs must be accompanied by data on toxicity in experimental animals. Acute toxicity is the initial toxic test to see the lethal dose 50 (LD50) in the population (Akhila et al. 2007; Dewoto 2007). Acute toxicity studies have indicated that water suspensions of selected herbal aqueous extracts (five medicinal plants: *Euphorbia hirta*, *Solanum torvum*, *Zingiber officinale*, *Curcuma longa*, and *Zingiber zerumbet*) are not toxic when administered orally to experimental birds at 2,000 mg kg[-1]. The ethanol extract of white oyster mushrooms (*Pleurotus ostreatus*) in the study revealed that up to a dose of 5,000 mg/kgBB did not show changes in behavior or death in experimental animals. This result supports that *P. ostreatus* compared to other herbal ingredients is a safe herbal ingredient, so these data are very supportive of the development of white oyster mushrooms into standardized herbs (Hashemi et al. 2008).

5 CONCLUSION

These results indicate that the ethanol extract of the white oyster mushroom is relatively safe and does not show any acute toxicity, because it has a wide dose range. The white oyster mushroom can be developed into an herbal medicine because of its efficacy and safety in animal trials.

6 CONFLICT OF INTEREST

The author has no conflict of interest with the publication of this paper.

REFERENCES

Akhila, J. S., Shyamjith, D., & Alwar, M. C. 2007. Acute toxicity studies and determination of median lethal dose. *Current science* 93(7): 917–920.

Bhekti Rahimah, S., Djunaedi, D. D., Soeroto, A. Y., & Bisri, T. 2019. The phytochemical screening, total phenolic contents and antioxidant activities in vitro of white oyster mushroom (*Pleurotus ostreatus*) Preparations. *Open access Macedonian journal of medical sciences* 7(15): 2404–2412.

Chinedu, E., Arome, D., & Ameh, F. S. 2013. A new method for determining acute toxicity in animal models. *Toxicology international* 20(3): 224–226.

Dewoto, H. R. 2007. Pengembangan obat tradisional Indonesia menjadi fitofarmaka. *Majalah Kedokteran Indonesia* 57(7): 205–211.

Hashemi, S. R., Zulkifli, I., Hair Bejo, M., Farida, A., & Somchit, M. N. 2008. Acute toxicity study and phytochemical screening of selected herbal aqueous extract in broiler chickens. *International journal of pharmacology* 4(5): 352–360.

Ifeoma, O. & Oluwakanyinsola, S. 2013. *New insights into toxicity and drug testing* Rijeka: InTech, 63–88.

Iwalokun, B. A., Usen, U. A., Otunba, A. A., & Olukoya, D. K. 2007. Comparative phytochemical evaluation, antimicrobial and antioxidant properties of *Pleurotus ostreatus*. *African journal of biotechnology* 6(15): 1732–1739.

Jayakumar, T., Thomas, P. A., & Geraldine, P. 2009. In-vitro antioxidant activities of an ethanolic extract of the oyster mushroom, *Pleurotus ostreatus. Innovative food science & emerging technologies* 10(2): 228–234.

Jayakumar, T., Thomas, P. A., Sheu, J. R., & Geraldine, P. 2011. In-vitro and in-vivo antioxidant effects of the oyster mushroom *Pleurotus ostreatus. Food research international* 44(4): 851–861.

Kähkönen, M. P., Hopia, A. I., Vuorela, H. J., Rauha, J. P., Pihlaja, K., Kujala, T. S., & Heinonen, M. 1999. Antioxidant activity of plant extracts containing phenolic compounds. *Journal of agricultural and food chemistry* 47(10): 3954–3962.

Validity and reliability of anatomy examination of organ structure and topography

Y. Yuniarti, R. Perdana, A.R. Putera & F.A. Yulianto
Universitas Islam Bandung, Bandung, Jawa Barat, Indonesia

ABSTRACT: Knowledge of the structure of the human body from the macro level to the molecular level is fundamental to understanding body functions and how the structure and function of body organs change due to disease. An appropriate instrument is needed to measure student knowledge of anatomy, especially of the structure and topography of organs. The purpose of this study was to find out which questions are valid and reliable to measure medical students' knowledge regarding organ structure and topography. The research method used was a survey approach, where 88 samples were collected randomly. Validity of the questions was analyzed by Pearson's correlation coefficient, while Kappa's coefficient agreement was chosen to analyze their reliability. The result showed that analysis of nine organ systems (dermatomuscular, endocrine, neurological, genitourinary, reproductive, special senses, cardiovascular, genitourinary, and respiratory) consisted of structure and topography questions. Most of the questions met validity and reliability criteria according to statistical analysis. In conclusion, the instrument was valid and reliable to determine the level of knowledge pertaining to organs' structure and topography. Any additional questions should also be analyzed for validity and reliability.

1 INTRODUCTION

Professional education and medical competency standards reinforced and developed in accordance with advances in medical science and technology and as an effort to answer the community's needs for medical quality assurance as the earliest part of achieving patient safety in the implementation of medical practice (Cahyaningrum et al. 2016).

Anatomy is a basic science that is very important for medical students (Nugraha et al. 2019). Anatomy is the science of body structures and the relationships among them (Gerard et al. 2009). Doctors without an education in anatomy are like moles working in the dark (Hegazy & Minhas 2015). Knowledge of the structure of the human body at a macro level and down to the molecular level is the basis for understanding bodily functions and their changes due to disease. Anatomy studies the normal structure of the body starting from the shape, size, and location of body parts and organs and their support of and relationship with surrounding structures. Anatomy is also the basis for knowing physiology and pathology or changes in the structure of the bodily relationships due to disease (Nugraha et al. 2019).

Anatomical knowledge supports the examination of a patient, the formation of a diagnosis, and the communication of these findings to the patient and other medical professionals. It provides a platform of knowledge suitable to all medical careers (Hegazy & Minhas 2015).

Gross anatomy is one of the fundamental topics in medical education. By learning gross anatomy, medical students get a first impression about the structure of the normal human body, which is the basis for understanding pathologic and clinical problems (Hegazy & Minhas 2015).

Based on this explanation, we need a measuring tool to assess the knowledge of medical students about anatomy, especially about the structure and topography of organs. In epidemiological research, the most commonly used measurement tool is a questionnaire. The questionnaire is a list of written questions addressed to respondents. A questionnaire must have several conditions including relevancy to the purpose of the study, easy-to-ask and easy-to-answer questions, and data that are easily processed. The questionnaire can be used as a measurement tool if it has been tested for validity and reliability (Bolarinwa 2015; Noviyani et al. 2016; Widi 2011). Presenting the value of reliability and validity of a questionnaire is important so that other researchers are confident with the quality of the data they gain later (Ghazali 2016).

Psychometricians have established criteria for declaring measuring devices, like an instrument, good measurement tools. These criteria include validity, reliability, standardization, economy, and practicality (Arifin 2017).

Fourth-year students at the medical faculty of the Islamic University of Bandung (UNISBA), Indonesia, are expected to have a good level of anatomical knowledge because after graduating with a medical degree, they will continue their education in the medical profession and then work face to face with patients. During this education, an understanding of anatomy is needed to help them understand the process of a disease and its handling principles (Munawaroh et al. 2018).

The purpose of this study was to determine the validity and reliability of the questionnaire as the main measuring instrument in research regarding the knowledge of medical students about the structure and topography of organs so the questionnaire can be used for further research.

2 METHOD

This research method utilized a survey approach, where 23 samples were collected randomly from fourth grade students of the faculty of medicine, UNISBA, in November–December 2018. The questions contained in the questionnaire were closed single-choice questions. The questionnaire contained questions about the structure and topography of organs in nine organ systems – namely the dermatomuscular, endocrine, neurological, genitourinary, reproductive, special senses, cardiovascular, genitourinary, and respiratory systems. Validation and reliability tests were carried out twice with a range of data collection for 2 weeks. Validity of the questions was analyzed by Pearson's correlation coefficient, while Kappa's coefficient agreement was chosen to analyze their reliability.

3 RESULTS AND DISCUSSION

Tables 1 and 2 show the test results indicating valid, invalid, reliable, and non-reliable questions.

Validity basically means "measure[ing] what is intended to be measured" (Taherdoost 2016). Validity is defined as the extent to which a concept is accurately measured in a quantitative study (Heale & Twycross 2015). The validity of an instrument specifies how well the instrument met the criteria of quality (Oktavia et al. 2018). Validity measures the accuracy and preciseness of the questionnaire results. Validity goes beyond where the measuring instrument really assesses what must be accepted (Hendryadi 2017; Puspitawati & Herawati 2018).

Reliability concerns the extent to which a measurement of a phenomenon provides stable and consist results. Testing for reliability is important as it refers to the consistency across the parts of a measuring instrument (Taherdoost 2016). It also measures the precision, repeatability, and trustworthiness of research (Chakrabartty 2013).

Reliability is the extent to which test scores are not affected by chance factors – the luck of the draw (Livingston et al. 2018). A measurement result can be trusted if in several measurements in the same group of subjects, the results obtained are relatively the same, as long as the aspects measured in the subject have not change (Matondang 2009). Reliability tests determine whether the instrument in terms of this questionnaire can be used more than once, at

Table 1.　Validity and reliability test of structure organ questionnaire.

Group question	Number of questions	Validity and reliability test results				Used questions
		Valid	Not valid	Reliable	Not reliable	
Dermatomuscular system (DMS)	15	6	9	9	6	9
Endocrine metabolic system (EMS)	14	7	7	11	3	11
Neurobehavioral system (NBS)	14	5	9	13	1	13
Genitourinary system (GUS)	13	5	8	10	3	10
Reproductive system (RPS)	11	10	1	10	1	10
Special sense (SS)	5	0	5	5	0	5
Cardiovascular system (CVS)	9	1	8	6	3	6
Gastrointestinal system (GIS)	10	3	7	8	2	8
Respiration system (RS)	10	5	5	6	4	6

Table 2.　Validity and reliability test of topography organ questionnaire.

Group question	Number of questions	Validity and reliability test results				Used questions
		Valid	Not valid	Reliable	Not reliable	
Dermatomuscular system (DMS)	11	5	6	10	1	10
Endocrine metabolic system (EMS)	9	3	6	6	3	6
Neurobehavioral system (NBS)	12	5	7	9	3	9
Genitourinary system (GUS)	9	6	3	7	2	7
Reproductive system (RPS)	9	5	4	9	0	9
Special sense (SS)	5	5	0	4	1	4
Cardiovascular system (CVS)	11	5	6	8	3	8
Gastrointestinal system (GIS)	10	5	5	7	3	7
Respiration system (RS)	10	5	5	7	3	7

least by the same respondents, and consistent data will be obtained. In other words, the reliability of the instrument characterizes the level of consistency (Janti 2004).

A question becomes invalid due to material aspects, construction aspects, and language/cultural aspects (Retnawati 2016). Material aspects that need to be considered in making questions include the contents of the questions in accordance with the indicators of expected learning achievement, the distractor functions (for multiple choice items), and the availability of answers for each question.

Nine aspects of construction need to be considered in drafting questions – that is, the questions are written briefly, clearly, and firmly; the formulation of the main questions and the answer choices are appropriate for the questions; the questions do not give clues to the answers; the questions are free from multiple negatives and pictures, graphs, tables, diagrams, discourses, and similar things contained in the question are displayed clearly; the length of the answer choices is relatively the same; the answers do not include the statement "all of the answers above are wrong" or "all of the answers above are correct"; similar answers in the form of numbers or times are arranged in the order of the size of the numbers or chronologically; and the answer to a question does not depend on the answer to the previous question.

Viewed from the aspect of language/culture, a question should use language in accordance with Indonesian language rules, use communicative language, avoid local language, and not repeat the same words or groups of words.

The results of the study revealed some questions were not reliable; this might be caused by the instrument being too long, unclear instructions given before the test, strict supervision when collecting data that caused respondents to feel afraid or less comfortable or free to respond to the instrument. If supervision is lacking, however, participants will work together, making the results unreliable; the environment for filling out the questionnaire was therefore uncomfortable (Retnawati 2016).

Three important reliability criteria of great interest for researchers include: (1) stability, (2) internal consistency, and (3) equivalence. Stability measures how similar the results are when measured at two different times. Internal consistency shows whether all subparts of an instrument measure the same characteristic. Equivalence is the concordance degree of two or more observers regarding instrument scores (Souza et al. 2017).

In this research questionnaire, the questions taken certainly met the criteria for validity and reliability, but questions with invalid but reliable results were used with the consideration that invalid questions must be tested by improving the structure of the question text based on the nine aspects of construction. A study has observed that a valid tool must be reliable, but a reliable tool may not necessarily be valid (Mohajan 2017; Oluwatayo 2012).

Validity is closely related to reliability. Reliability or consistency of measurements is needed to obtain valid results, but reliability can be obtained without validity. Validity does not guarantee reliability since they are two different things in interrelated conditions (Handani & Ambar 2015; Pujihastuti 2010).

4   CONCLUSION

The instrument was valid and reliable; thus, it can be used to determine the level of knowledge pertaining to organ structure and topography.

ACKNOWLEDGMENT

We would like to thank the Faculty of Medicine, UNISBA, for facilitating this study.

REFERENCES

Arifin, Z. 2017. Kriteria Instrumen dalam suatu penelitian. *Jurnal THEOREMS (The original research of mathematics)* 1(2): 92–100.

Bolarinwa, O. A. 2015. Principles and methods of validity and reliability testing of questionnaires used in social and health science researches. *Nigerian postgraduate medical journal* 22(4): 195.

Cahyaningrum, Y. D., Mulyaningrum, U., & Pravitasari, P. 2016. Validasi Kuesioner Evaluasi Progress Test pada Mahasiswa Tahap Sarjana Kedokteran Universitas Islam Indonesia. *Jurnal Kedokteran Universitas Lampung* 1(2): 319–323.

Chakrabartty, S. N. 2013. Best split-half and maximum reliability. *IOSR journal of research & method in education* 3(1): 1–8.

Gerard J., Tortora, T., & Derrickson, B. 2009. *Principles of anatomy and physiology: Organization, support and movement, and control systems of the human body. Vol. 2: Maintenance and continuity of the human body.* New York: John Wiley & Son.

Ghazali, N. H. M. 2016. Reliability and validity of an instrument to evaluate the school-based assessment system: A pilot study. *International journal of evaluation and research in education* 5(2): 148–157.

Handani, T. & Ambar, H. 2015. Validitas Dan Reliabilitas Soal Tengah Semester Genap Kaitannya Dengan Ketercapaian Tujuan Pembelajaran Bahasa Indonesia Kelas VIIIA SMP Negeri 2 Banyudono Tahun Pelajaran 2013/2014 (Doctoral dissertation, Universitas Muhammadiyah Surakarta).

Heale, R. & Twycross, A. 2015. Validity and reliability in quantitative studies. *Evidence-based nursing* 18(3): 66–67.

Hegazy, A. M. & Minhas, L. 2015. Reflection of the type of medical curriculum on its anatomy content: Trial to improve the anatomy learning outcomes. *International journal of clinical and developmental anatomy* 1(3): 52.

Hendryadi, H. 2017. Validitas isi: tahap awal pengembangan kuesioner. *Jurnal Riset Manajemen dan Bisnis (JRMB) Fakultas Ekonomi UNIAT* 2(2): 169–178.

Janti, S. 2014. Analisis validitas dan reliabilitas dengan skala likert terhadap pengembangan si/ti dalam penentuan pengambilan keputusan penerapan strategic planning pada industri garmen. *Prosiding Seminar Nasional Aplikasi Sains & Teknologi (SNAST)* 15: 155–160.

Livingston, S. A., Carlson, J., Bridgeman, B., Golub-Smith, M., & Stone, E. 2018. *Test reliability: Basic concepts. Research Memorandum No. RM-18-01.* Princeton, NJ: Educational Testing Service.

Matondang, Z. 2009. Validitas dan reliabilitas suatu instrumen penelitian. *Jurnal Tabularasa* 6(1): 87–97.

Mohajan, H. K. 2017. Two criteria for good measurements in research: Validity and reliability. *Annals of Spiru Haret University. Economic Series* 17(4): 59–82.

Munawaroh, S., Kartikasari, M. N. D., & Hermasari, B. K. 2018. Konsensus Pakar Anatomi Indonesia mengenai Materi Inti Anatomi Sistem Pencernaan. *Jurnal Biomedik: JBM* 10(1).

Noviyani, R., Ketut, T., Ayu, I., & Nyoman, G. B. 2016. Uji validitas dan reliabilitas kuesioner EORTC QLQ C-30 untuk menilai kualitas hidup pasien kanker ginekologi di RSUP Sanglah Denpasar. *Jurnal Farmasi Klinik Indonesia* 5(2): 106–114.

Nugraha, Z. S., Khadafianto, F., & Fidianingsih, I. 2019. Refleksi pembelajaran anatomi pada mahasiswa kedokteran fase ketiga melalui applied and clinical question. *Refleksi Pembelajaran Inovatif* 1(1).

Oktavia, R., Mentari, M. & Mulia, I. S. 2018. Assessing the validity and reliability of questionnaires on the implementation of Indonesian curriculum K-13 in STEM education. *Journal of physics: Conference series* 1088(1): 012014.

Oluwatayo, J. A. 2012. Validity and reliability issues in educational research. *Journal of educational and social research* 2(2): 391–400.

Pujihastuti, I. 2010. Prinsip Penulisan KuesionerPenelitian. *CEFARS: Jurnal Agribisnis dan Pengembangan Wilayah* 2(1): 43–56.

Puspitawati, H. & Herawati, T. 2018. Reliabilitas dan Validitas Indikator Ketahanan Keluarga di Indonesia. *Jurnal Kependudukan Indonesia* 13(1): 1–14.

Retnawati, H. 2016. *Analisis kuantitatif instrumen penelitian.* Yogyakarta: Parama Publishing.

Souza, A. C. D., Alexandre, N. M. C., & Guirardello, E. D. B. (2017). Psychometric properties in instruments evaluation of reliability and validity. *Epidemiologia e Serviços de Saúde* 26: 649–659.

Taherdoost, H. 2016. Validity and reliability of the research instrument: How to test the validation of a questionnaire/survey in a research *International Journal of Academic Research in Management (IJARM)* 5(3): 28–36.

Widi, R. 2011. Uji validitas dan reliabilitas dalam penelitian epidemiologi kedokteran gigi. *Stomatognatic (JKG Unej)* 8(1): 27–34.

The role of FTO gene polymorphism in weight loss: An evidence-based case report

M. Nathania & L.I. Octovia
Universitas Indonesia, Depok, Indonesia
Ciptomangunkusumo National Hospital, Jakarta Pusat, Indonesia

ABSTRACT: Obesity and its comorbidities are major health problems throughout the world that continue to increase. The "fat mass and obesity–associated" (FTO) gene is the strongest genetic predictor of body weight, which alters the regulation of *IRX3* and *IRX5* homeobox gene expression and affects mitochondrial fat metabolism. It is hoped that nutrigenomic findings can answer the urgent need to develop more effective anti-obesity strategies, especially in solving various individual responses to treatment. We conducted a literature survey on PubMed, Science Direct, and Cochrane according to the clinical question to determine the effect of lifestyle modification on weight loss in patients with FTO gene polymorphisms. Screening for title, abstract, and full text was based on eligibility criteria (systematic review or meta-analysis of clinical trials, suitability with the clinical question, English language, full-text availability, and human studies), followed by a duplication filter and critical appraisal. One meta-analytical article was obtained showing the FTO AA genotype producing more significant weight loss compared to TT genotypes with a weighted mean difference of −0.44 kg (95% CI: 0.09 to 0.79; P = 0.015). This shows individuals carrying the FTO homozygous AA genotype predisposition can lose more weight through lifestyle modification than noncarriers, but these differences might be too small to be clinically important. Further consideration is needed to determine the benefit of FTO gene testing over the cost. In carriers of the TT genotype, more aggressive anti-obesity treatment is needed.

1 INTRODUCTION

Obesity and its comorbidities are major health problems that have increased substantially over the past 40 years. This epidemic disease has almost quadrupled among men and almost tripled among women, and it also has had a significant impact on mortality, morbidity, and health care costs (Jaacks et al. 2019).

Long noted in weight loss trials, lifestyle modification as the primary modality for obesity management shows individual variability in responding to the same interventions (Fenwick et al. 2019). With the development of the Human Genome Project, genes have been found to play an important role in obesity pathogenesis. In particular, more than 100 loci associated with body mass index have been identified (Akiyama et al. 2017), with the "fat mass and obesity–associated" (FTO) gene being the strongest genetic predictor (Ehrlich & Friedenberg 2016). Nutrigenomics, introduced in 2001, is now often performed with the aim of applying individualized interventions based on genetic background, although there is no definite evidence that this test contributes to the strategy of providing therapy (Pavlidis et al. 2015).

Therefore, the role of genetic testing in obesity treatment needs further exploration. It is hoped that this research can contribute to the development of more effective obesity management and provide answers to whether genetic testing influences the success of obesity management.

2 CLINICAL SCENARIO

Mr. N, 55 years old, who works as an accountant, presented with a complaint of body weight that could not be reduced despite his diet. He used to eat fatty foods and snacks while working in the office, but he had replaced his diet with fruits and vegetables for the previous 2 months. The patient had no exercise habit. His height and weight were 155 centimeters and 65 kilograms, respectively, with a waist circumference of 95 centimeters. His calculated body mass index was 27.1 kg/m^2.

The patient stated that his friend advised him to carry out a genetic examination of the FTO gene, and asked if the test was necessary. Based on this clinical scenario, a question arose whether the FTO gene plays a role in the effectiveness of lifestyle modification in obese patients.

3 METHODS

A literature search was performed on PubMed Clinical Query, Science Direct, and Cochrane according to these clinical questions on October 13, 2019. The keywords used were "genetic," "gene," "FTO," "nutrigenomic," "nutrigenetic," "fat loss," and "weight loss." Screening for title, abstract, and full text was based on the duplication filters and eligibility criteria (systematic review or meta-analysis of clinical trials, suitability with the clinical question, English language, full-text availability, and human studies) shown in Figure 1. Critical appraisal was done by consensus of the authors using several criteria from the Centre of Evidence-Based Medicine, University of Oxford, for systematic review and meta-analysis.

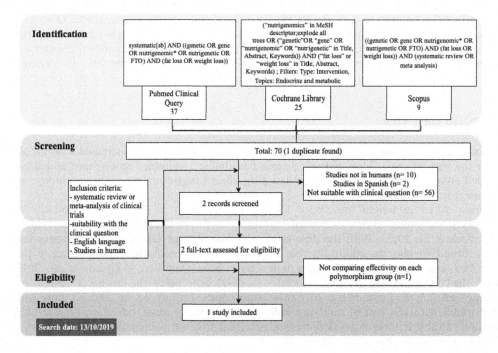

Figure 1.   Flowchart of search strategy.

4 RESULTS

Following the search strategy, one original article was eligible for this evidence-based case report. The design and summary of the selected article are available in Table 1.

4.1 *Critical appraisal*

Studies conducted by Xiang and colleagues (2016) presented clear clinical questions and were used in the search as indicated in the methods section. Database searches were performed on PubMed and Embase, but not on other major databases such as Cochrane or on unpublished studies, and also through manual searching or via expert contacts. Weaknesses of the study included language limitations, and the quality assessment methods of each study were not mentioned even though there appears to be a further selection of articles after the eligibility criteria. The results of the study have been correctly described in the characteristics table between studies as well as on a complete forest plot. A heterogeneity test was conducted with the use of two different methods, Cochrane's Q test and the I2 statistics, in accordance with the *Cochrane Handbook for Systematic Reviews of Interventions*. No significant heterogeneity was detected in the analysis.

From the aspects of importance and applicability, these studies had various weighted mean difference (WMD) with the results showing significant weight loss difference in the group AA genotype compared to the TT genotype of 0.44 kg (95% CI: 0.09, 0.79 kg, p = 0.015), consistent with sensitivity and several subgroup analyses. Analysis comparing the TA genotype to the TT genotype resulted in an insignificant result of 0.18 kg (95% CI: −0.45, 0.09; p = 0.19). The confidence interval can be considered narrow, showing a precise result. The result of this study can be applied to daily clinical practice because the characteristics of the sample, diagnostic criteria, and guidelines for lifestyle modification in obese patients were similar.

Table 1. Literature characteristics.

Article	Design	Population	Exposure	Results
Xiang et al. 2016	Systematic review and meta-analysis of randomized controlled trials	Obese adults aged ≥ 18 years old who had undergone lifestyle modification (without drug intervention)	Group AA vs. Group TT	FTO AA genotype group had greater weight loss compared to the TT genotype (−0.44 kg; 95% CI: 0.09, 0.79 kg, p = 0.015). Adjustment for baseline body mass index (BMI) or body weight (−0.70 kg; 95% CI: 1.16, 0.23 kg, p = 0.003) Stratified analysis: Participants who were > 50 years old (−0,44 kg; 95% CI: 1.12, 0.28 kg; p = 0.033) BMI < 35 (−0.70 kg; 95% CI: −1.16, −0.23 kg; p = 0.003) Studies with men and women combined (mixed) (−0.46 kg; 95% CI −0.83, −0.10 kg; p = 0,013)
			Group TA vs. Group TT	In individuals with the TA genotype, there was also a decrease in body weight, but not as significant as in the AA genotype, which was −0.18 kg (95% CI: −0.45, 0.09; P = 0.19).

5   DISCUSSION

The results of the study suggest that the FTO gene polymorphism may play a role in the effectiveness of lifestyle modification for weight loss in obese patients, although the effect size may be too small to be clinically important. However, the true effect might be underestimated because of the different methods of intervention. In the subgroup analysis, it was found that the effect size became larger when the intervention method was only a diet, but the results were insignificant. This is possibly because of the insufficient number of studies and participants. Apart from that, stratified analysis of other intervention methods could not be carried out due to small sample sizes.

Weaknesses of the study were due to language restriction and literature source limitation, as well as an assessment of study quality that was not explained. Nevertheless, nutrigenomics is indeed a new technique, and researchers have not carried out many studies. The appraised study by Xiang and colleagues (2016) was the first meta-analysis study to assess weight loss in response to diet/lifestyle interventions, where previous studies generally only linked FTO polymorphism to the incidence of obesity (Yeo 2014).

Weight loss is a complex process that can be achieved only if there is an energy deficit, either by modulation of energy intake or energy expenditure. FTO gene expression is known to play a role in the arcuate nucleus (ARC) of the hypothalamus, which is primary to the regulation of hunger and satiety (Yang et al. 2017). Individuals with FTO polymorphism showed an increase in high-calorie, high-fat diets and sugar intake (Crovesy & Rosado 2019). Recent evidence suggests that lipid metabolism is dependent on N6-methyladenine (m6A) methylation, which is under-methylated in FTO gene polymorphism. This dysregulation in the nuclear level (mRNA) causes abnormal gene expression, decreases ghrelin mRNA (the only orexigenic hormone) and peptide levels (Wu et al. 2020). FTO expression also regulates *IRX3* and *IRX5* homeobox gene expression, affecting the mitochondrial metabolism of fat (de Araujo et al. 2019).

For this patient, a more thorough medical examination is needed, which includes a lipid profile and blood sugar level, uric acid, liver and kidney function, and body composition examination using bioelectrical impedance analysis to determine the best diet plan. Calorie reduction in pediatric obesity ranged from 200 to 500 calories per day, with a weight loss target of 0.5 kilograms per week (Park et al. 2012). Efforts should be made to increase energy expenditure. Physical activity with recommendations of a minimum of 20 minutes of moderate to vigorous activity daily and reducing sedentary lifestyle by limiting nonacademic screen time to a maximum of 2 hours a day are recommended (Styne et al. 2017). Pharmacotherapy can be applied if the patient does not respond to conventional therapies (Park et al. 2012).

This evidence-based case report shows nutrigenomic testing in already obese patients requires further consideration. Future public health strategies for obesity management should focus on developing lifestyle modification programs to achieve sustainable weight loss regardless of the FTO genotype.

6   CONCLUSIONS

Individuals carrying the FTO gene with the AA genotype are capable of losing more weight through lifestyle modification than noncarriers, although these differences might be too small to be clinically important and higher-quality studies are needed. Further considerations are needed to determine the benefit of FTO gene testing as an obesity management strategy.

REFERENCES

Akiyama, M., Okada, Y., Kanai, M., Takahashi, A., Momozawa, Y., Ikeda, M., ... & Iwasaki, M. 2017. Genome-wide association study identifies 112 new loci for body mass index in the Japanese population. *Nature genetics* 49(10): 1458–1467.

Crovesy, L. & Rosado, E. L. 2019. Interaction between genes involved in energy intake regulation and diet in obesity. *Nutrition*.

de Araujo, T. M., Razolli, D. S., Correa-da-Silva, F., de Lima-Junior, J. C., Gaspar, R. S., Sidarta-Oliveira, D., . . . & Velloso, L. A. (2019). The partial inhibition of hypothalamic IRX3 exacerbates obesity. *EBioMedicine* 39: 448–460.

Ehrlich, A. C. & Friedenberg, F. K. 2016. Genetic associations of obesity: The fat-mass and obesity-associated (FTO) gene. *Clinical and translational gastroenterology* 7(1): e140.

Fenwick, P. H., Jeejeebhoy, K., Dhaliwal, R., Royall, D., Brauer, P., Tremblay, A., . . . & Mutch, D. M. 2019. Lifestyle genomics and the metabolic syndrome: A review of genetic variants that influence response to diet and exercise interventions. *Critical reviews in food science and nutrition* 59(13): 2028–2039.

Jaacks, L. M., Vandevijvere, S., Pan, A., McGowan, C. J., Wallace, C., Imamura, F., . . . & Ezzati, M. 2019. The obesity transition: Stages of the global epidemic. *Lancet diabetes & endocrinology*.

Park, M. H., Falconer, C., Viner, R. A., & Kinra, S. 2012. The impact of childhood obesity on morbidity and mortality in adulthood: A systematic review. *Obesity reviews* 13(11): 985–1000.

Pavlidis, C., Patrinos, G. P., & Katsila, T. 2015. Nutrigenomics: A controversy. *Applied & translational genomics* 4: 50–53.

Styne, D. M., Arslanian, S. A., Connor, E. L., Farooqi, I. S., Murad, M. H., Silverstein, J. H., & Yanovski, J. A. 2017. Pediatric obesity: Assessment, treatment, and prevention: An Endocrine Society clinical practice guideline. *Journal of clinical endocrinology & metabolism* 102(3): 709–757.

Wu, J., Frazier, K., Zhang, J., Gan, Z., Wang, T., & Zhong, X. (2020). Emerging role of m6A RNA methylation in nutritional physiology and metabolism. *Obesity reviews* 21(1): e12942.

Xiang, L., Wu, H., Pan, A., Patel, B., Xiang, G., Qi, L., . . . & Qi, Q. (2016). FTO genotype and weight loss in diet and lifestyle interventions: A systematic review and meta-analysis. *American journal of clinical nutrition* 103(4): 1162–1170.

Yang, Q., Xiao, T., Guo, J., & Su, Z. 2017. Complex relationship between obesity and the fat mass and obesity locus. *International journal of biological sciences* 13(5): 615.

Yeo, G. S. 2014. The role of the FTO (fat mass and obesity related) locus in regulating body size and composition. *Molecular and cellular endocrinology* 397(1–2): 34–41.

Medical Technology and Environmental Health – Abdullah, Widiaty & Abdullah (eds)
© 2020 Taylor & Francis Group, London, ISBN 978-0-367-86053-0

Investigating the hepatoprotective potential of *Ocimum americanum L.* ethanol extract in rifampicin-induced hepatotoxicity mice

D. Renovaldi, E. Multazam & Y. Safitri
Universitas Muhammadiyah Jakarta, South Tangerang, Indonesia

ABSTRACT: Rifampicin (RIF) is broadly used in the world for the treatment of tuberculosis, but the hepatotoxicity is still a major concern during clinical therapy related to oxidative stress activity in liver. *Ocimum americanum* (OA) contains antioxidant molecules and provides protection against free radicals caused by oxidative damage. This study attempts to investigate the hepatoprotective activity of *Ocimum americanum* L. ethanol extract against RIF–induced hepatotoxicity mice. Thereafter, the level serum of alanine transaminase (ALT) and aspartate transaminase (AST) were investigated. Treatment groups (RIF 300 mg/kg BW+OA 2.8 mg/20gr BW and RIF 300 mg/kg BW+OA 5.6 mg/20gr BW) were compared to RIF and to normal control group for 14 days. Treatment with RIF significantly increased the level of ALT and AST serum while OA-treatment reduced these activities in both extract dosage groups. These findings indicate that the ethanol extract of *Ocimum americanum* L. exerted significant hepatoprotector effects, likely related to its antioxidant compounds.

1 INTRODUCTION

Tuberculosis (TB) remains a major global health problem despite decades of highly effective treatment. The World Health Organization (WHO) declared TB as a global public health emergency in 1993, when an estimated 7-8 million new cases and 1.3-1.6 million deaths occurred each year (Ramappa & Aithal 2013). According to World Health Organization data (WHO) published in 2016, 9.6 million people suffered from tuberculosis and 1.5 million died due to it in 2014 (Kim et al. 2017). There are four main TB treatment agents, namely, isoniazid, rifampicin (RIF), pyrazinamide, and ethambutol. Typical regimens in clinical practice for tuberculosis include isoniazid, RIF, and pyrazinamide for two months, followed by isoniazid and RIF for four additional months (Lian et al. 2013). Apart from its ability to fight TB, anti-TB drugs, such as isoniazid, RIF, and pyrazinamide are also reported to often cause hepatotoxicity (Saukkonen et al. 2006). At present, RIF is one of the main drugs of choice for tuberculosis, and RIF can induce hepatotoxicity. RIF is known to cause hepatocellular dysfunction in the early stages of the administration, but this symptom disappears after the administration is stopped. In addition, RIF affects bilirubin excretion temporarily by causing hyperbilirubinemia. Besides, RIF also causes liver injury in the form of cholestasis, it is also associated with typical liver lesions resulting from hepatocellular changes accompanied by centrilobular necrosis. Previous studies have shown that oxidative stress occurs in patients taking anti-TB drugs that are induced by hepatotoxicity (Singh et al. 2011). Several studies have shown that activities that can fight against oxidative stress may play a role in liver protection. Endogenous lipid peroxidation has been shown to be a major factor in RIF cytotoxic action. This mechanism is generally associated with the formation of highly reactive oxygen species (ROS), which act as stimulators of lipid peroxidation and a source of destruction and damage to cell membranes (Lian et al. 2013).

Traditional medicine has been used by many people around the world for the treatment of liver disease for a long time without significant toxic effects. Therefore, it is necessary to look for complementary and alternative medicine, especially herbal medicines for the treatment of liver diseases for better efficacy and safety to replace the drugs that are currently used (Wati et al. 2015). *Ocimum americanum* was formerly known as *Ocimum canum*, was a wild species in India, but was commonly cultivated in Indonesia for its own essential oils for commercial purposes. The Ocimum plant contains a large number of antioxidants besides vitamin C, vitamin E, flavonoids and carotenoids (Pandey et al. 2014). The presence of many pharmacologically active compounds in the Ocimum species makes this species was believed to provide protection against oxidative damage induced free radicals from cellular components including hepatotoxicity (Aluko et al. 2013). Our literature survey reveals that the hepatoprotective activity of *Ocimum americanum* leaves has not been widely studied scientifically. The potential of this plant in Indonesia is considered to be very large because of the number of cultivation and is quite commonly used as a complement to food and herbal medicine. Based on this phenomenon, the aim of this study was to determine the hepatoprotective effect of ethanol extract of *Ocimum americanum* leaves on rifampicin-induced hepatotoxicity mice.

2   METHOD

2.1   *Chemical material*

All chemicals were of an analytical grade. Kits for aspartate aminotransaminase (AST) and alanine amino transaminase were purchased from ELITech Clinical System SAS Z.I. 61500 SEES, France.

2.2   *Procedure of Ocimum americanum L. ethanol extraction*

Standardized and certified plant leaves are obtained from the Bogor Spices and Medicinal Plant Research Institute (BALITTRO) Bogor, Indonesia. Basil leaves as much as 0.5-1 kg are washed thoroughly with water, dried using an oven with a temperature of 60°C. Then mashed using a blender until it becomes smoother and dried, then sifted using a sieve until a fine powder is obtained. Furthermore, 128 grams of basil leaf powder was put into the macerator, after that added 90% ethanol solvent as much as 1 L (ratio 1: 7.5), allowed to stand for 24 hours. After that, the filtrate shelter was carried out. The pulp obtained from the filtering results is then macerated (repeated a maximum of 3 times). Maceration method by soaking using 90% ethanol and stirring continuously. After the filtrate is obtained, evaporation is carried out using a rotary evaporator at a temperature of 50°C until the 90% ethanol solvent evaporates and the semi-ethanol basil extract is obtained.

2.3   *Experimental animal*

28 male mice (Strain Deutschland, Denken, & Yonken/DDY) were purchased from the Non-Ruminent and Animals Laboratory, Faculty of Animals Husbandry, Bogor Agricultural Institute. The mice weighed about 30-40 gr. Animals were provided with a standard diet that has been recommended by the breeder. After 7 days of acclimatization period, they were randomly assigned into four groups, each comprising seven mice. Group I (normal control group) mice received 0,2 ml of carboxymethyl cellulose 1% (CMC1%). Group II (RIF group) mice received 300 mg/kg BW. Group III and IV (treatment groups) mice received 2.8 mg/20gr BW extract ethanol of *Ocimum americanum* + RIF 300 mg/kg BW and 5.6 mg/20gr BW extract ethanol of *Ocimum americanum* + RIF 300 mg/kg BW respectively. All treatments were given intragastrically once a day for 14 days.

2.4   *Sample analysis and biochemical parameters*

At the end of the experimental period, mice remained fasting for 10 h before sacrifice. Blood samples were collected for biochemical studies. The serum was separated by centrifugation for 10 min at 3000 rpm. AST and ALT activities were estimated in serum using the respective kits. The comparison of the reagents was R1 4: R2 1 based on the kits instruction (ELITech, France) and estimated using a clinical chemistry analyzer for quantifications (Microlab 300, ELITechGroup, France).

2.5   Statistical analysis

All results were expressed as mean value ± SD. The statistical analysis was performed using analysis of variance followed by student's t-test with $p < 0.05$ considered statistically significant.

3   RESULTS

3.1   Rifampicin induces hepatotoxicity

In this study, it was found that there were significant differences in the activity of serum transaminases ALT (p-value 0.00) and AST (p-value 0.00) in the RIF group compared to the normal control group. Figure 1 shows a striking increase in serum ALT and AST in the RIF group and a striking increase from day 1, day 7 to day 14. While in the normal control group there was no significant increase in serum ALT and AST which tended to be stable until the 14th day.

The results of this analysis indicate that the administration of 300 mg/kg BW rifampicin can have a hepatotoxic effect characterized by elevated levels of serum ALT and AST liver function biomarkers that are significant in the study sample.

3.2   Ocimum americanum reduce the activity of serum transaminase enzymes

ANOVA test analysis was carried out to determine the mean differences in the population of serum ALT levels and the population of AST serum levels in the treatment and control groups after administration of *Ocimum americanum* ethanol extract. ANOVA test results showed that there were statistically significant mean differences in serum ALT levels and in serum AST levels between each group of study subjects (p-value 0.00). The next test used was the student's t-test to find out which groups that have a significant difference to another group. In ALT serum level, it was found that a significant difference was only seen in the

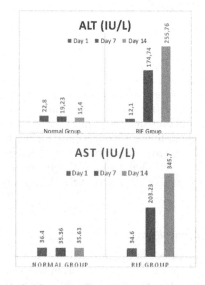

Figure 1.   Rifampicin administration increases the activity of ALT and AST serum in mice.

Table 1. Effect of OA treatment on serum ALT and AST of RIF-induced hepatotoxicity mice.

Groups	ALT (IU/L)	AST (IU/L)
Normal Control	13.7 ± 3.99	33.3 ± 4.26
Rifampicin	272.7 ± 51.1	287.2 ± 98.0
Rifampicin + OA 2,8 mg/20gr BW	151.6 ± 39.7	242.7 ± 86.8
Rifampicin + OA 5,6 mg/20gr BW	158.7 ± 54.2	240.1 ± 79.4

treatment group with a dose of *Ocimum americanum* ethanol extract of 2.8 mg/20gr BW (p-value 0.002), while in the treatment group at a dose of 5.6 mg/20gr BW was not significantly different (p-value 0.0079). In AST serum level, no significant difference was found in the treatment groups with the dosage of the *Ocimum americanum* ethanol extract 2.8 mg/20gr BW (p-value 0.80) or at a dose of 5.6 mg/20gr BW (p-value 0.75).

In this analysis, some of the treatment groups were found to have no significant decrease in ALT and AST levels. But on a mean difference, each treatment group was able to reduce ALT and AST levels compared to the RIF group (Table 1).

The results of this study indicated that protection by *Ocimum americanum* extract had a potential effect in decreasing liver function damage markers (ALT and AST levels) in both the extract dosage of 2.8 mg/20gr BW and 5.6 mg/20gr BW. However, a statistically significant decrease was only found in serum ALT levels, whereas AST was considered to be an insignificant decrease. Furthermore, the results of this study also showed that the decrease in serum ALT and AST levels were not affected by the number of administered dosage of *Ocimum americanum* extract.

4 DISCUSSION

In this study, the hepatotoxic effect of rifampicin was observed in oral administration at a dose of 300 mg/kg BW for 14 days. We evaluated liver injury in experimental animals through clinical biochemical examinations such as ALT and AST in sample serum. Increased serum ALT and AST are indicators of hepatotoxicity. This study showed that in the RIF group there was a significant and striking increase in serum ALT and AST compared to the normal control group. These results were consistent with research conducted by Kim et al. (2017) which found that liver injury was characterized by an increase in serum ALT and AST as well as pathological changes at the molecular level in a sample of rats given rifampicin treatment. A study conducted by Lian et al. (2013) also found that rifampicin can cause an increase in hepatic oxidative stress conditions, histopathological injury to the liver, increase in serum AST and ALT, and increase in plasma malondialdehyde (MDA) in rifampicin-administered rats compared to the normal control group.

ALT is the biomarker that is most often relied upon to detect hepatoxicity. This is a liver enzyme that plays an important role in amino acid metabolism and gluconeogenesis. It catalyzes the reductive transfer of amino groups from alanine to α-ketoglutarate to produce glutamate and pyruvate. The normal level of this enzyme is in the range of 5-50 U/L. Increased levels of this enzyme are released in circulation during damages to liver cells. In other hands, AST is another liver enzyme that helps in producing/forming proteins. This enzyme catalyzes the reductive transfer of amino groups from aspartate to α-ketoglutarate to produce oxaloacetate and glutamate. In addition to the liver, AST is also found in other organs such as the heart, muscles, and kidneys. Injury to these tissues can cause an increase in AST levels in the blood. Normal levels in the range of 7-40 U/L. AST also helps in detecting hepatocellular necrosis but it is considered as a biomarker enzyme that is less specific for hepatocellular injury because it can also indicate abnormalities in the heart, muscles, brain or kidneys (Pandit et al. 2012, Singh et al. 2011)

The administration of rifampicin is known to produce many metabolic and morphological deviations in the liver because it is the main detoxification site for this antituberculosis drug. Rifampicin induces cytochrome P450 enzyme, causing an increase in the production of toxic metabolites from acetyl hydrazine (AcHz) (Pandit et al. 2012). In addition, many studies have shown that one of the main pathways that induce hepatotoxicity by rifampicin is the presence of toxic metabolites that can induce oxidative stress in the liver. In biological systems, the balance between pro-oxidants and antioxidants is very important for cellular homeostasis. Research conducted by Lian et al. (2013) supports this hypothesis that RIF causes a decrease in the status of antioxidant levels in the liver, as evidenced by a significant reduction in Glutathione Peroxidase (GPx) activity, one of the primary antioxidant molecular enzyme.

The hypothesis in this study is the potential for hepatoprotective action by antioxidant activity derived from *Ocimum americanum*. This study found that the administration of *Ocimum americanum* extract can reduce serum ALT and AST levels in mice induced to hepatotoxicity by rifampicin. These results are in accordance with studies conducted by Aluko et al. (2013) which found that administration of *Ocimum americanum* extract can provide hepatoprotective effect by decreasing levels of ALT, AST, Alkaline phospatase/ALP, and total bilirubin (TBIL) to normal values and improvement in liver histopathological structure in rats that were induced hepatotoxicity using paracetamol. A similar study was also conducted by Radhika et al. (2017) that found that administration of *Ocimum americanum* extract can mediate restoration of serum marker enzymes (ALT, AST, ALP, and Gamma-glutamyl transferase/GGT) to normal levels and increase in pro-antioxidant enzymes (lactoperoxidase/LPO, superoxide dismutase/SOD, glutathione/GSH, and glutathione peroxidase/GPx) in isoniazid-induced hepatotoxic samples. In addition, administration of *Ocimum americanum* extract can also significantly reduce the lipid profile (LDL, VLDL and cholesterol) in isoniazid-induced sample groups compared to sample groups without administration of *Ocimum americanum* extract. This was considered to have a beneficial effect in which studies conducted by Ghadir et al. (2010) showed that serum levels of HDL, LDL, and total cholesterol have a positive correlation with the severity of liver cell damage and administering *Ocimum americanum* would be able to prevent this damage.

Significant reductions were found particularly striking in ALT serum levels, compared to AST serum levels in the treatment group given *Ocimum americanum* extract. ALT is a standard clinical biomarker of hepatotoxicity. This enzyme measurement is more specific to detect abnormalities in the liver because this enzyme is primarily found in the liver. While AST is considered less specific as a biomarker enzyme for liver injury because the increase of AST level can also be found in heart, brain and kidney disorders (Singh et al. 2011).

The results of this study indicate that the *Ocimum americanum* extract has a protective effect on damage that occurs in liver cells. Ocimum plants are known to contain large amounts of antioxidants such as vitamin C, vitamin E, flavonoids, and carotenoids. This antioxidant composition is pharmacologically active in producing a protective effect against free radicals induced by oxidative damage from cell components (Pandey et al. 2014). Bunrathep et al. (2007) reported a comparison of antioxidant activity in extracts from 4 species of ocimum using DPP bioassay and found that *Ocimum americanum* ranked second as the most antioxidant ocimum species after Ocimum free maximum, and was followed by Ocimum sanctum, and Ocimum basilicum. Ntonga et al. (2014) study determined that the antioxidant effect of *Ocimum americanum* is mainly due to phenolic compounds while some aldehydes and monoterpene ketones such as citral, citronellal, isomenthone, and menthone are also strong scavenger compounds. This statement supports the results of this study to suggest that *Ocimum americanum* has the potential to be a hepatoprotector against rifampicin-induced hepatotoxicity.

5 CONCLUSION

This study showed that rifampicin can cause hepatotoxicity status in the liver of experimental animal samples characterized by significant increases in serum ALT and AST levels. Treatment with *Ocimum americanum* ethanol extract was considered to have a hepatoprotective effect marked by the decrease in ALT and AST serum levels in the hepatotoxic-induced sample mice. These results imply that the administration of *Ocimum americanum* ethanol extract can have a hepatoprotective effect on hepatotoxicity caused by rifampicin administration.

ACKNOWLEDGEMENT

We would like to thank the Faculty of Medicine and Health, Universitas Muhammadiyah Jakarta, Indonesia, for funding this study. We also thank the Faculty of Pharmacy, Pancasila University, Indonesia, for providing us with all the necessary facilities to complete this work.

REFERENCES

Aluko, B. T., Oloyede, O. I. & Afolayan, A. J. 2013. Hepatoprotective activity of Ocimum americanum L Leaves against paracetamol–induced liver damage in rats. *Am J Life Sci* 1: 37–42.

Bunrathep, S., Palanuvej, C. & Ruangrungsi, N. 2007. Chemical compositions and antioxidative activities of essential oils from four ocimum species endemic to Thailand. *J Health Res.* 21(3): 201–206.

Ghadir, M. R., Riahin, A. A., Havaspour, A., Nooranipour, M. & Habibinejad, A. A. 2010. The relationship between lipid profile and severity of liver damage in cirrhotic patients. *Hepatitis monthly* 10(4): 285.

Kim, J. H., Nam, W. S., Kim, S. J., Kwon, O. K., Seung, E. J., Jo, J. J., ... & Lee, H. S. 2017. Mechanism investigation of rifampicin-induced liver injury using comparative toxicoproteomics in mice. *International journal of molecular sciences* 18(7): 1417.

Lian, Y., Zhao, J., Xu, P., Wang, Y., Zhao, J., Jia, L., ... & Peng, S. 2013. Protective effects of metallothionein on isoniazid and rifampicin-induced hepatotoxicity in mice. *PloS one* 8(8).

Ntonga, P. A., Baldovini, N., Mouray, E., Mambu, L., Belong, P. & Grellier, P. 2014. Activity of Ocimum basilicum, Ocimum canum, and Cymbopogon citratus essential oils against Plasmodium falciparum and mature-stage larvae of Anopheles funestus ss. *Parasite* 21.

Pandey, A. K., Singh, P. & Tripathi, N. N. 2013. Chemistry and bioactivities of essential oils of some Ocimum species: an overview. *Journal of Coastal Life Medicine* 1(3): 192–205.

Pandit, A., Sachdeva, T. & Bafna, P. 2012. Drug-induced hepatotoxicity: a review. *J Appl Pharm Sci* 2(5): 233–43.

Radhika, J., Jothi, G., Nivethetha, M., Vijayaroslin, L. & Yuvarani, S. 2017. Hepatoprotective efficacy of ocimum canum sims. *On isoniazid induced hepatotoxicity.*

Ramappa, V. & Aithal, G. P. 2013. Hepatotoxicity related to anti-tuberculosis drugs: mechanisms and management. *Journal of clinical and experimental hepatology* 3(1): 37–49.

Saukkonen, J. J., Cohn, D. L., Jasmer, R. M., Schenker, S., Jereb, J. A., Nolan, C. M., ... & Bernardo, J. 2006. An official ATS statement: hepatotoxicity of antituberculosis therapy. *American journal of respiratory and critical care medicine* 174(8): 935–952.

Singh, A., Bhat, T. K. & Sharma, O. P. 2011. Clinical biochemistry of hepatotoxicity. *J Clin Toxicol* 4(0001): 1–9.

Wati, Aulia, and Dessysehati Masawoy. "Activity of Ethanol Extract of Purpleleaves." 7.3 (2015): 497–501. Print.

Medical Technology and Environmental Health – Abdullah, Widiaty & Abdullah (eds)
© 2020 Taylor & Francis Group, London, ISBN 978-0-367-86053-0

Micronucleus assay and oral hygiene index in smokers

M.M. Damayanti, Y. Kharisma, I.M. Nur, M. Rachmawati, A.H. Hasan, F.A. Yulianto, S.B. Rahimah & W. Maharani
Universitas Islam Bandung, Bandung, Indonesia

ABSTRACT: Smoking can lead to oral disease. Many people are unaware of the relationship between smoking habits and oral hygiene index (OHI), up to potential oral diseases. Micronuclei (MN) is used as an indicator to look at the cytotoxic effects of smoking and as a biomarker for assessing DNA damage. This study aimed to identify a correlation between the amount of MN and the OHI on smoking habits. Sixty-two subjects with more than 5 years of smoking habits were examined. Oral mucosal cells were collected from both sides of the cheeks and slides were prepared and examined for cells with MN where the presence of MN was assessed under 1000x magnification. The OHI was measured to assess dental debris and calculus. The results showed no significant correlation between MN and OHI of smoking statistically ($P = 0.61$). The genotoxic effect of smoking caused chromosomal damage in epithelial cells of the oral mucosa and was reflected in the occurrence of MN and smokers' poor OHI.

1 INTRODUCTION

Smoking has been identified as the second greatest risk factor causing global death and disability. Smoking is a known cause of cardiovascular disease, lung disease, cancer, and other systemic pathologies; the area of the human body that is directly exposed to the effects of smoking is the oral cavity. Smoking has impacts on the oral cavity ranging from aesthetic changes to fatal diseases such as oral cancer. Oral health behavior can influence patients' oral status and oral hygiene index (OHI) levels (Tatullo et al. 2016; Sabounchi 2016).

Smoking changes the structure of oral tissue. Many researchers have reported a significant correlation between smoking and an increase in the frequency of micronuclei and other nuclear abnormalities. The nicotine in cigarettes is one of the genotoxic compounds that will convert nitrosation into nitrosamine, which can damage the DNA. DNA damage due to genotoxic exposure can manifest as micronuclei, a second cell nucleus that is smaller than the real nucleus cell (Farhadi et al. 2016; Rahmah et al. 2016).

The displaced chromosome or chromosome fragment is eventually covered by a nuclear membrane; except for the smaller size, it is morphologically similar to the nuclei after conventional nuclear staining (Fenech et al. 2011; Vinodkumar et al. 2016).

The buccal cell micronucleus (MN) test was first performed in 1983 and continues to gain popularity as a biomarker of genetic damage in various applications. The MN test provides information about cytogenetic damage in tissues, which is a target of human carcinogens and where carcinomas can develop (Palaskar & Jindal 2010).

The calculation through the MN is regarded as a fast, efficient, and economical technique that provides a quantitative analysis of reliable genotoxicity. Significantly high MN frequencies have been observed in people exposed to organic solvents, antineoplastic agents, diesel derivatives, polycyclic aromatic hydrocarbons, paints containing solvents, arsenic-contaminated drinking water, food/vitamin deficiencies, smoking, and alcohol. Thus, MN buccal cell testing can be used as a biomarker of genetic damage in various applications, especially in smokers (Dindgire et al. 2012; Kharisma et al. 2019; Kiran et al. 2018).

Smoking can cause changes in the composition of microbial plaque and alter the host response to the plaque. Cigarette smoke potentially contains at least 500 toxins, including hydrogen cyanide, carbon monoxide (resulting in carboxyhemoglobin), free radicals, nicotine, nitrosamines (potential carcinogens), and various oxidant gases (causing platelet activation and endothelial dysfunction) (Hilgers & Kinane 2004). Previous research has shown that smoking has a direct influence on periodontal tissue. Smokers have a greater chance of suffering periodontal diseases such as alveolar bone loss, increasing tooth pocket depth, and tooth loss compared to nonsmokers. Plaque and calculus scores have also been shown to be higher in smokers compared to nonsmokers. Pathological conditions in the oral cavity also often found in smokers include root caries, halitosis, periimplantitis, reduction of the taste function, tooth staining or restoration, and periodontal disease. Periodontal disease includes plaque and calculus index, periodontal pocket, gingival inflammation, gingival recession, and alveolar bone loss (Affrin 2015; Kusuma 2019).

The number of micronuclei as genotoxic biomarkers and oral hygiene as an index of the healthy oral cavity are indicative of a smoker. Therefore, the purpose of this study was to see whether a correlation exists between the number of micronuclei and oral hygiene index in smokers.

2 METHODS

The study was conducted in the biomedical laboratory of the Faculty of Medicine, Islamic University of Bandung (UNISBA), Indonesia. Sixty-two subjects were collected based on research criteria (age, gender, experience smoking, socioeconomic status, employment). The research was quantitative and the method used was observational analytics with cross-sectional findings. The tools and materials used in this study included cytobrushes, a glass object, a glass lid, a light microscope, a mouth glass, a mouth mirror, a dental probe, an informed consent sheet, and hematoxylin-eosin (HE) staining material.

Explanation regarding the procedures involved in this study was given to all subjects. All subjects were required to sign a consent form before participating. Before sampling, all individuals washed their mouths using clean water, then the right and left buccal mucosa swabs were taken. The results of the buccal mucosa swab on the cytobrush were then rubbed into the glass object. The epithelial cells in the glass object were then fixed, and the next step was HE staining. The preparats were then viewed on 100 cells and evaluated under a light microscope at 1000x magnification (Kharisma et al. 2019).

The OHI examination was carried out using a mouth mirror and a dental probe. The Simple Oral Hygiene Index (OHI-S) was used to check the subjects' oral hygiene. The OHI-S consists of a debris index and a calculus index. Six tooth surfaces – 16, 11, 26, 36, 31, and 46 – were examined for debris and calculus. Scores of 0, 1, 2, and 3 were given according to the number of debris or calculus on the tooth surface. The debris and calculus indexes were then combined to obtain the OHI-S, which was then recorded for the analysis (Jiun et al. 2015).

This study was approved by the Health Research Ethics Committee of the Faculty of Medicine of UNISBA with the ethical clearance number 379/Komite Etik.FK/X/2018.

3 RESULTS

The effect of smoking on OHI within the good category was quite high; 29 subjects were categorized as poor and moderate with a percentage of 46.77% (Figure 1).

The study involved evaluating 100 buccal cells from each subject. The micronuclei are shown by a black arrow with one or more MN. The MN are round or ovoid with the same color as real nuclei (Figure 2).

The null hypothesis that a zero correlation exists and that the relationship is not linear was accepted (P 0.61 is greater than the significance determined alpha 0.05) with a correlation value of −0.07 (Table 1).

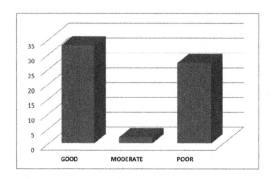

Figure 1.   Distribution of oral hygiene index.

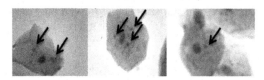

Figure 2.   Micronuclei with 1000x magnifications.

Table 1.   Correlation between OHI-S and MN.

Pearson r	P
−0.07	0.61

4 DISCUSSION

This study was conducted with a sample of 62 smokers with more than 5 years of a smoking habit. No correlation was found between the number of micronuclei and the oral hygiene index in smokers, but the research shows an increase in the average number of micronuclei in active smokers. This finding is in line with several previous studies; those by Dr. Nikhil Solomon and colleagues and Kewan Kamal Ahmad and colleagues state that the mean micronucleus count in buccal cells of smokers was significantly higher than in nonsmokers (Ahmad et al. 2015; Raj et al. 2019).

The smokers demonstrated poorer oral hygiene and greater calculus accumulation compared to the nonsmokers (Pereira et al. 2013). Other research found that smokers have worse oral hygiene habits than nonsmokers (Santos et al. 2015). The OHI in this research was quite good; the factors that involved oral hygiene status are oral health education, socioeconomic status, behavioral beliefs, perceived power, and subjective norms. Oral health behaviors can affect oral health and efforts to build good oral health behaviors can affect an individual's general health. Smoking is a bad habit that will result in decreased oral hygiene because it will leave stains and calculus. Another factor that is very instrumental in assessing the OHI is oral health instruction such as how to brush the teeth (Bozorgmehr et al. 2013).

Smoking and breathing the smoke of cigarettes can damage the DNA due to genotoxic substances such as benzopyrene and nicotine. In addition, tobacco also has carcinogenic ingredients and the heat generated from smoking can add to the aggressive action of the carcinogenic material on the oral mucosa. Titrated nicotine is a dangerous nitrosamine compound, according to the International Agency for Research on Cancer (IARC), in producing DNA adducts either through methylation or hydroxylation pathways. DNA adducts can then induce the formation of a micronucleus, a mass with a structure like a nucleus that has a small size and that is located close to the real nucleus, in the cytoplasm. It is formed due to chromosomal abnormalities in the cell division phase – namely anaphase. Chromosome damage caused by genotoxic substances in basal cells can be seen in the form of micronuclei that migrate to the upper epithelial layer until they are exfoliated. Micronucleus frequency is a measure of chromosomal damage at the beginning of cell division and the number of micronuclei related to carcinogenic stimuli, before the development of clinical symptoms – namely premalignant and malignant events.

The genotoxic effect of smoking causes chromosomal damage in epithelial cells of the oral mucosa and is reflected in the occurrence of micronuclei and smokers' poor oral hygiene index. For this reason, micronuclei can be used as an indicator of the early detection of potential oral diseases.

5 CONCLUSION

The number of micronuclei was not linear with a category of oral hygiene index regarding the smoking habit. Based on this study. the evaluation of micronuclei and oral hygiene index may be a new bio-monitoring tool to use as a simple biomarker for the screening of premalignant changes in cells.

CONFLICT OF INTEREST

There is no conflict of interest in this research.

ACKNOWLEDGMENTS

We acknowledge all biomedical laboratory staff and the pathology anatomy department of the Medical Faculty of the Islamic University of Bandung who contributed to this research.

REFERENCES

Affrin, J. H. 2015. Assessment of oral hygiene status in smokers, nonsmokers and alcoholics. *Research journal of pharmacy and technology* 8(8): 1167–1170.

Ahmad, K. K., Mustafa, S. K., & Karim, J. K. 2015. Prevalence of micronucleated cell in buccal smears among smokers and non-smokers. *International journal of advanced research* 3(4): 972–977.

Bozorgmehr, E., Hajizamani, A., & Malek Mohammadi, T. 2013. Oral health behavior of parents as a predictor of oral health status of their children. *ISRN dentistry* 2013: 1–5.

Dindgire, S. L., Gosavi, S., Kumawat, R. M., Ganvir, S., & Hazarey, V. 2012. Comparative study of exfoliated oral mucosal cell micronucleus frequency in potentially malignant and malignant lesions. *International journal of oral and maxillofacial pathology* 3(2): 15–20.

Farhadi, S., Jahanbani, J., Jariani, A., & Ghasemi, S. 2016. Bio-monitoring of the nuclear abnormalities in smokers using buccal exfoliated cytology. *Advanced biomedical research* 7(4): 128–133.

Fenech, M., Kirsch-Volders, M., Natarajan, A. T., Surralles, J., Crott, J. W., Parry, J., & Thomas, P. 2011. Molecular mechanisms of micronucleus, nucleoplasmic bridge and nuclear bud formation in mammalian and human cells. *Mutagenesis* 26(1): 125–132.

Hilgers, K. K. & Kinane, D. F. 2004. Smoking, periodontal disease and the role of the dental profession. *International journal of dental hygiene* 2(2): 56–63.

Jiun, I. L., Siddik, S. N., Malik, S. N., Tin-Oo, M. M., Alam, M. K., & Khan, M. M. 2015. Association between oral hygiene status and halitosis among smokers and nonsmokers. *Oral health and preventive dentistry* 13(5): 395–405.

Kharisma, Y., Damayanti, M. M., & Yulianto, F. A. 2019. Folic acid usual doses decrease the buccal micronucleus frequency on smokers.

Kiran, K., Agarwal, P., Kumar, S., & Jain, K. 2018. Micronuclei as a predictor for oral carcinogenesis. *Journal of cytology* 35(4): 233.

Kusuma, A. R. P. 2019. Pengaruh merokok terhadap kesehatan gigi dan rongga mulut. *Majalah Ilmiah Sultan Agung* 49(124): 12–19.

Palaskar, S. & Jindal, C. 2010. Evaluation of micronuclei using Papanicolaou and May Grunwald Giemsa stain in individuals with different tobacco habits: A comparative study. *Journal of clinical and diagnostic research* 4: 3607–3613.

Pereira, A. D. F. V., Castro, A. C. S., de Lima Ramos, Q., Alves, C. M. C., & Pereira, A. L. A. 2013. Effects of cigarette smoking on oral hygiene status. *Revista Odonto Ciência* 28(1): 04–07.

Rahmah, N., Dewi, N., & Raharja, S. D. (2016). Analisis Sitogenik Mikronukleus Mukosa Bukal Pada Perokok Aktif Dan Pasif. *Dentino* 1(1): 15–20.

Raj, N. S. S. & Ramdas, A. (2019). Micronucleus assay of buccal mucosal cells in smokers and non-smokers. *Indian journal of applied research* (04).

Sabounchi, S. S., Torkzaban, P., Sabounchi, S. S., & Ahmadi, R. 2016. Association of oral health behavior-related factors with periodontal health and oral hygiene. *Avicenna journal of dental research* 8(2): 1–7.

Santos, A., Pascual, A., Llopis, J., Giner, L., Kim, D. M., Levi Jr, P., & Ramseier, C. A. 2015. Self-reported oral hygiene habits in smokers and nonsmokers diagnosed with periodontal disease. *Oral health and preventive dentistry* 13(3): 245–251.

Tatullo, M., Gentile, S., Paduano, F., Santacroce, L., Marrelli, M., 2016. Crosstalk between oral and general health status in e-smokers. *Medicine (Baltimore)*, 95(49): 1–7.

Vinodkumar, M. P., Vanaki, S. S,. & Puranik, R. S. 2016. Assessment of cytotoxicity of chlorhexidine containing mouth rinses by micronucleus test in exfoliated buccal epithelial cells. *National journal of integrated research in medicine* 7(3).

The vascular risk factors of ischemic stroke in young adults

A. Tursina
Universitas Islam Bandung, Bandung, Jawa Barat, Indonesia
Department of Neurology, Rumah Sakit Angkatan Udara Dr. M. Salamun, Bandung, Indonesia,

R.A. Indrianti
Universitas Islam Bandung, Bandung, Jawa Barat, Indonesia

W. Nurruhyuliawati
Rumah Sakit Umum Daerah Al Ihsan, Bandung, Indonesia

ABSTRACT: The incidence of stroke in young adults has increased in various countries due to a decrease in quality of life and reduced productivity. Disadvantages of stroke are more severe in younger patients than in older patients, placing a burden on the sufferers. The appropriate prevention is to reduce vascular risk factors that can cause ischemic stroke. This research aimed to determine the correlation of vascular risk factors with the prevalence of stroke in young adults. This research used a quantitative descriptive method and a cross-sectional design. Data analysis was done via frequency distribution. Hospitalized patients' medical records at the neurology ward in the hospital in Bandung, Indonesia, from August 2016 to August 2018 were consulted. These data were classified based on the patients' age, gender, and risk factors. The results of this research showed that women have the same risk factors as men aged 36–40 years old. The researchers concluded that vascular risk factors increasing blood sugar level, hypertension, and dyslipidemia were less significant in causing the occurrence of ischemic stroke in young adults while gout in young adults was the main risk factor for ischemic stroke. The early detection of risk factors is very important in preventing the occurrence of ischemic stroke in young adults.

1   INTRODUCTION

1.1   *Background*

Stroke is the third highest cause of death and also the highest cause of disability in the world (Department Kesehatan Indonesia 2013; Sacco et al. 2013; Van Alebeek et al. 2018). This condition also occurs in Indonesia. National data on stroke in Indonesia have shown that 15.4% of deaths are due to stroke. In addition to being one of the highest causes of death, stroke is also the highest cause of disability in Indonesia (Kusuima et al. 2009).

The prevalence of stroke in Indonesia has been estimated to be as many as 8.3 per 1,000 population, but health workers have stated that the prevalence of stroke in Indonesia is 6 per 1,000 population. This shows that about 72.3% of cases of stroke in the society have been diagnosed by health workers. Data health workers have provided have revealed that the prevalence of stroke in 11 provinces in Indonesia, including West Java, is higher than the national prevalence (Department Kesehatan Indonesia 2013; Kusuima et al. 2009).

Classification of stroke is based on anatomic pathology, and stroke classifications consist of ischemic stroke and bleeding stroke. Research at Denmark Hospital showed that the incidence of ischemic stroke is higher, 89.9%, than the incidence of bleeding stroke, which is 10.1%. Research in China and Japan also showed that ischemic stroke (70–80%) occurs more often than bleeding stroke (20–30%) (Hoy et al. 2013). Research into stroke by Misbach in

Southeast Asia revealed that the incidence of ischemic stroke was 74% while the incidence of bleeding stroke was 26.6% (American Stroke Association 2013).

The incidence of stroke increases with age. Stroke occurs less in young adult patients than in older patients (Mozaffarian et al. 2016). Data show stroke rarely happens in various countries but that the incidence of stroke in young adults continues to increase (Smajlović 2015). Stroke patients aged about and over 60 years old comprise the second largest group of stroke patients in Asia while stroke patients aged 15–59 years form the fifth largest group.

Young adult patients with stroke suffer greater losses than elderly stroke patients. Stroke that occurs in younger patients causes patients to become disabled at a productive age (Smajlović 2015). About half of stroke patients have physical or cognitive disabilities. They need help to do daily activities. This causes a decrease in their quality of life and has negative psychological and economic effects for themselves and their country (Mozaffarian et al. 2016; Smajlović 2015). Prevention should be target at reducing the incidence of stroke in young adults (American Stroke Association 2013). Appropriate prevention of stroke risk factors is expected to reduce mortality and morbidity due to stroke in young adults. Researchers have not conducted many studies regarding stroke in young adults in West Java or Indonesia. A study on risk factors for stroke in young adults was conducted in 2015–2016 in the neurological unit of RSUP Dr. Wahidin Sudirohusodo. Two types of samples were taken in this research – patients with stroke and patients who did not suffer with stroke aged 18–40 years old. The sample consisted of a total of 174 patients. It can be concluded from this research that a history of diabetes mellitus and hypertension are the main risk factors for the incidence of stroke in young adults aged 18–40 years (Burhanuddin 2013). Further research was conducted at the Dr. Hasan Sadikin General Hospital in Bandung from April to May 2017. The population of this research comprised patients with stroke in West Java while the subject of this research included the medical records of patients treated in this hospital from January 1, 2011, to December 31, 2016. Patients with stroke aged 18–45 years old were included in this research. A total of 450 samples was obtained.

The highest number of cases of stroke occurred in patients aged 42–45 years old (45.11%). The incidence of ischemic stroke (50.44%) was higher than the incidence of bleeding stroke (49.56%). The greatest risk factor was hypertension (42.06%) (Syifa et al. 2017). The incidence of stroke in young adults examined in the preliminary study at RS Salamun Bandung in 2016 was as many as 45 people, and it increased in 2017 to 58 people while in RSUD Al Ihsan, there was an increase every year from 35 to 48 patients.

Researchers have identified modifiable and non-modifiable risk factors for stroke. The modifiable risk factors for stroke include hypertension, heart disease, diabetes mellitus, dyslipidemia, and hyperuricemia while the non-modifiable risk factors for stroke consist of age, sex, and genetics. The incidence of stroke in young adults is related to lifestyle choices such as consuming lots of delicious fatty foods and being too lazy to move or do sport. This is one of the risk factors for stroke that can certainly be modified. Cholesterol levels below 200 mg/dl are considered safe, whereas cholesterol levels over 240 mg/dl are considered a dangerous condition resulting in greater risk of heart disease and stroke (Del Mar Bibiloni et al. 2016; MedicineNet 2018). Excessive sugar consumption can cause diabetes mellitus, which is one of the risk factors for stroke in young adults. If diabetes is accompanied by high cholesterol levels, high triglycerides, and high blood pressure, the risk of stroke is four times greater (Bell et al. 2015; Ferrier 2014). Hypertension, both systolic and diastolic blood pressure, is the biggest cause of stroke (Grace et al. 2016). Increased uric acid production also plays a role in increasing the incidence of stroke. Hyperuricemia is elevated uric acid levels in the blood resulting from the consumption of foods with high fat, purines. and fructose, obesity, and some kidney disorders (Ferrier 2014).

Based on various facts about the risk of stroke at a young age, the theories of experts, and the results of previous studies, the researchers were interested in studying the correlation of stroke vascular risk factors – namely hypertension, diabetes mellitus, hypercholesterolemia, and hyperuricemia – with the prevalence of stroke in young adults at RSAU Dr. M Salamun and RSUD Al Ihsan Bandung. Preventative measures should be taken to reduce the incidence of stroke in young adults. This research is expected to discover the vascular risk factors so that the mortality and morbidity of stroke in young adults can be reduced.

2 METHOD

2.1 Research stages

The preparation and implementation of the research were carried out in two stages. In the first stage, the researchers consulted the license department of the Education and Neurology Division, Dr. RSAU. M. Salamun Bandung and RSUD Al Ihsan Bandung, in order to obtain research objectives and to complete technical preparation, after which the researchers signed an informed consent. The second stage consisted of collecting complete medical records, taking data according to inclusion, and processing and analyzing these data.

2.2 Research materials

Research materials included the complete medical records of hospitalized patients in the neurology wards of RSAU Dr. M. Salamun and RSUD Al Bandung, Indonesia, from August 2016 to August 2018 who met the inclusion criteria.

2.3 Inclusion criteria

1. Diagnosis of ischemic stroke
2. Stroke patients aged 18–45 years old
3. Blood pressure examination
4. Head CT scan without contrast
5. Laboratory examination of lipid profile, GDP, GD2JPP, hyperuricemia

2.4 Exclusion criteria

1. Incomplete medical records
2. Diagnosis of stroke accompanied by other diseases such as kidney disorders and heart disease

The selection of research samples was based on total sampling. The hospitals consulted for this research saw 60 patients from January 2016 to December 2018.

2.5 Research design

This research study used a descriptive method with a cross-sectional design. The measurement employed both independent and dependent variables examined at the same time.

2.6 Statistical analysis

The data analysis and hypothesis testing design used the SPSS 20 software. Frequency distribution and the assumptions of the normal distribution of age, systole, diastole, and glucose were performed for each sex group. The assumption of a normal distribution was met if the P-value of the Shapiro-Wilk test was greater than the specified significance ($\alpha = 0.05$). Several variables contained the assumption that the normal distribution was not fulfilled so that a nonparametric (Kruskal-Wallis) different test was conducted. The results indicated a statistically significant relationship with a significance level of 0.05 or $\alpha = 5\%$.

3 RESULTS AND DISCUSSION

This study was carried out at RSAU Dr. M. Salamun Bandung and RSUD Al Ihsan West Java from August 2016 to August 2018 by collecting samples taken from patients' medical records. The total population sample comprised 60 patients with ischemic stroke at a young age.

3.1 *Distribution of stroke patients by gender*

Table 1 shows a comparison of the incidence of stroke in women and men. The table reveals that women (53.33%) are more susceptible to the risk factors for stroke than men (46.73%). These results were consistent with the results of previous research suggesting that women have more of the risk factors for stroke than men (Valery & Krishnamurthi 2016).

3.2 *Distribution of stroke patients by age*

Table 2 shows the incidence of ischemic stroke by age. The data, taken from the research of Alchuriyah and Wahjuni (2016), show that the youngest adult patients with stroke were aged 43 years old.

According to Sitorus (2008), a family history of stroke has a significant influence on the risk for stroke of 3.91 times those without a family history of stroke. A relationship exists between family history of stroke and incidence of stroke at a young age; offspring of stroke sufferers are known to experience changes in early atherosclerotic markers – namely fat deposits under the lining of blood vessel walls that can trigger stroke (Sitorus 2008).

Table 3 shows that the incidence of ischemic stroke in patients aged 36–40 years old is the same between women and men. In addition, women and men aged 18–40 years old have the same risk of stroke. This proves that the risk factors for stroke in young adult women and men are the same (Burhanuddin 2013).

3.3 *Distribution of stroke patients based on risk factors for increased blood sugar*

Diabetes mellitus can cause changes in the vascular system (blood vessels and heart). Diabetes mellitus also accelerates the occurrence of heavier, more diffuse arteriosclerosis so that it can increase the risk factors for ischemic stroke. Table 4 shows that blood sugar levels in women ischemic stroke patients with an average fasting blood sugar of 118 mg/dl and blood

Table 1. Distribution of stroke patients by gender.

Gender	Percentage Frequency (%)	Cumulative
Women	32	53.33
Men	28	46.73
Total	60	100.00

Table 2. Distribution of stroke patients by age.

Age (year) Frequency	Percentage %	Age (year) Frequency
21–25	1	1
26–30	3	5
31–35	7	12
36–40	15	25
41–45	34	57
	60	100

Table 3. Distribution of ischemic stroke patients by age and gender.

Variable	Frequency	Women	Men	P (K-Wallis)
Age	Minimum	23.00	33.00	0.42
	Maximum	45.00	45.00	
	Median	39.50	40.50	
	Mean	37.94	40.07	
	SD	6.16	3.34	

Table 4. Distribution of ischemic stroke patients based on risk factors for increased blood sugar.

Variable	Frequency	Men	Women	P(K-Wallis)
Fasting	Minimum	69.00	19.00	0.99
Glucose	Maximum	291.00	166.00	
(mg/dl)	Median	91.50	96.00	
	Mean	118.78	99.57	
Glucose 2JPP	Minimum	75.00	79.00	0.82
(mg/dl)	Maximum	626.00	212.00	
	Median	115.50	115.50	
	Mean	160.06	128.07	
	SD	112.58	38.70	

sugar 2 hours postprandial of 160 mg/dl were higher than those in men. Although the results of the P-value of 0.99 and 0.82 indicate an increase, blood sugar is not significant as a risk factor for ischemic stroke. From these results, it can be seen that diabetes mellitus is not the main factor of the history of young adult patients with stroke.

Researchers carried out studies on risk factors for stroke in young adults in the neurological department of RSUP Dr. Wahidin Sudirohusodo from 2015 to 2016. The sample included a total of 174 patients with stroke and patients who did not suffer stroke aged 18–40 years. A history of diabetes mellitus is a risk factor for stroke in young adults. Diabetes mellitus is genetically inherited and is also triggered by unhealthy lifestyle choices such as consuming food containing high glucose and junk food and not doing regular exercise (Burhanuddin 2013).

3.3.1 Distribution of stroke patients based on risk factors for hypertension

Hypertension is often cited as the main cause of stroke. Hypertension causes damage to blood vessel walls due to blood pressure that exceeds normal limits and the release of collagen. A peeled endothelium causes a positively charged basement membrane to attract negatively charged platelets, resulting in platelet aggregation. Also, thrombokinase is released, causing stable blood clots, and if the blood vessels are no longer strong enough to withstand high blood pressure, it will result in a fatal rupture of blood vessels in the brain – a stroke. Hypertension can cause ischemic stroke resulting in atherosclerosis and degenerative processes of blood vessels that can cause microaneurysms, a part of ischemic stroke (Eshak et al. 2017).

Table 5 shows that, of 60 patients, the average blood pressure indicated first-degree hypertension in both men and women. This is consistent with various theories stating that a history of hypertension is a major risk factor for stroke. Hypertension is a major cause of complications in several cardiovascular diseases and one of the most common health problems in the world. Data obtained from Wahidin Sudirohusodo in 2015–2016 imply that a history of hypertension is a risk factor for stroke in young adults aged 18–40 years old (Burhanuddin 2013).

Table 5. Distribution of ischemic stroke patients based on risk factors for hypertension.

Variable	Frequency	Men	Women	(K-Wallis)
Systolic (mmHg)	Minimum	70.00	100.00	0.86
	Maximum	233.00	230.00	
	Median	150.00	150.00	
	Mean	147.53	150.96	
	SD	34.66	33.68	
Diastolic (mmHg)	Minimum	40.00	60.00	0.39
	Maximum	38.00	160.00	
	Median	90.00	90.00	
	Mean	88.59	95.11	
	SD	21.04	20.69	

Table 6. Distribution of ischemic stroke patients based on risk factors for dyslipidemia.

Variable	Frequency	Men	Women	(K-Wallis)
Cholesterol total (mg/dl)	Minimum	106.00	79.00	0.8
	Maximum	336.00	338.00	
	Median	199.50	206.50	
	Mean	201.62	198.43	
	SD	54.01	58.41	
Triglycerides (mg/dl)	Minimum	11.00	39.000	0.53
	Maximum	510.00	878.00	
	Median	127.00	164.00	
	Mean	135.06	174.86	
	SD	88.95	177.41	
HDL (mg/dl)	Minimum	21.00	21.00	0.09
	Maximum	85.00	59.00	
	Median	46.00	39.00	
	Mean	44.47	38.00	
	SD	13.00	11.12	
LDL (mg/dl)	Minimum	224.00	40.00	0.93
	Maximum	224.00	237.00	
	Median	131.00	132.50	
	Mean	125.91	124.57	
	SD	44.38	45.16	

3.3.2 *Distribution of stroke patients based on risk factors for dyslipidemia*

Table 6 shows that most women patients have dyslipidemia because their total cholesterol and LDL levels increase while their HDL decreases. In turn, for men, only LDL levels have increased. When the HDL level drops, it shows that a significant value can cause an increase in risk factors for ischemic stroke.

This happens because cholesterol levels in normal circumstances increase in young adults at an average of 200 mg. Hypercholesterolemia is a disease that is chronic or prolonged while hypocholesterolemia in young adults is less frequent. Patients who have a history of hyper-cholesterolemia suffer from stroke because their cholesterol levels increase due to their unhealthy lifestyle choices such as consuming foods high in cholesterol and saturated fat. High cholesterol levels can cause atherosclerosis, which results in narrowing of the walls of blood vessels, disrupting the blood supply to the brain. This is what will cause strokes (Alchuriyah & Wahjuni 2016; Burhanuddin 2013; Eshak et al. 2017; Sitorus 2008; Syifa et al. 2017).

The incidence of stroke in young adults aged 18–40 years old has an OR value of 3.92 (95% CI 1.939–7.928). This means that patients with a history of hypercholesterolemia have risk of

Table 7. Distribution of ischemic stroke patients based on increased uric acid.

Variable	Frequency	Men	Women	(K-Wallis)
Uric acid (mg/dl)	Minimum	2.10	3.30	0.00
	Maximum	11.10	11.70	
	Median	4.60	6.55	
	Mean	5.27	6.72	
	SD	2.39	1.88	

stroke that is 3.92 times greater than that in young adults aged 18–40 years old compared to patients who have no history of hypercholesterolemia. The results of this study were in line with research conducted by Setyowati in (Sitorus 2008), who found that a history of hypercholesterolemia had a 4.64 times greater risk of suffering from stroke compared to someone who did not have a history of hypercholesterolemia (Burhanuddin 2013; Eshak et al. 2017; Sitorus 2008). Research conducted at the Dr. Hasan Sadikin General Hospital in Bandung found that the high incidence of ischemic stroke in the group aged 35–45 years old can be associated with an increased percentage of risk factors for dyslipidemia. Dyslipidemia contributes to the formation of atherosclerosis, which is one mechanism of ischemic stroke (Syifa et al. 2017).

3.4 *Distribution of stroke patients based on risk factors for increased gout*

Table 7 shows that the average uric acid of women and men was normal with a significant P result.

Research conducted by Storhaug and colleagues (2013) also showed that gout is a risk factor for ischemic stroke.

4 CONCLUSIONS

Based on the results of this research, the risk factors for stroke for women are the same as those for men aged 36–40 years old. Vascular risk factors increase blood sugar levels while hypertension and dyslipidemia are less significant in causing ischemic strokes in young adults. Gout is also a risk factor for ischemic stroke in young adults.

REFERENCES

Alchuriyah, S. & Wahjuni, C. U. 2016. Faktor risiko kejadian stroke usia muda pada pasien rumah sakit Brawijaya Surabaya. *Jurnal Berkala Epidemiologi* 4(1): 62–73.
American Stroke Association. 2013. *Stroke statistics* [Online]. Retrieved from Stroke.org.uk.
Bell, K., Twiggs, J., Olin, B. R., & Date, I. R. 2015. Hypertension: The silent killer. Updated JNC-8 guideline recommendations. *Alabama Pharmacy Association* 334: 1–8.
Burhanuddin, M. 2013. Faktor risiko kejadian stroke pada dewasa awal (18–40 tahun) di kota Makassar tahun 2010–2012 1-14.
Del Mar Bibiloni, M., Salas, R., Yolanda, E., Villarreal, J. Z., Sureda, A., & Tur, J. A. 2016. Serum lipid profile, prevalence of dyslipidaemia, and associated risk factors among northern Mexican adolescents. *Journal of pediatric gastroenterology and nutrition* 63(5): 544–549.
Departemen Kesehatan Indonesia, Lap Nas 2013. Badan Penelitian dan Pengembangan Kesehatan. *Proporsi Penyebab Kematian pada Kelompok Umur 55–64 tahun menurut Tipe Daerah*. Jakarta: Badan Penelitian dan Pengembangan Kesehatan. Riset Kesehatan Dasar (RISKESDAS), 1–384.
Eshak, E. S., Honjo, K., Iso, H., Ikeda, A., Inoue, M., Sawada, N., & Tsugane, S. 2017. Changes in the employment status and risk of stroke and stroke types. *Stroke* 48(5): 1176–1182.
Ferrier, D. R. 2014. *Biochemistry*. Philadelphia: Lippincott Williams & Wilkins.
Grace, M., Jacob, K. J., Kumar, A. V., & Shameer, V. K. 2016. Role of dyslipidemia in stroke and comparison of lipid profile in ischemic and hemorrhagic stroke: A case control study. *International journal of advances in medicine* 3(3): 694–698.

Hoy, D. G., Rao, C., Hoa, N. P., Suhardi, S., & Lwin, A. M. M. 2013. Stroke mortality variations in South-East Asia: Empirical evidence from the field. *International journal of stroke* 8: 21–27.

Kusuima, Y., Venketasubramanian, N., Kiemas, L. S., & Misbach, J. 2009. Burden of stroke in Indonesia. *International journal of stroke* 4(5): 379–380.

MedicineNet. 2018. *The medical definition of lipid profile* [Online]. Retrieved from: MedicineNet.

Mozaffarian, D., Benjamin, E. J., Go, A. S., Arnett, D. K., Blaha, M. J., Cushman, M., & Howard, V. J. (2016). Executive summary: Heart disease and stroke statistics – 2016 update: A report from the American Heart Association. *Circulation* 133(4): 447–454.

Sacco, R. L., Kasner, S. E., Broderick, J. P., Caplan, L. R., Connors, J. J., Culebras, A., & Hoh, B. L. 2013. An updated definition of stroke for the 21st century: A statement for healthcare professionals from the American Heart Association/American Stroke Association. *Stroke* 44(7): 2064–2089.

Sitorus, R. J. 2008. Faktor-Faktor Risiko Yang Mempengaruhi Kejadian Stroke Pada Usia Muda Kurang Dari 40 Tahun (Studi Kasus Di Rumah Sakit Di Kota Semarang). *Jurnal Epidemiologi*. (unpublished).

Smajlović, D. 2015. Strokes in young adults: Epidemiology and prevention. *Vascular health and risk management* 11: 157–164.

Storhaug, H. M., Norvik, J. V., Toft, I., Eriksen, B. O., Løchen, M. L., Zykova, S., & Jenssen, T. 2013. Uric acid is a risk factor for ischemic stroke and all-cause mortality in the general population: A gender specific analysis from the Tromsø study. *BMC cardiovascular disorders* 13(115): 1–10.

Syifa, N., Amalia, L. & Bisri, D. Y. 2017. Gambaran Epidemiologi Pasien Stroke Dewasa Muda yang Dirawat di Bangsal Neurologi RSUP Dr. Hasan Sadikin Bandung Periode 2011–2016. *Jurnal Neuroanestesi Indonesia* 6(3): 143–150.

Valery, L. F. & Krishnamurthi, R. V. 2016. Global burden of stroke. Ed. J. C. Grotta, G. W. Albers, J. P. Broderick, S. E. Kasner, E. H. Lo, A. D. Mendelow, & L. K.S. Wong *Stroke: Pathophysiology, diagnosis and management*. 6th edn. New York: Elsevier, 207–216.

Van Alebeek, M. E., Arntz, R. M., Ekker, M. S., Synhaeve, N. E., Maaijwee, N. A., Schoonderwaldt, H., & de Leeuw, F. E. 2018. Risk factors and mechanisms of stroke in young adults: The FUTURE study. *Journal of cerebral blood flow & metabolism* 38(9): 1631–1641.

Medical Technology and Environmental Health – Abdullah, Widiaty & Abdullah (eds)
© 2020 Taylor & Francis Group, London, ISBN 978-0-367-86053-0

Antihepatotoxic activity of ethanol extract of *Ocimum americanum L.* on isoniazid-induced hepatotoxicity mice

E. Multazam, D. Renovaldi & Y. Safitri
Universitas Muhammadiyah Jakarta, South Tangerang, Indonesia

ABSTRACT: Isoniazid (INH) remains a fundamental drug for tuberculosis (TB) treatment despite the fact that it causes hepatotoxicity due to bioactivation of the metabolites and oxidative stress. Ocimum americanum (OA) is Lamiaceae herb that contain antioxidants and potential to protect against free radicals and reactive oxygen species (ROS) molecules. This study aims to investigate the hepatoprotector potential of Ocimum americanum L. ethanol extract against INH-induced hepatotoxicity mice. The liver biomarker serum, alanine transaminase (ALT) and aspartate transaminase (AST) were investigated. Treatment groups (INH 100mg/kg BW + OA 2,8 mh/20gr BW and INH 100mg/kg BW + OA 5,6 mg/20gr BW) were compared to INH group (INH 100 mg/Kg BW) and to normal control group for 14 days. Treatment with INH significantly increased the ALT and AST levels, meanwhile OA-treatment reduced these activities (p-value<0.05). These results suggest the potential of antihepatotoxic role of Ocimum americanum L. ethanol extract related to its antioxidant activity.

1 INTRODUCTION

Due to its high efficacy rate, isoniazid (INH) remains the drug of choice for the treatment of latent TB (TB) despite the fact that it can cause liver injury (Wang et al. 2016). Drug-induced liver injury (DILI) caused by INH has the clinical characteristics of liver damage that are quite typical, including malaise effects, jaundice, nausea, and vomiting. The duration of therapy before manifestation can vary between 1-25 weeks with an average of 12 weeks. In most cases, liver injury cannot be noticeable and only diverts by analyzing liver biomarkers such as alanine aminotransferase (ALT) and aspartate aminotransferase (AST). This is important for mild cases of liver damage, which occurs in 20% of patients who require it. In more than half of patients, an increase in ALT occurs between 1–6 months (Sotsuka et al. 2011). The sudden increase in ALT that causes liver failure is idiosyncratic and is not clearly related to medication, drug dosage, fever or eosinophil count. When liver injury is identified, the first treatment is to stop the drug and monitor patient's recovery. However, the re-challenge of patients with more severe liver injuries can lead to a rapid onset of symptoms (within a few hours) and is contraindicated (Metushi 2016).

DILI is defined as liver damage or disease caused by drugs, plants, or other related substances. This is the most common cause of acute liver failure in the United States (US). DILI accounts for 7% of the side effects of drugs, 2% of jaundice in hospitals, and about 30% of fulminant liver failure. DILI has switched the hepatitis virus as a major cause of acute liver failure. A brief search of a commercial pharmacology database showed that there are more than 700 drugs with hepatotoxicity which were launched and used for use in the United States (Saukkonen et al. 2006). In particular, INH side effects such as liver injury is seen as controversial in the treatment for tuberculosis. The noteworthy is the using of full dose or discontinuation of anti-tuberculosis drugs because hepatotoxicity can lead to prolonged treatment time and potential to occur as drug-resistant Mycobacterium tuberculosis, so this can cause inhibition to TB treatment (Gourishankar et al. 2014).

Herbal medicines have been used by many people around the world for treatment including liver disease for using in a long time without significant toxic effects. Therefore, the use of herbal medicines can be complementary and alternative medicine, for the treatment of acute liver disease for better efficacy and safety (Mohamed et al. 2015). Ocimum americanum, also known as Ocimum canum, is a wild species in India but is generally cultivated in Indonesia for processing in the form of essential oils and food supplements. Ocimum plants contain many antioxidants such as vitamin C, vitamin E, flavonoids and carotenoids (Pandey et al. 2014). The presence of many active pharmacological compounds in the Ocimum species makes this species can have the potential to provide protection against oxidative-induced free radicals from cellular components including hepatotoxicity (Aluko et al. 2013). Our literature survey reveals that the hepatoprotective activity of Ocimum americanum leaves has not been widely discussed scientifically and extensively. The potential of this plant in Indonesia is considered very large because of the amount of cultivation and is very commonly used as a food supplement and utilization as herbal medicine. Based on this phenomenon, the purpose of this study is to prove the existence of hepatoprotective activity of ethanol extract of Ocimum americanum leaves in isoniazid-induced hepatotoxicity mice.

2 METHOD

2.1 *Chemical material*

All chemicals were of an analytical grade. Kits for aspartate aminotransaminase (AST) and alanine amino transaminase were purchased from ELITech Clinical System SAS Z.I. 61500 SEES, France.

2.2 *Preparation of Ocimum americanum L. ethanol extract*

Standardized and certified plant leaves are obtained from the Bogor Spices and Medicinal Plant Research Institute (BALITTRO) Bogor, Indonesia. Basil leaves as many as 0.5-1 kg are washed thoroughly with water, dried using an oven with a temperature of 60°C. Then mashed using a blender until it becomes smoother and dried, then sifted using a sieve until a fine powder is obtained. Furthermore, 128 grams of basil leaf powder was put into the macerator, after that added 90% ethanol solvent as much as 1 L (ratio 1: 7.5), allowed to stand for 24 hours. After that, the filtrate shelter was carried out. The pulp obtained from the filtering results is then macerated (repeated a maximum of 3 times). Maceration method by soaking using 90% ethanol and stirring continuously. After the filtrate is obtained, evaporation is carried out using a rotary evaporator at a temperature of 50°C until the 90% ethanol solvent evaporates and the semi-ethanol basil extract is obtained.

2.3 *Experimental animals*

28 male mice (Strain Deutschland, Denken, & Yonken/DDY) were purchased from the Non-Ruminent and Animals Laboratory, Faculty of Animals Husbandry, Bogor Agricultural Institute. The mice weighed about 30-40 gr. Animals were provided with a standard diet that has been recommended by the breeder. After 7 days of acclimatization period, they were randomly assigned into four groups, each comprising seven mice. Group I (normal control group) mice received 0,2 ml of carboxymethyl cellulose 1% (CMC1%). Group II (INH group) mice received 100 mg/kg BW. Group III and IV (treatment groups) mice received 2.8 mg/20gr BW extract ethanol of *Ocimum americanum* + INH 100 mg/kg BW and 5.6 mg/20gr BW extract ethanol of *Ocimum americanum* + INH 100 mg/kg BW respectively. All treatments were given intragastrically once a day for 14 days.

2.4 Sample analysis and biochemical parameters

At the end of the experimental period, mice remained fasting for 10 h before sacrifice. Blood samples were collected for biochemical studies. The serum was separated by centrifugation for 10 min at 3000 rpm. AST and ALT activities were estimated in serum using the respective kits. The comparison of the reagents was R1 4: R2 1 based on the kits instruction (ELITech, France) and estimated using a clinical chemistry analyzer for quantifications (Microlab 300, ELITechGroup, France).

2.5 Statistical analysis

All results were expressed as mean value ± SD. The statistical analysis was performed using a non-parametric test of analysis of variance (Kruskal Wallis) followed by Mann-Whitney test with p-value < 0.05 considered statistically significant.

3 RESULTS

3.1 Isoniazid increased ALT and AST levels

In this study, it was found that there were significant differences in the activity of serum transaminases ALT (p-value 0.001) and AST (p-value 0.00) in the INH group compared to the normal control group. Figure 1 shows an increase in serum ALT and AST in the INH group and a striking increase from day 1, day 7 to day 14. While in the normal control group there was no significant increase in serum ALT and AST which tended to be stable until the 14th day.

The results of this analysis indicate that the administration of 100 mg/kg BW isoniazid can have a hepatotoxic effect characterized by elevated levels of serum ALT and AST liver function biomarkers that are significant in the study sample.

3.2 Ocimum americanum reduce the activity of serum transaminase enzymes

Kruskal-Wallis test analysis was carried out to determine the mean differences in the population of serum ALT and AST levels in the treatment and control groups after administration of

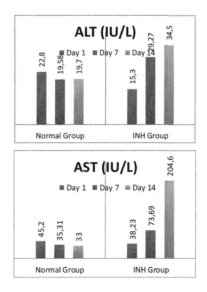

Figure 1. Isoniazid administration increases the activity of ALT and AST serum in mice.

Table 1. Effect of OA treatment on serum AST and ALT of INH-induced hepatotoxicity mice.

Groups	AST (IU/L)	ALT (IU/L)
Normal Control	38.78 ± 4.35	12.7 ± 2.09
Isoniazid	207.0 ± 20.8	22.46 ± 2.87
Isoniazid + OA 2,8 mg/20gr BW	124.4 ± 18.65	13.83 ± 3.42
Isoniazid + OA 5,6 mg/20gr BW	137.3 ± 28.4	14.19 ± 4.57

Ocimum americanum ethanol extract. Kruskal-Wallis test results showed that there were statistically significant mean differences in serum AST levels and in serum ALT levels between each group of study subjects. The next test used was the Mann Whitney test to find out which groups that have a significant difference to another group. In the ALT serum level, it was found that a significant difference was seen in both treatment groups with a dose of Ocimum americanum ethanol extract of 2.8 mg/20gr BW (p-value 0.001) and in the dose of 5.6 mg/20gr BW (p-value 0.002). These findings also found in the AST serum level, there were significant differences found in the treatment groups with the dosage of the Ocimum americanum ethanol extract 2.8 mg/20gr BW (p-value 0.001) and at a dose of 5.6 mg/20gr BW (p-value 0.001).

Table 1 shows additional results which explain that the administration of *Ocimum americanum* ethanol extract can significantly reduce the activity of the ALT and AST enzymes in the serum of experimental animals given hepatotoxic induction using isoniazid.

The results of this study indicated that protection by Ocimum americanum extract had a potential effect in decreasing liver function damage markers (ALT and AST levels) in both the extract dosage of 2.8 mg/20gr BB and 5.6 mg/20gr BB. However, a statistically significant decrease was only found in serum ALT levels, whereas in AST was considered to be an insignificant decrease. Furthermore, the results of this study also showed that the decrease in serum ALT and AST levels were not affected by the number of administered dosage of Ocimum americanum extract.

4 DISCUSSION

The isoniazid hepatotoxic effect was observed in oral administration at a dose of 100 mg/kg for 14 days of treatment. We evaluated the presence of liver injury in experimental animals through clinical biochemical examination of ALT and AST levels in serum samples. Increased ALT and serum AST are markers of hepatotoxicity. This study showed that in the INH group there was a significant and significant increase in serum ALT, especially on AST examination compared to the normal control group. These results are similar to research conducted by Kim et al. (2017) who found that liver injury was characterized by an increase in serum ALT and AST and followed by pathologic changes at the molecular level in samples of rats given isoniazid treatment. Research conducted by Lian et al. (2013) also found that isoniazid can cause an increase in hepatic oxidative stress conditions, histopathological injury to the liver, increase in serum AST and ALT, and increase in plasma malondialdehyde (MDA) in mice given isoniazid compared to normal control groups.

ALT is the biomarker most often used to detect hepatoxicity. This is a liver enzyme that plays an important role in amino acid metabolism and gluconeogenesis. This enzyme catalyzes the reductive transfer of amino groups from alanine to α-ketoglutarate to produce glutamate and pyruvate. Increased levels of this enzyme will be released in circulation during damage to liver cells. On the other hand, AST is another liver enzyme that helps in producing/forming proteins. This enzyme catalyzes the reductive transfer of amino groups from aspartate to α-ketoglutarate to produce oxaloacetate and glutamate. Aside from the liver, AST is also found in other organs such as the heart, muscles, and kidneys. Injury to this organ tissue can cause an increase in AST levels in the blood. AST also helps in detecting hepatocellular necrosis and acute injury to the liver. (Pandit et al. 2012, Singh et al. 2011).

Isoniazid administration is known to produce many metabolic and morphological disorders in the liver. this is believed because the liver is the main detoxification site for this antituberculosis drug. INH is an easily oxidized hydrazide molecule (Metushi et al. 2016). Three metabolites have been reported to be responsible for liver injury induced by INH, namely acetyl hydrazine (AcHz), hydrazine (Hz) and more recently a metabolite that results from bioactivation of INH itself (Pandey 2014). Experiments involving AcHz and Hz as hepatotoxic species were conducted several decades ago, mostly in rats where acute liver injury correlated with cHzal-AcHz binding and with blood Hz levels (Wang et al. 2016). In this connection, many studies have shown that one of the main pathways reported to be involved in hepatotoxicity by isoniazid is the presence of toxic metabolites that can induce oxidative stress in the liver. Many studies have proven that the balance between pro-oxidants and antioxidants is very important for cellular homeostasis. Research conducted by Lian et al. (2013) support this hypothesis that INH causes a decrease in the status of antioxidant levels in the liver, as evidenced by a significant decrease in the activity of Glutathione Peroxidase (GPx), one of the main antioxidant molecular enzymes.

The hypothesis in this study is the potential for hepatoprotective activity by antioxidant compounds derived from Ocimum americanum. This study found that the administration of Ocimum americanum extract can reduce levels of ALT and AST in mice induced by isoniazid hepatotoxicity. This result was also found in a study conducted by Aluko et al. (2013) who found that administration of Ocimum americanum extract can provide antihepatotoxic effects by reducing levels of ALT, AST, ALO, and TBIL to normal values and helps improve liver histopathological structure in rats induced by hepatotoxicity using paracetamol. Similar research was also carried out by Radhika et al. (2017) who found that administration of Ocimum americanum extract can mediate the recovery of enzymes that show liver damage (SGOT, SGPT, ALP, and GGT) to normal levels and also found an increase in pro-antioxidant enzymes (LPO, SOD, GSH, and GPx) in Hepatotoxic samples. In addition, the administration of Ocimum americanum extract can also significantly reduce the lipid profile (LDL, VLDL, and cholesterol) in the isoniazid-induced sample group compared to the sample group without administration of Ocimum americanum extract. This is considered to have a positive effect where research conducted by Ghadir et al. (2010) showed that serum HDL, LDL, and total cholesterol levels had a positive correlation with the severity of liver cell damage and administration Ocimum americanum have ability to prevent this injury.

The results of this study indicate that the Ocimum americanum extract has a protective effect on damage that occurs in liver cells. The mechanism of hepatoprotection is believed to originate from the presence of phytochemical composition in the Ocimum americanum, especially those that act as antioxidants. Antioxidants are substances that can prevent cell oxidation due to exposure to toxins and free radicals. Drug metabolism can produce metabolites that are toxic, especially in the liver which is the main organ that performs metabolism and drug detoxification, including isoniazid. Reactive Oxygen Species (ROS) including single oxygen, superoxide ion, hydroxyl ion, and hydrogen peroxide are reactive molecules and toxins produced by cells during normal metabolism. However, exposure to toxin metabolites from drugs can cause an increase in ROS production so that it can cause damage to proteins, lipids, enzymes, and DNA, and is often associated with the pathogenesis of oxidative diseases. Ocimum plants are known to contain large amounts of antioxidants such as vitamin C, vitamin E, flavonoids, and carotenoids. This antioxidant composition is pharmacologically active in producing a protective effect against free radicals induced by oxidative damage from cell components (Pandey et al. 2014). This statement supports the results of this study to declare that Ocimum americanum has an effect as an antihepatotoxic agent against isoniazid-induced hepatotoxicity.

5 CONCLUSION

This study showed that isoniazid is able to cause hepatotoxicity in the liver of mice characterized by significant increases in ALT and AST serum levels. Treatment with Ocimum americanum ethanol extract was considered to have an antihepatotoxic effect marked by the decrease in ALT and AST serum levels in the hepatotoxic-induced sample mice. These results suggest that the administration of *Ocimum americanum* ethanol extract has an effect in providing a protector against hepatotoxicity caused by the administration of isoniazid.

ACKNOWLEDGEMENT

We would like to thank the Faculty of Medicine and Health, Universitas Muhammadiyah Jakarta for funding us to conduct this study. We also thank Faculty of Pharmacy, Pancasila University for providing us with all the necessary facilities to complete this work.

REFERENCES

Aluko, B. T., Oloyede, O. I. & Afolayan, A. J. 2013. Hepatoprotective activity of Ocimum americanum L Leaves against paracetamol–induced liver damage in rats. *Am J Life Sci.* 1: 37–42.

Ghadir, M. R., Riahin, A. A., Havaspour, A., Nooranipour, M. & Habibinejad, A. A. 2010. The relationship between lipid profile and severity of liver damage in cirrhotic patients. *Hepatitis monthly* 10(4): 285.

Gourishankar, A., Navarro, F., DebRoy, A. N. & Smith, K. C. (2014). Isoniazid hepatotoxicity with clinical and histopathology correlate. *Annals of Clinical & Laboratory Science* 44(1): 87–90.

Kim, J. H., Nam, W. S., Kim, S. J., Kwon, O. K., Seung, E. J., Jo, J. J., ... & Lee, H. S. 2017. Mechanism investigation of rifampicin-induced liver injury using comparative toxicoproteomics in mice. *International journal of molecular sciences* 18(7): 1417.

Lian, Y., Zhao, J., Xu, P., Wang, Y., Zhao, J., Jia, L., ... & Peng, S. 2013. Protective effects of metallothionein on isoniazid and rifampicin-induced hepatotoxicity in mice. *PloS one* 8(8).

Metushi, I., Uetrecht, J. & Phillips, E. 2016. Mechanism of isoniazid-induced hepatotoxicity: then and now. *British journal of clinical pharmacology* 81(6): 1030–1036.

Mohamed, A. H., Elhusain, B. A. & Abuelgasim, A. I. 2015. Hepatoprotective activity of ethanol extract of Ocimum basilicum against CCl4–induced hepatotoxicity in albino rats. *Journal of Science and Technology* 16(2).

Pandey, A. K., Singh, P. & Tripathi, N. N. 2014. Chemistry and bioactivities of essential oils of some Ocimum species: an overview. *Asian Pacific Journal of Tropical Biomedicine* 4(9): 682–694.

Pandit, A., Sachdeva, T. & Bafna, P. 2012. Drug-induced hepatotoxicity: a review. *J Appl Pharm Sci.* 2(5): 233–43.

Radhika, J., Jothi, G., Nivethetha, M., Vijayaroslin, L. & Yuvarani, S. 2017. Hepatoprotective Efficacy of Ocimum Canum Sims. *on Isoniazid Induced Hepatotoxicity.*

Saukkonen, J. J., Cohn, D. L., Jasmer, R. M., Schenker, S., Jereb, J. A., Nolan, C. M., ... & Bernardo, J. 2006. An official ATS statement: hepatotoxicity of antituberculosis therapy. *American journal of respiratory and critical care medicine* 174(8): 935–952.

Singh, A., Bhat, T. K. & Sharma, O. P. 2011. Clinical biochemistry of hepatotoxicity. *J Clin Toxicol* 4(0001): 1–9.

Sotsuka, T., Sasaki, Y., Hirai, S., Yamagishi, F. & Ueno, K. 2011. Association of isoniazid-metabolizing enzyme genotypes and isoniazid-induced hepatotoxicity in tuberculosis patients. *In vivo* 25(5): 803–812.

Wang, P., Pradhan, K., Zhong, X. B. & Ma, X. (2016). Isoniazid metabolism and hepatotoxicity. *Acta pharmaceutica sinica B* 6(5): 384–39`2.

Medical Technology and Environmental Health – Abdullah, Widiaty & Abdullah (eds)
© 2020 Taylor & Francis Group, London, ISBN 978-0-367-86053-0

Ameliorative effects of ethanol extract of sea cucumber (*Holothuria edulis* spp.) in alloxan-induced rats

Y. Andriane, R.A. Indriyanti, R. Damailia & U.A. Lantika
Universitas Islam Bandung, Bandung, Indonesia

ABSTRACT: Oxidative stress is a common cause of organ damage, including liver injury. The purpose of this study was to observe the hepato-protective effects of sea cucumber (*Holothurian edulis* spp.) on rats suffering from alloxan-induced liver injury. This was an experimental study using Wistar strain rats. The rats were randomly divided into six groups, each containing six rats (a control group, an alloxan-only group, a glibenclamide group, and ethanol extract of sea cucumber groups divided into three doses). After the rats were given alloxan, researchers intervened with ethanol extract of sea cucumber and glibenclamide for 12 days. At the end of the study, the rats were sacrificed and their liver histopathologies were determined. The results revealed sinusoid dilatation and congestion, central veins that were not intact, dilated portal veins and bile ducts in the portal area, and a histiocytic appearance of the cellular architecture in the alloxan group. However, in the ethanol extract of sea cucumber groups, the sinusoid began narrowing and the central venous was intact, especially in the 200 mg/200 gBW/day dose group. This study showed that ethanol extract of sea cucumber has potential as a hepato-protective agent that might be due to reduction of oxidative damage by decreased free radical generation.

1 INTRODUCTION

The liver is the main organ of metabolism and has important functions to protect the body against the accumulation of harmful substances such as drugs and other xenobiotics (Tso & McGill 2003). Despite these functions, hepatic diseases remain a problem. No completely effective drugs have yet been developed that repair liver function, protect the liver, or regenerate hepatic cells. Hence, it is necessary to find alternative substances or natural products for the treatment of liver diseases (Abdulkadir & Tungadi 2018; Madrigal-Santillán et al. 2014; Rachmawati & Ulfa 2018).

The human body is continuously exposed to different types of agents that result in the production of reactive species called free radicals (reactive oxygen species), which by the transfer of their free, unpaired electrons cause the oxidation of the cellular machinery. In order to counter the deleterious effects of such species, the body has endogenous antioxidant systems or it obtains exogenous antioxidants from diet that neutralize such species and keeps the homeostasis of the body. Any imbalance between the reactive species and antioxidants leads to a condition known as "oxidative stress" that results in the development of a pathological condition (Asmat et al. 2016).

Oxidative stress can be defined as any disturbance in the balance of antioxidants and prooxidants in favor of the latter due to different factors such as aging, drug interactions and toxicity, inflammation, and/or addiction. Biomarkers of oxidative stress include lipid peroxidation (LPO), vitamins, glutathione, catalase (CAT), and superoxide dismutase (SOD) (Asmat et al. 2016).

Sea cucumbers have long been used as a food and traditional medicine in Asian countries with *Stichopus hermanni*, *Thelenota ananas*, *Thelenota anax*, *Holothuria fuccogilva*, and *Actinopyga mauritiana* the most highly valued species. These organisms are potential sources of high value-added compounds with therapeutic properties such as triterpene glycosides, carotenoids, bioactive peptides, vitamins, minerals, fatty acids, collagens, gelatins, chondroitin sulfates, and amino acids. Health benefit effects of sea cucumbers have been validated through scientific research and have shown medicinal value such as wound healing, neuroprotection, and antitumor, anticoagulant, antimicrobial, and antioxidant properties (Pangestuti & Arifin 2018).

Sea cucumber mixed extract contains physiologically active phenolic compounds with antioxidant activity, which affords potential hepato-protective activity against thioacetamide-induced liver injury in a rat model (Esmat et al. 2013). Research conducted by Abdulkadir and Tungadi (2018) has shown that sea cucumber can improve hepatic damage induced by paracetamol in mice.

Holothuria edulis has low values, but abundant sea cucumber species live in the Red Sea in Saudi Arabia (Hasan 2009). The current study aimed to investigate the protective effects of the sea cucumber *Holothuria edulis* extract against alloxan-induced liver injury in male Wistar rats.

Alloxan is a pyrimidine derivate. Alloxan-induced diabetes triggered morphological and ultrastructural liver changes that closely resembled human disease, ranging from steatosis to steatohepatitis and liver fibrosis (Lucchesi et al. 2015).

Based on this description, the aim of this study was to evaluate the hepato-protective activity of sea cucumber in treating alloxan-induced liver injury in rats.

2 MATERIALS AND METHODS

2.1 *Chemicals*

Alloxan was purchased from Aldrich Chemistry.

2.2 *Animals*

Male Wistar rats weighing 200–300 g were obtained from the animal laboratory of Pusat Pendidikan Antar Universitas (PAU) Bandung Institute of Technology, Bandung, Indonesia. Animals were given food and water ad libitum. Rats were maintained in a good environment with a 12 h/12 h light-dark cycle. Rats were acclimatized to laboratory conditions for 7 days.

2.3 *Sample collection*

Sea cucumbers (*Holothuria edulis* spp.) were collected from the Makassar coast of Indonesia. The taxonomic identity of the samples was confirmed by the biology department at Padjadjaran University. The animals were transported to our laboratory in an icebox containing ice cubes and a pinch of table salt. All visceral organs were removed and then washed until clean, then cut into small pieces and dried using an oven at 70°C. The dried samples were then mashed using a blender. The mashed sea cucumber was then put into a container for maceration and soaked with ethanol for 24 hours, then filtered using filter paper. The extracted solution was evaporated using an evaporator at 70°C. The extract was concentrated using a water bath at 70°C until it produced a viscous sea cucumber extract/paste that was ready to use.

2.4 *Preparation of Na-Carboxymethyl cellulose (CMC) 0.5% suspension*

An Na-CMC suspension was prepared by gradually dissolving 5 g of CMC into 100 ml of distilled water (heated at 70°C) and stirring with an electric stirrer until homogenous suspension was performed.

2.5 *Preparation of glibenclamide suspension dose 0.1 mg/200 gBW*

Glibenclamide was suspended into 0.5% Na-CMC prepared by diluting 5 mg of glibenclamide into 100 ml of Na-CMC 0.5%.

2.6 Preparation of extract of sea cucumber into Na-CMC 0.1% suspension

The extract/paste of sea cucumber was suspended into an Na-CMC 0.5% suspension made by dissolving the extract, which had been weighed according to the desired concentrations of 100 mg/200 g BW, 200 mg/200 g BW, and 400 mg/200 g BW.

2.7 Work design

After a period of adaptation, 36 rats were initially randomized into six groups for treatment with a single intraperitoneal (i.p.) injection of 150 mg/BW alloxan dissolved in a normal saline (alloxan group (n = 6), glibenclamide group (n = 6), ethanol extract of sea cucumber groups 100 mg/200 gBW/day dose (n = 6), 200 mg/200 gBW/day (n = 6), and 400 mg/200 gBW/day (n = 6)) versus an equal volume of CMC 10% alone (n = 6) and served as a control group (Kusumawati et al. 2012; Rohilla & Ali 2012).

The rats receiving alloxan were randomly assigned to five groups (n = 6 for each group): the control group (treated with N-CMC only), the alloxan group (treated with alloxan only), the glibenclamide group (treated with alloxan followed by glibenclamide 0.1 mg/200 gBW), and the ethanol extract of sea cucumber groups (treated with alloxan followed by ethanol extract of sea cucumber divided into three doses: 100 mg/200 gBW/day, 200 mg/200 gBW/day, and 400 mg/200 gBW/day). Glibenclamide and ethanol extract of sea cucumber were administered in a divided dose orally for 12 consecutive days starting from day 3 after the alloxan treatment (Kusumawati et al. 2012; Rohilla & Ali 2012).

All animals were kept separately in the cage. At the time of euthanasia, liver tissues from all animals were collected and slices of liver tissue were processed for histopathological studies.

2.8 Ethical considerations

All experiments on laboratory animals were performed in accordance with the protocol approved by the Health Ethics Committee, the Faculty of Medicine, Islamic University of Bandung, No. 378/Komite Etik.FK/X/2018.

2.9 Histopathological preparation from the rat liver

Fresh slices of liver collected from all animals in all groups at the time of euthanasia were fixed in 10% formalin and embedded in paraffin wax blocks, then cut in sections and stained by hematoxylin and eosin and examined under a light microscope in order to determine pathological changes.

3 RESULTS

3.1 Liver histopathological examination

The liver sections obtained from control rats showed the characteristic normal histology of the liver, which exhibited the well-organized lobular architecture with the normal central vein (CV) and surrounding hepatocytes distributed radialis surrounded by normally sized sinusoids and apparently healthy liver parenchyma (Figure 1). The liver sections obtained from the group treated only with alloxan exhibited sinusoid dilatation, the central veins were not intact, the portal veins and the bile ducts in the portal area were dilated and congestion had occurred in the blood sinusoids, and they presented a histiocytic appearance (Figure 2). The liver sections prepared from the group receiving a combination of alloxan and glibenclamide showed that the sinusoids were still dilated and the central veins were not intact (Figure 3). The liver sections prepared from the group receiving a combination of alloxan and sea cucumber extract showed that the liver tissue had restored its structure to almost normal (Figures 4–6).

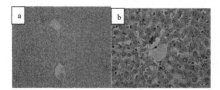

Figure 1. Photomicrograph of rat livers from the control group 100x (a), 400x (b) showing normal histology of the liver, which exhibits a well-organized lobular architecture with the normal central vein (CV) and surrounding hepatocytes distributed radialis surrounded by normally sized sinusoids and apparently healthy liver parenchyma.

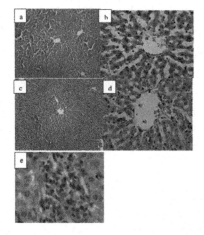

Figure 2. Photomicrograph of rat livers from the alloxan group (100x (a and c), 400x (b, d, e)) showing sinusoid dilatation. The central veins are not intact, the portal veins and the bile ducts in the portal area are dilated and congestion has occurred in the blood sinusoids, and there is a histiocytic appearance (e).

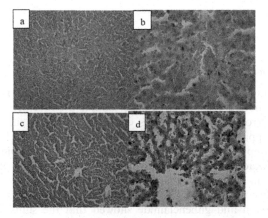

Figure 3. Photomicrograph of rat livers from the glibenclamide group (100x (a and c), 400x (b and d)) showing that the sinusoids still dilated and the central veins are not intact.

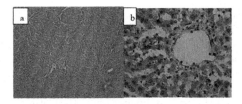

Figure 4. Photomicrograph of rat livers from the ethanol extract of sea cucumber 100 mg/200 g BW dose group (100x (a), 400x (b)) showing that the liver tissue restored its structure to almost normal, but the sinusoids are still dilated.

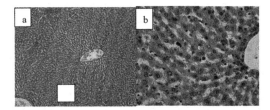

Figure 5. Photomicrograph of rat livers from the ethanol extract of sea cucumber 200 mg/200 g BW dose group (100x (a), 400x (b)) showing that the liver tissue restored its structure to almost normal.

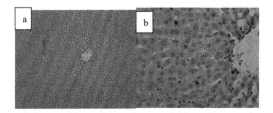

Figure 6. Photomicrograph of rat livers from the ethanol extract of sea cucumber 400 mg/200 g BW dose group (100x (a), 400x (b)) showing that the liver tissue restored its structure to almost normal.

4 DISCUSSION

Traditional medicine is focused on finding treatments for as many diseases as possible. The present study was designed to investigate the potential protective effect of sea cucumbers against liver injury in rats. Liver dysfunction in the rats manifested in liver tissue inflammation and abnormalities in the microstructure of the liver tissue (Figure 1). Histopathologic observations in this research found that alloxan caused liver damage and the effective dose of sea cucumber extract to prevent liver damage starts at 200 mg/200 g BW. This result may be due to the high protein content of sea cucumber. Amino acids, glutamate, and glycine are the main components of glutathione (GSH). These proteins can reuse GSH that is exhausted due to toxicity in cytoplasm and mitochondria. GSH is a robust peroxy-nitrite eliminator and it prevents the effective formation of nitro-tyrosine, reduces the necrosis of liver cells, and activates the cell regeneration cycle (Abdulkadir & Tungadi 2018; Asmat et al. 2016).

Hyperglycemia increases mitochondrial reactive oxygen species (ROS) production (Circu & Aw 2010). The generation of ROS such as superoxide anion rapidly induces apoptotic cell death (Circu & Aw 2010; Redza-Dutordoir & Averill-Bates 2016). The rise in blood glucose is accompanied by disturbance of the lipid profile, which makes patients at high risk for several complications, including fatty liver degeneration (Schofield et al. 2016; Zhou et al. 2008). Alloxan can induce diabetes and trigger liver morphological and ultrastructural changes (Lucchesi et al. 2015). LPO leads to a cascade reaction, generating endogenous toxicants that react with membrane proteins, which may lead to hepatic damage (Sahreen et al. 2011; Schofield et al. 2016).

The results of this research showed an ameliorating effect of sea cucumber extract on liver insult in rats. This effect is indicated by an improvement of the liver histological picture. The hepato-protective effect of sea cucumber extract manifested in this study might be due to antioxidant activity.

This is related to previous studies finding that extract of sea cucumber can improve hepatic damage because of the high protein content of sea cucumber (Abdulkadir & Tungadi 2018).

5 CONCLUSION

The conclusion reached from this study was that the extract of sea cucumber at a dose of 200 mg/200 gBW can improve alloxan-induced hepatic damage in rats.

ACKNOWLEDGMENT

We would like to thank the Faculty of Medicine, Islamic University of Bandung (UNISBA), Indonesia, for financial support for this study.

CONFLICT OF INTEREST

The authors claim no conflict of interest.

REFERENCES

Abdulkadir, W. S. & Tungadi, R. 2018. The effect of sea cucumber (*Holothuria scabra*) extract as hepatoprotective: Histopathological study. *Asian journal of pharmaceutical and clinical research* 11(9): 391–393.

Asmat, U., Abad, K., & Ismail, K. 2016. Diabetes mellitus and oxidative stress: A concise review. *Saudi pharmaceutical journal* 24(5): 547–553.

Circu, M. L. & Aw, T. Y. 2010. Reactive oxygen species, cellular redox systems, and apoptosis. *Free radical biology and medicine* 48(6): 749–762.

Esmat, A. Y., Said, M. M., Soliman, A. A., El-Masry, K. S., & Badiea, E. A. 2013. Basic nutritional investigation: Bioactive compounds, antioxidant potential, and hepatoprotective activity of sea cucumber (*Holothuria atra*) against thioacetamide intoxication in rats. *Nutrition* 29: 258–267.

Hasan, M. H. 2009. Stock assessment of holothuroid populations in the Red Sea waters of Saudi Arabia. *SPC Beche-de-mer information bulletin* 29: 31–37.

Kusumawati, I., Sukardiman, H., Studiawan, H., Rahman, A., Santosa, M. H., & Susandi, E. 2012. Hypoglycemic activity of the herbal tea combination of bitter melon (*Momordica charantia L.*) and *Lagerstroemia speciosa* leaves in alloxan-induced mice. *Group* 121(131.00): 10–12.

Lucchesi, A. N., Cassettari, L. L., & Spadella, C. T. 2015. Alloxan-induced diabetes causes morphological and ultrastructural changes in rat liver that resemble the natural history of chronic fatty liver disease in humans. *Journal of diabetes research* 2015: 1–11.

Madrigal-Santillán, E., Madrigal-Bujaidar, E., Álvarez-González, I., Sumaya-Martínez, M. T., Gutiérrez-Salinas, J., Bautista, M., ... & Morales-González, J. A. 2014. Review of natural products with hepatoprotective effects. *World journal of gastroenterology: WJG* 20(40): 14787–14804.

Pangestuti, R., & Arifin, Z. 2018. Medicinal and health benefit effects of functional sea cucumbers. *Journal of traditional and complementary medicine* 8(3), 341–351.

Rachmawati, E. & Ulfa, E. U. (2018). Uji Toksisitas Subkronis Ekstrak Kayu Kuning (Arcangelisia flava Merr) terhadap Hepar dan Ginjal. *Global medical & health communication* 6(1): 1–6.

Redza-Dutordoir, M. & Averill-Bates, D. A. 2016. Activation of apoptosis signalling pathways by reactive oxygen species. *Biochimica et Biophysica Acta (BBA) – Molecular Cell Research* 1863(12): 2977–2992.

Rohilla, A. & Ali, S. 2012. Alloxan induced diabetes: Mechanisms and effects. *International journal of research in pharmaceutical and biomedical sciences* 3(2): 819–823.

Sahreen, S., Khan, M. R., & Khan, R. A. 2011. Hepatoprotective effects of methanol extract of *Carissa opaca* leaves on CCl 4-induced damage in rat. *BMC complementary and alternative medicine* 11(48): 1–9.

Schofield, J. D., Liu, Y., Rao-Balakrishna, P., Malik, R. A. & Soran, H. 2016. Diabetes dyslipidemia. *Diabetes therapy* 7(2): 203–219.

Tso, P., & McGill, J. 2003. *The Physiology of the Liver. Medical Physiology*, RA Rhoades, GA Tanner, eds. Baltimore: Lippincott Williams and Wilkins, 514–525.

Zhou, J. Y., Zhou, S. W., Zhang, K. B., Tang, J. L., Guang, L. X., Ying, Y., ... & Li, D. D. 2008. Chronic effects of berberine on blood, liver glucolipid metabolism and liver PPARs expression in diabetic hyperlipidemic rats. *Biological and pharmaceutical bulletin* 31(6): 1169–1176.

Medical Technology and Environmental Health – Abdullah, Widiaty & Abdullah (eds)
© 2020 Taylor & Francis Group, London, ISBN 978-0-367-86053-0

The chronic effects of lemon aqueous fraction administration on body weight and visceral fat mass

A.R. Furqaani, R. Ekowati, A.B. Yulianti, M. Tejasari, H. Heriansyah & M.K. Dewi
Universitas Islam Bandung, Bandung, Jawa Barat, Indonesia

ABSTRACT: Flavonoids contained in lemon have beneficial effects in modulating lipid profile and obesity. This study aimed to determine the chronic effects of aqueous fraction of lemon administration on body weight and visceral fat mass of old mice fed by a high-fat diet. Twenty-six female mice were divided into 5 groups, group 1 (K1) were fed by high-fat diet (PTL); group 2 (K2) were fed by standard diet; and three other groups were fed by PTL and got aqueous fraction of lemon with 3 different doses, 0.20 g/kgBB (K3), 0.40 g/kgBB (K4), and 0.80 g/kgBB (K5) for 75 days. Initial body weight, final body weight, and visceral fat mass were measured. The results showed that the final weight and visceral fat mass of K3 were significantly lower among the groups ($p<0.05$). Therefore, chronic administration of lemon aqueous fraction, in a dose-dependent manner, has the potential to reduce visceral fat.

1   INTRODUCTION

Aging has been defined as a progressive and generalized impairment of function resulting in a loss of adaptive response to stress and a growing risk of age-associated disease (Samarakoon et al. 2011). Aging is characterized by a deterioration in maintaining metabolism homeostatic, leading to functional decline, increasing the risk for health deterioration and death. The aging process is characterized metabolically by insulin resistance, changes in body composition, and physiological declines in growth hormone (GH), insulin-like growth factor-1 (IGF-1), and sex steroids. Associated with aging are a dysregulation of metabolic homeostasis usually manifested as age-related obesity, diminished insulin sensitivity and impaired glucose and lipid homeostasis. Aged humans similar to rodents, marked by increased adiposity, develop increased fat mass, with a disproportionate increase in visceral fat (VF) compared with subcutaneous fat (SF) mass (Barzilai et al. 2012).

Metabolic deterioration contributes to the aging phenotype and metabolic pathologies are thought to be one of the main factors limiting the potential for lifespan extension (Bettedi & Foukas 2017). Metabolism process will decrease in line with increasing age. As aging occurs, muscle mass decreases thereby slowing down the metabolic process. Implementing a healthy-active lifestyle and consuming good nutrition consistently can slow down the aging process or promote "healthy aging". unfortunately, the modern lifestyle tends to lack physical activity and unhealthy instant diets with high-fat contribute to causing serious problems, including metabolic syndrome, cardiovascular disease and gastrointestinal issues (Atallah et al. 2018, Villareal et al. 2011).

Calorie restriction is the most potent environmental intervention known to increase lifespan and healthspan in some species including primates (Colman et al. 2014). The various mechanisms underlying the beneficial effects of calorie restriction on lifespan and healthspan, such as enhanced stress resistance and improved proteostasis (Mitchell et al. 2016), induces browning of the adipose tissue so that more insulin sensitive and glucose tolerant (Fabbiano et al. 2016), induces DNA-repair (Vermeij et al. 2016), or by modulating the Insulin/IGF-1 Signalling (IIS) and mechanistic Target of Rapamycin (mTOR) (Barzilai et al. 2012, Fontana et al. 2010, Kenyon 2010). The IIS and mTOR pathway are the most extensively studied pathways shown to regulate lifespan and healthspan in several model organisms (Fontana et al. 2010, Kenyon 2010). Pharmacological agents are widely developed to interfere with targeting components of these mechanisms to mimic the beneficial effects of calorie restriction.

On the other hand, various herbs and fruits also have potential in promoting health and providing protection against health problems, including dyslipidemia, cardiovascular diseases, or obesity (Kou et al. 2017, Kim et al. 2015, Karimah et al. 2014, Olukanni et al. 2013, Khan et al. 2010). Citrus has long been considered as valuable fruit for a healthy and nutritious diet. It is well established that some of the nutrients in citrus promote health and provide protection against chronic disease. The citrus genus is the most important fruit tree crop in the world and lemon is the third most important Citrus species. Lemon contains many chemical compounds such as phenolic compounds, flavon-oids, pectin, and other nutrients (vitamins, minerals, dietary fiber, essential oils, and carotenoids) (Kou et al. 2017, Kim et al. 2015, Al-Juhaimi 2014, Olukanni et al. 2013, Khan et al. 2010). Although the benefit of lemons on health has been widely studied, the effects of the lemon aqueous fraction are still limited. Therefore, this research aims to determine the chronic effects of lemon aqueous fraction administration on body weight and visceral fat mass of old mice fed by a high-fat diet.

2 METHOD

This study is pure in vivo laboratory experimental research with a completely randomized design. The experimental animals used in this study were mice (Mus musculus L.) DDY strains with inclusion criteria: adult females aged 45-50 weeks, weight 40-50 g, health, and responsive. Before the treatment period, an acclimatization period was done for one week. During the research period, animals placed in cages (size: 40x28x25 cm). Each cage filled with ten experimental animals. Cages are also equipped with food and drink so that animals can get access to food and drinks ad libitum. Wood sawdust is placed on the base of the cage to absorb animal urine and waste. Twenty-six female mice were divided into 5 groups, group 1 (K1) were fed by high-fat diet (PTL); group 2 (K2) were fed by standard diet; and three other groups that fed by PTL and got aqueous fraction of lemon with 3 different doses, 0.20 g/kg BW (K3), 0.40 g/kg BW (K4), and 0.80 g/kg BW (K5) for 75 days. The standard diet used in this study was CP-551. While the high-fat diet (PTL) was made from 1 kg of soluble cow fat and 20 duck eggs for every 5 kg CP-551.

Weight measurements were taken before and at the end of the treatment period. After that, the mice were sacrificed to be measured visceral fat mass. At the end of the study, all experimental animals were sacrificed and parts of the body of the mice that were not used in the study were collected and buried. Data of initial and final body weight obtained from the study were statistically analyzed using ANOVA Oneway followed by Fisher's Posthoc Test with confidence interval 95% ($\alpha=0.05$). While data of visceral fat mass were not normally distributed so that those data analyzed using Kruskal Wallis non-parametric test. Data are presented in mean $\pm$ SD (standard deviation). This study had been reviewed and approved by the Ethical Committee of Medical Faculty, Universitas Islam Bandung, with ethical approval number: 062/Komite Etik.FK/III/2018.

3 RESULTS AND DISCUSSION

Bodyweight and visceral fat mass of the research subject are presented in Table 1. The data show that there is no difference in the initial body weight of all groups. However, at the end of the treatment period, the average of final body weight of K3 was significantly lower than the entire group except for K2. The final body weight of K2 was also significantly lower compared to K1 which received the high-fat diet. The visceral fat mass of K3 is also significantly lower than that among the other groups. The visceral fat mass of K2 was significantly lower than that of K5.

Table 1. Body weight and visceral fat mass.

Parameters	K1 (N=5)	K2 (N=6)	K3 (N=5)	K4 (N=5)	K5 (N=5)	p Value
Initial Body Weight (g)	43,60±4,51	41,00±1,26	40,00±0,00	44,40±4,04	41,80±2,68	0.142
Final Body Weight (g)	43,80±1,92[a]	38,33±1,75[bc]	33,60±2,51[c]	41,60±5,90[ab]	40,60±6,43[ab]	0.010*
Visceral Fat Mass (g)	3,62±3,19	1,95±0,48*	0,84±0,62**	4,65±4,13	8,22±2,84	0.004*

Note: the* sign indicates significant difference. The letters a, b, c: parameters are not labeled with the same letter show significant differences; N: number of samples.

Data of K1 and K2 showed that a high-fat diet tends to increase visceral fat mass. Then by comparing K1 and the three treatment groups that received lemon aqueous fraction (K3, K4, K5), we can conclude that the effect this treatment on final body weight and visceral fat of animal research. The final body weight and visceral fat mass OF K3 were significantly lower than other groups, except for K2. These results indicate that lemon aqueous fraction has the potential to modulate lipid profile and controlling obesity. The bioactive compounds of lemon, such as phenolic compounds are involving as antioxidants, lowering and preventing obesity, and altering secretion of adipokine (Al-Juhaimi 2014). Citrus limon also consists of high vitamin C that plays a role as co-factor towards Enzyme 7-a-Hydroxylase so that it limits the synthesis of bile enzyme so that it will further decrease of low-density lipoprotein-cholesterol (LDL-C) in the blood (Charlton-Menys, & Durrington 2008). The various flavonoids of citrus, such as nobiletin, tangeretin, and hesperetin, are involved in lowering LDL-C level by reducing apolipoprotein B (apo B) and the triglycerides (TAG). As known that apo B and TAG are building blocks of the low-density lipoprotein (LDL) formation (Kou et al. 2017). This condition will lead to hypercholesterolemia (Karimah et al. 2014). The limonoid of citrus has a beneficial effect in lowering inflammation makers' tumor necrosis factor-α (TNF-α) and matrix metallopeptidase 9 (MMP-9). TNF-α contributes to inducing acute phase response along with insulin resistance. On the other hand, an increase of MMP-9 level is associated with instability of the atherosclerotic plaque structure (Kelley et al. 2015, Robertson et al. 2007).

Furthermore, pectin compounds can reduce and slow the increase in plasma triacylglycerol concentrations. The interaction between pectin and substrate emulsion results in inhibition absorption of lipase to the surface of substrate emulsion. Therefore, lemon has the potential to reduce fat mass and weight within a certain period (Kim et al. 2015, González-Molina et al. 2010). TAG, as in the case of hyperlipidemia, can lead to greater destabilization of lipoprotein particles and also affect the high-density lipoprotein (HDL) structure and their antiathero-genic properties (Kou et al. 2017). Olukanni et al. (2013) reported that 5 weeks of lemon juice intake improved the lipid profile by the significant reduction of total cholesterol, LDL-C and a significant increase in high-density lipoprotein-cholesterol (HDL-C). Olukanni et al. (2013) also showed that the elevation of serum glutathione (GSH) could be responsible in part for the reported antioxidant effects of lemon juice. Results suggest that the hypocholesterolemic effects of citrus lemon juice may be due to its antioxidant activities. Khan et al. (2010) also reported the hypolipidemic effect of citrus lemon, their research showed that the administration of citrus lemon juice (1ml/kg/day) for 30 days revealed a significant reduction in serum cholesterol, triglycerides, LDL levels and resulted in an increase in HDL. Another study showed that the lemon detox program significantly decreased the ratio of trunk fat to total fat, decreased serum IGF-1 levels through reducing circulating insulin levels, decreased serum high-sensitive C-reactive protein (hs-CRP) level. Therefore, the lemon detox program contributed to reducing body fat and improving insulin resistance through caloric restriction. Besides, the reduction of circulating hs-CRP and changes in adipokine production might have a potential beneficial effect on risk factors for cardiovascular disease (Kim et al. 2015).

In line with those researches, our study also shows that lemon aqueous fraction has the potential to reduce body weight and visceral fat mass. This might be related to several lemon bioactive compounds that have hypolipidemic activity as previously described. However, our results also showed that the mechanism of action of bioactive compounds in the lemon aqueous fraction works depending on the dose. The effective dose in reducing body weight and visceral body mass is the lowest dose, which is 0.2 g/kg BW. Otherwise, the chronic administration of lemon aqueous fraction for higher doses has the opposite effect.

4 CONCLUSION

We can conclude that the chronic administration of lemon aqueous fraction has a potential effect in controlling obesity by reducing visceral fat mass, but its mechanism of action depends on the dose of administration. Administration of lowest dose (0.20 g/kg BW) aqueous fraction of lemon significantly reduce visceral fat mass so that it also reduces the final bodyweight of the animal research.

ACKNOWLEDGEMENT

We would like to thank the Ministry of Research, Technology and Higher Education of the Republic of Indonesia for funding the research. We also thank Biomedical Laboratory, Faculty of Medicine, Universitas Islam Bandung (UNISBA), Indonesia, for facilitating this research study.

REFERENCES

Al-Juhaimi, F. Y. 2014. Citrus fruits by-products as sources of bioactive compounds with antioxidant potential. *Pak. J. Bot* 46(4): 1459–1462.

Atallah, N., Adjibade, M., Lelong, H., Hercberg, S., Galan, P., Assmann, K. E. & Kesse-Guyot, E. 2018. How healthy lifestyle factors at midlife relate to healthy aging. *Nutrients* 10(7): 854.

Barzilai, N., Huffman, D. M., Muzumdar, R. H. & Bartke, A. (2012). The critical role of metabolic pathways in aging. *Diabetes* 61(6): 1315–1322.

Bettedi, L., & Foukas, L. C. 2017. Growth factor, energy and nutrien sensing signalling pathways in metabolic ageing. *Biogerontology* 18(6): 913–929.

Charlton-Menys, V. & Durrington, P. N. 2008. Human cholesterol metabolism and therapeutic molecules. *Experimental physiology* 93(1): 27–42.

Colman, R. J., Beasley, T. M., Kemnitz, J. W., Johnson, S. C., Weindruch, R. & Anderson, R. M. 2014. Caloric restriction reduces age-related and all-cause mortality in rhesus monkeys. *Nature communications* 5(1): 1–5.

Kelley, D. S., Adkins, Y. C., Zunino, S. J., Woodhouse, L. R., Bonnel, E. L., Breksa III, A. P., & Mackey, B. E. 2015. Citrus limonin glucoside supplementation decreased biomarkers of liver disease and inflammation in overweight human adults. *Journal of functional foods* 12: 271–281.

Fabbiano, S., Suárez-Zamorano, N., Rigo, D., Veyrat-Durebex, C., Dokic, A. S., Colin, D. J. & Trajkovski, M. 2016. Caloric restriction leads to browning of white adipose tissue through type 2 immune signaling. *Cell metabolism* 24(3): 434–446.

Fontana, L., Partridge, L. & Longo, V. D. 2010. Extending healthy life span—from yeast to humans. *science* 328(5976): 321–326.

González-Molina, E., Domínguez-Perles, R., Moreno, D. A. & García-Viguera, C. 2010. Natural bioactive compounds of Citrus limon for food and health. *Journal of pharmaceutical and biomedical analysis* 51(2): 327–345.

Karimah, F., Achmad, S. & Prawiradilaga, R. S. 2014. Efek Jus Buah Naga Super Merah (Hylocereus costaricensis) dan Simvastatin terhadap Kadar Kolesterol Total Darah dan Bobot Badan Tikus Jantan Galur Wistar Hiperkolesterolemia. *Global Medical & Health Communication* 2(2): 79–84.

Kenyon, C. J. 2010. The genetics of ageing. *Nature* 464(7288): 504–512.

Khan, Y., Khan, R. A., Afroz, S. & Siddiq, A. 2010. Evaluation of hypolipidemic effect of Citrus lemon. *Journal of Basic and Applied Sciences* 6(1): 39–43.

Kim, M. J., Hwang, J. H., Ko, H. J., Na, H. B. & Kim, J. H. 2015. Lemon detox diet reduced body fat, insulin resistance, and serum hs-CRP level without hematological changes in overweight Korean women. *Nutrition Research* 35(5), 409–420.

Kou, G., Zhao, Z., Dong, X. Y., Zhang, Y., Guo, L. Y. & Zhou, Z. Q. 2017. Effects of Citrus Fruits on Blood Lipid Levels: A Systematic Review and Meta-Analysis. *Acta Medica Mediterranea* 33(6): 1143–1150.

Mitchell, S. J., Madrigal-Matute, J., Scheibye-Knudsen, M., Fang, E., Aon, M., González-Reyes, J. A., ... & Wahl, D. 2016. Effects of sex, strain, and energy intake on hallmarks of aging in mice. *Cell metabolism* 23(6): 1093–1112.

Olukanni, O. D., Akande, O. T., Alagbe, Y. O., Adeyemi, O. S., Olukann, A. T. & Daramola, G. G. 2013. Lemon juice elevated level of reduced glutathione and improved lipid profile in Wistar rats. *American-Eurasian J. Agric. & Environ. Sci.* 13(9): 1246–1251.

Robertson, L., Grip, L., Mattsson Hulten, L., Hulthe, J. & Wiklund, O. 2007. Release of protein as well as activity of MMP-9 from unstable atherosclerotic plaques during percutaneous coronary intervention. *Journal of internal medicine* 262(6): 659–667.

Samarakoon, S. M. S., Chandola, H. M. & Ravishankar, B. 2011. Effect of dietary, social, and lifestyle determinants of accelerated aging and its common clinical presentation: A survey study. *Ayu* 32(3): 315.

Vermeij, W. P., Dollé, M. E. T., Reiling, E., Jaarsma, D., Payan-Gomez, C., Bombardieri, C. R., ... & Youssef, S. A. 2016. Restricted diet delays accelerated ageing and genomic stress in DNA-repair-deficient mice. *Nature* 537(7620): 427–431.

Villareal, D. T., Chode, S., Parimi, N., Sinacore, D. R., Hilton, T., Armamento-Villareal, R., ... & Shah, K. 2011. Weight loss, exercise, or both and physical function in obese older adults. *New England Journal of Medicine* 364(13): 1218–1229.

Medical Technology and Environmental Health – Abdullah, Widiaty & Abdullah (eds)
© 2020 Taylor & Francis Group, London, ISBN 978-0-367-86053-0

The impact of tender coconut water on preventing lipid peroxidation and increasing antioxidant enzymes in lead-induced rats

S.T. Zulaikhah & J. Wahyuwibowo
Universitas Islam Sultan Agung (Unissula), Semarang, Indonesia

ABSTRACT: Plumbum (Pb) is a heavy metal that can trigger the formation of reactive oxygen species (ROS), increasing the generation of free radicals and lipid peroxidation, and lowering antioxidant enzymes such as glutathione peroxidase (GPx). Tender coconut water contains antioxidants, L-arginine, and other compounds that can reduce the effect of Pb exposure. The purpose of this research was to examine the effect of tender coconut water on prevention of lipid peroxidation and increasing antioxidant enzymes in lead-induced rats. This experimental study used a posttest-only control group design, with 18 white male Wistar strain rats randomly divided into three groups: K1 (standard feed only), K2 (standard feed + Pb), and K3 (standard feed + Pb + tender coconut water). The induction of Pb was performed by inhalation at a dose of 10 mg/day/rat, while tender coconut water (8 mL/200 gr BW rats/day) was given orally for 4 weeks. Rats' blood from the ophtalmicus venous was analyzed using ELISA to measure malondialdehyde (MDA) levels and GPx. Data were analyzed using ANOVA tests. Results showed that the average rate of MDA in Group 2 was increased compared to Group 1, but in Group 3, it decreased compared to Group 2. The GPx average in Group 2 decreased compared to Group 1, but in Group 3, it increased compared to Group 2. The statistical analysis obtained p-values of < 0.05. Tender coconut water administration was proven to be able to prevent lipid peroxidation and to increase the antioxidant enzymes characterized by increased GPx levels in lead-induced rats.

1 INTRODUCTION

Indonesia has the world's third highest air pollution level. Of the many sources of air pollution, motor vehicles (transportation) are the largest source (85%) (Ardillah 2016). In the past decade, NT exposure has expanded and become a health problem in society. The use of plumbum (Pb) in the workplace today must be controlled and restricted because it has a detrimental effect on health (Assi et al. 2016). Lead can be detected in the blood because more than 90% of lead metals are absorbed by red blood cells. Continuous Pb exposure will trigger the formation of reactive oxygen species (ROS), increase lipid peroxidation, cause oxidative stress in cells or tissues, inhibit enzymes, break down protein structures, damage DNA, and interfere with the body's antioxidant metabolism, particularly superoxide dismutase (SOD) and glutathione peroxidase (GPx), lowering glutathione (GSH) and vitamin C and inhibiting thiamin (B1) and pyridoxine (B6) (Flora et al. 2012). Malondialdehyde (MDA) is a highly reactive compound and is the end product of lipid peroxidation, usually used as a sign of the occurrence of lipid peroxidation in order to assess oxidative stress (Winarsi 2007). Oxidative stress due to Pb exposure can occur through depletion of antioxidants and increased ROS. Pb forms a covalent bond with sulfhydryl groups in antioxidants such as

GSH, glutathione reductase (GR), and glutathione S-transferase (GST). Pb also reduces the absorption of GST so that it can cause a decrease in the activity of antioxidant enzymes such as GPx, superoxide dismutase (SOD), and catalase (CAT), so that glutathione levels are reduced (GSH) and H_2O_2 is built up, which ultimately leads to oxidative stress (Abadin 2005).

In normal physiological conditions, the amount of endogenous antioxidants (antioxidant enzymes) is sufficient to ward off free radicals, but in conditions of increased free radicals caused by environmental pollution, an imbalance occurs between free radicals with antioxidants so that it can cause oxidative stress. This is where the exogenous antioxidant roles of food, fruits, or vegetables are essential to prevent effects such as heavy metal toxicity (Flora et al. 2012). Foods containing natural antioxidants can be used to reduce morbidity and mortality rates in particular due to oxidative stress. Natural antioxidants can protect the body against cell damage caused by ROS, inhibit the occurrence of degenerative diseases, and inhibit lipid peroxidation (Winarsi 2007).

Tender coconut water is a natural drink that contains beneficial compounds such as iron, vitamin C, vitamin B6, folic acid, L-arginine, and fatty acids. L-arginine is known to act as an antioxidant and can reduce the formation of free radicals. Vitamin C also acts as an antioxidant that prevents the occurrence of lipid peroxidase. Vitamin B6 plays a role in enhancing the production of GSH, and Vitamin B1 (thiamine) can prevent the occurrence of lipid peroxidation on the liver and kidneys of a rat induced with Pb (Abadin 2005). Earlier research results stated that tender coconut water administration proved to lower MDA levels and increase GPx levels in people exposed to mercury (Zulaikhah 2018). This research aimed to prove the effect of tender coconut water in preventing lipid peroxidation and increasing GPx levels in lead-induced rats.

2 MATERIAL AND METHODS

This study was designed as experimental research with a posttest control group.

2.1 *Ethical clearance*

This study received ethical clearance from the Bioethics Committee of Medical Research/Faculty of Medicine, Islam Sultan Agung Semarang University (42/II/2019/Bioethics Commission).

2.2 *Tender coconut water*

Tender coconut water was obtained from green coconut aged 5–7 months. At this age, coconut has tender, thin, jelly-like endosperm, and it is edible using a spoon. The dosage administered was 8 ml/200 grBW/day for 4 weeks (Loki & Rajamohan 2003).

2.3 *Experimental animals*

2.3.1 *Lead (Pb) induction*
The dosage of lead given was 10 mg/day by inhalation; 10 mg of lead powder was diluted with 1 mL aquadest, poured into an open container, then placed in the cage. In this research, one group consisted of six rats; thus one cage had one 60 mg lead container diluted with 6 mL aquadest (Baker 2007). Male white Wistar strain rats were used that met the following criteria: 2 months old, weighed 180–220 g, healthy looking, active motion, normal feeding and drinking, no injuries, and no disabilities. A total of 18 rats was adapted for 1 day and then randomly divided into three groups of six rats each and fed the following diet:

After 4 weeks, blood was drawn to measure the MDA and GPx levels.

Table 1. Lead (Pb) induction.

Group 1 (K1)	Fed standard diet + distilled water ad libitum for 4 weeks
Group 2 (K2)	Fed standard diet + distilled water ad libitum + Pb inhalation (10 mg Pb + 1 mL aquadest/day) for 4 weeks
Group 3 (K3)	Fed standard diet + distilled water ad libitum + Pb inhalation (10 mg Pb + 1 mL aquadest/day) + 8 mL/200 g body weight/day tender coconut water for 4 weeks

2.3.2 *Blood-drawing procedure*

The equipment used was sterile microhematocrit tubes, blood vials, and sterile cotton. Blood was taken by inserting a microhematocrit tube in the ophthalmic vein in the corner of the rats' eyes' periorbita then slowly rotated until the blood came out. Blood coming out was contained in an ependrof, as much as 2 cc. The microhematocrit tube was plugged once sufficient blood was obtained, and the remaining blood in the corner of the rat's eye was cleaned using sterile cotton. The MDA and GPx levels were examined using ELISA (Enzyme-Linked Immunosorbent Assay).

2.3.3 *Research location*

The treatment of experimental animals and the examination of MDA and GPx levels was carried out in at PAU Gadjah Mada University Yogyakarta.

2.4 *Statistical analysis*

All statistical analyses were performed using SPSS software version 22.0. Data were tested for normality via a Shapiro-Wilk test and homogeneity testing was performed with Leuvene's test. Data on MDA and GPx levels were normally distributed and homogenous so that they were analyzed by a parametric statistical test, which was the One Way ANOVA followed by a post hoc LSD to determine the difference between groups. P-values < 0.05 were considered statistically significant.

3 RESULTS

Tender coconut water can lower MDA levels as a parameter to lipid peroxidation and increase the antioxidant GPx enzyme, as shown in Table 2.

Table 2. The average of MDA (nmol/mL) and GPx (u/mg) in three groups (K1, K2, K3).

Variable	Group			p-Value
	1 Means ± SD	2 Means ± SD	3 Means ± SD	
MDA levels (nmol/mL)	3.30 ± 0.50	9.38 ± 0.37	4.48 ± 0.86	> 0.05*
Shapiro Wilk	0.981	0.861	0.918	0.339**
Leuvene's test				0.000***
One Way ANOVA				
GPx levels (u/mg)	66.75 ± 3.26	23.02 ± 2.03	59.93 ± 2.56	> 0.05*
Shapiro Wilk	0.871	0.867	0.956	0.341**
Leuvene's test				0.000***
One Way ANOVA				

information: Significant *> 0.05
**> 0.05
***< 0.05

Table 2 indicates that the average rate of MDA in Group 2 increases when compared to Group 1, but decreases in Group 3 when compared to Group 2. Average GPx in Group 2 decreases compared to Group 1, but increases in Group 3 when compared to Group 2. Statistical analysis results with the ANOVA test showed that tender coconut water administration at a dose of 8 ml/200 grBW/day for 4 weeks can lower MDA levels and increase antioxidant levels of GPx enzymes (p-value < 0.05).

4 DISCUSSION

Continuous lead (Pb) exposure will trigger the formation of reactive oxygen species (ROS), increase lipid peroxidation, cause oxidative stress in cells or tissues, inhibit enzymes, break down protein structures, damage DNA, and interfere with the body's antioxidant metabolism, particularly superoxide dismutase (SOD) and glutathione peroxidase (GPx), lowering glutathione (GSH) and vitamin C as well as inhibiting thiamin (B1) and pyridoxine (B6) (Flora et al. 2012). Pb exposure results in increased ROS production through inhibition of biosynthesis and activation of NAD (P) H oxidase. Increased lipid peroxidation and inactivation and/or depletion of antioxidant enzymes can reduce the absorption of glutathione S-transferase (GST), thinning GSH and thiol proteins so that GPx activity decreases (Abadin 2005). There is a positive correlation between the blood Pb levels and the decrease in glutathione (GSH). Pb exposure can also potentially reduce the absorption of selenium (the compound needed by GPx) and/or thiol binding in antioxidant proteins. The higher the level of Pb in the blood, the higher the level of free radicals in the body. Pb will bind to the thiol cluster found in antioxidant proteins so that antioxidant activities and antioxidant enzymes such as GPx will decrease. Deficiency of one of the antioxidant components causes a thorough decline of antioxidant status (Maslachah et al. 2008). This condition causes the depletion of GSH and increases H_2O_2. Hydrogen peroxide, when it reacts with transitional metals such as Fe^{++} and Cu^+ in a Fenton reaction, will produce radical hydroxyl ($\cdot OH$), which is dangerous because of its very high reactivation (Suryohudoyo 2007; Winarsi 2007). Hydroxyl radicals can damage cell membranes that are rich in source poly unsaturated fatty acids (PUFA) and can cause lipid peroxidation; one of its parameters is increased MDA levels.

Coconut water used in this research study was tender coconut water from a group of coconut cultivars, green coconut aged 5–7 months; the moderate antioxidant status measured was the antioxidant enzyme that is glutathione peroxidase (GPx).The total content of sugar and water in coconut reaches its maximum when the coconut is 5–7 months old, so at this time coconut water has a very sweet taste that decreases in sweetness until the coconut grows old, which is aged ± 12 months (Zulaikhah 2019). This study proved that administration of tender coconut water at a dose of 8 ml/200 gr BW/day in lead-induced rats for 4 weeks can prevent the lipid peroxidation being marked with decreased MDA levels and increased antioxidant levels of GPx enzymes (p < 0.05). These results are in line with the research of Zulaikhah and colleagues (2018), which proves that tender coconut water can lower MDA levels and increase the antioxidant levels of GPx enzymes in mercury-exposed gold miners (Zulaikhah et al. 2018). Loki and Rajamohan reinforce this research, stating that tender coconut water can increase GPx levels in mice that are induced with CCl_4 (Loki & Rajamohan 2003). Tender coconut water contains compounds that are very beneficial for health such as the amino acids L-arginine, methionine, vitamin C, B vitamins, selenium, and others (Zulaikhah 2019).

Components of amino acids such as methionine contained in tender coconut water are a source of sulfide for the sestein. Methionine is an essential amino acid that cannot be synthesized by the body alone; it must be supplied from outside of the body. Methionine serves as a precursor to the formation of cysteine, which is the main compound in glutathione (GSH) synthesis. Methionine will be synthesized into S-adenosilmethionine (SAM) with the help of the catalysts. Furthermore, SAM will be changed to S-adenosilhomocysteine (SAH), SAH will be converted into homocysteine with the enzyme adenosilhomosisteinase, and methionine synthase (MS) homocysteine can be changed back into methionine. In cysteine synthesis, homocysteine will be converted into a cystathionine with the help of the enzyme cystathionine β synthase (CBS) and vitamin B6, and subsequently converted into cysteine with the help of the enzyme cystathionine liase and vitamin B6. Finally, cysteine will be injected into glutathione (GSH) (Baker 2007). GSH is a kind of substrate for the antioxidant GPx in describing hydrogen peroxide. When the glutathione synthesis is interrupted, it will cause a decrease in GSH and a buildup of hydrogen peroxide, and eventually an increase of hydroxyl radicals will occur that are more harmful to the human body. GSH also serves as a kind of substrate in the radical regeneration reaction of vitamin C to vitamin C, so that a decrease in GSH will interfere with the reaction and will lead to lipid peroxidation. Methionine will work when synergized with vitamin B6. Vitamin B6 serves as a cofactor of cystathionine enzyme β synthesis and cystathionine liase in the process of cysteine and GSH synthesis. Tender coconut water can be used as a source of methionine and vitamin B6 (Zulaikhah & Sampurna 2016).

The content of other amino acids contained in the tender coconut water includes L-arginine. High L-arginine content in tender coconut water can be utilized to reduce radical generation of NO, increase antioxidant activity, lower MDA levels, and inhibit lipid peroxidation (Tripathi et al. 2009; Zulaikhah 2018). Increased levels of the examined group after L-arginine administration can increase the radical conversion of ascorbic acid into ascorbic acid (Tripathi & Misra 2009). Treatment with L-arginine can increase GPx activity in rats induced with PB (Tkachenko & Kurhalyuk 2011). The high content of L-arginine in tender coconut water can be used to reduce the generation of free radicals, increase antioxidant activity, and inhibit lipid peroxidation (Zulaikhah & Wahyuwibowo 2019). Tender coconut water contains selenium, an essential mineral that is very important for human health and plays an important role in protein synthesis and enzyme activity of glutathione peroxidase (GPx). GPx activity is strongly influenced by the presence of selenium; selenium deficiency in the body can decrease GPx activity by up to 90% (Zulaikhah 2018). Under normal conditions, selenium needs in the body may have been fulfilled from food, but for people exposed to Pb it is necessary to supply more selenium from other sources. Pb has a strong affinity for to bind to selenium, and both are antagonistic (Yang et al. 2008). Lipid peroxidation in this study was measured based on malondialdehyde (MDA) levels in plasma. MDA is one of the oxidation products of polyunsaturated fatty acids in cell membranes. Increased MDA levels are an important indicator of lipid peroxidation (Winarsi 2007).

In this study, tender coconut water was proven to lower MDA levels as a sign of the occurrence of lipid peroxidation reduction. The lipid peroxidation process can be inhibited by compounds contained in tender coconut water. For example, the L-arginine contained in coconut water can be used to decrease Fe and Cu, so as to inhibit the Fenton reaction, and the formation of hydroxyl radicals (OH$^{\bullet}$) will eventually inhibit lipid peroxidation (Suliburska et al 2014; Tkachenko & Kurhalyuk 2011). The vitamin C content contained in tender coconut water is also associated with decreasing lipid peroxidation. Vitamin C and vitamin E have a protection effect against heavy metal toxicity (Al-Azzawie et al. 2013). Vitamin C works when synergized with vitamin E in inhibiting lipid peroxidation. Vitamin E is an antioxidant chain breaker on membranes that can prevent cellular damage by lipid peroxidation and inhibit the formation of free radicals (Maslachah 2008). Vitamin E, which is oxidized by free radicals, can react with vitamin C; after getting a hydrogen ion, vitamin C will turn into vitamin E (Zulaikhah 2018). Vitamin E in membranes reacts with radical lipid (LOO$^{\bullet}$) to form radical vitamin E (vitamin E$^{\bullet}$). The radical vitamin E reacts with vitamin C to form radical vitamin C (vitamin C$^{\bullet}$). Radical vitamin C (vitamin C$^{\bullet}$) will experience regeneration into vitamin C by involving glutathione (GSH). GSH will be oxidized to oxidized glutathione (GSSG) by the enzyme glutathione peroxidase (GPx). GSSG will be reduced to a form of GSH by glutathione enzyme reductase (GRed) (Flora et al. 2012). GSH metabolism due to Pb exposure can be seen in Figure 1.

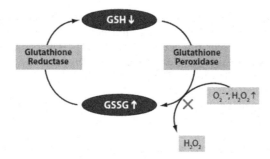

Figure 1.   Effect of lead on GSH metabolism (Flora et al. 2012).

5   CONCLUSION

The administration of tender coconut water at a dose of 8 mL/200 kgBW/day in lead-induced rats for 4 weeks can prevent lipid peroxidation, which is marked by a decrease in MDA levels and an increase in GPx antioxidant enzyme levels.

ACKNOWLEDGMENT

This study was funded by Faculty of Medicine Universitas Islam Sultan Agung Semarang Indonesia (2019).

CONFLICTS OF INTEREST

The authors declare no conflicts of interest.

REFERENCES

Abadin, H. 2005. *Draft toxicological profile for lead* Atlanta: Agency for Toxic Substances and Disease Registry.

Al-Azzawie, H. F., Umran, A., & Hyader, N. H. 2013. Oxidative stress, antioxidant status and DNA damage in mercury exposed workers. *British journal of pharmacology and toxicology* 4(3): 80–88.

Ardillah, Y. 2016. Faktor risiko kandungan timbal di dalam darah. *Jurnal Ilmu Kesehatan Masyarakat* 7(3): 150–155.

Assi, M. A., Hezmee, M. N. M., Haron, A. W., Sabri, M. Y. M., & Rajion, M. A. 2016. The detrimental effects of lead on human and animal health. *Veterinary world* 9(6): 660–671.

Baker, S. M. 2007. Who ignores individuality fails the patient. In *Proceedings from the 13th International Symposium of the Institute for Functional Medicine* 2007: 88–95.

Flora, G., Gupta, D.. & Tiwari, A. 2012. Toxicity of lead: A re-view with recent updates. *Interdisciplinary toxicology* 5(2): 47–58.

Loki, A. L. & Rajamohan, T. 2003. Hepatoprotective and antioxidant effect of tender coconut water on carbon tetrachloride induced liver injury in rats *Indian Journal of Biochemistry & Biophysics* 40(2003): 354–357.

Maslachah, L., Sugihartuti, R., & Rochmah Kurnijasanti, N. I. D. N. 2008. Hambatan Produksi React-ive Oxygen Species Radikal Superoksida (O2.-) oleh Antioksidan Vitamin E (α-tocopherol) pada Tikus Putih (Rattus norvegicus) yang Menerima Stressor Renjatan Listrik. *Media Kedokteran Hewan* 24(1): 21–26.

Suliburska, J., Bogdanski, P., Krejpcio, Z., Pupek-Musialik, D., & Jablecka, A. 2014. The effects of L-arginine, alone and combined with vitamin C, on mineral status in relation to its antidiabetic, anti-inflammatory, and antioxidant proper-ties in male rats on a high-fat diet. *Biological trace element research* 157(1): 67–74.

Suryohudoyo, P. 2007. *Kapita Selekta. Ilmu Kedokteran Molekuler.* Jakarta: Sagung Seto.

Tkachenko, H. & Kurhalyuk, N. 2011. Role of L-arginine against lead toxicity in liver of rats with different resistance to hypoxia. *Polish journal of environmental studies* 20(5): 1319–1325.

Tripathi, P., Chandra, M., & Misra, M. K. 2009. Oral administration of L-arginine in patients with angina or following myocardial infarction may be protective by increasing plasma superoxide dismutase and total thiols with reduction in serum cholesterol and xanthine oxidase. *Oxidative medicine and cellular longevity* 2(4): 231–237.

Tripathi, P. & Misra, M. K. 2009. Therapeutic role of L-arginine on free radical scavenging system in ischemic heart diseases *Indian Journal of Biochemistry & Biophysics* 46(6): 498–502.

Winarsi, H. 2007. *Antioksidan Alami & Radikal Bebas.* 3rd edn. Yogyakarta: Kanisius.

Yang, D. Y., Chen, Y. W., Gunn, J. M., & Belzile, N. 2008. Selenium and mercury in organisms: Interactions and mechanisms. *Environmental reviews* 16: 71–92.

Zulaikhah, S. T., Anies, A., Suwandono, A., Santoso, S., 2018. Effects of tender coconut water on antioxidant enzymatic superoxida dismutase (SOD), CATALASE (CAT), glutathione peroxidase (GPx) and lipid peroxidation in mercury exposed workers *International Journal of Science and Research* 4(12): 517–524.

Zulaikhah, S. T. 2019. Health benefits of tender coconut water (TCW). *International journal of pharmaceutical sciences and research* 10(2): 474–480.

Zulaikhah, S. T. & Sampurna, S. 2016. Tender coconut water to prevent oxidative stress due to mercury exposure. *IOSR journal of environmental science, toxicology, and food technology* 10(6): 35–38.

Zulaikhah, S. T. & Wahyuwibowo, J. 2019. Effect of tender coconut water to prevent anemia on Wistar rats induced by lead (plumbum). *Pharmacognosy Journal* 11(6): 1325–1330.

Digital and health technology

Medical Technology and Environmental Health – Abdullah, Widiaty & Abdullah (eds)
© 2020 Taylor & Francis Group, London, ISBN 978-0-367-86053-0

Perception and influences of smartphone use among primary school children in Muar district in Johor, Malaysia

R.R. Marzo, A. Ahmad, M.T. Win, T.L. Sheng & A.C.Y. Kung
Asia Metropolitan University, Johor, Malaysia

ABSTRACT: The general objective of this pilot study was to determine the extent of use and the influences of the smartphone among primary school children and the perception of its effect on their behavior and lifestyle. This study also specifically focused on the social behavior, academic performance, and health of these children with respect to smartphone use in their daily life. Smartphone addiction, its association with smartphone use, and its predictors had not yet been studied among primary school children in Muar, Johor, Malaysia. This was a cross-sectional study carried out using a convenient sampling method in three primary schools selected in Muar District. Each school was chosen from a multiracial mix of Malay, Chinese, and Tamil primary schools. The tool used was a questionnaire composed of 30 different questions distributed to 216 respondents. The questionnaire comprised questions that investigated the demographics, social behavior, academic performance, and health of the subjects (Kwon et al. 2013). We also took the opportunity of this pilot study to validate our questionnaires. More than half of the 216 schoolchildren who participated in the surveys, or more than 51% (110) of the schoolchildren between the ages of 10 and 12 years old, owned a smartphone. They admitted to not just owning but also actively using a smartphone on a daily basis (57.4% of the males and 44.4% of the females). Chinese students had the highest rate of smartphone ownership of 48/110 or 49%, compared to Malays (40%) and Indians (21%). Smartphone use was more rampant among lower-income students (34%) compared to students from higher-income families (28%). The majority admitted to using smartphones for entertainment purposes (57%) and social media (20%) while only a small proportion of the active users used their smartphones for educational purposes (13%). A portion of the school-children (106) did not own a smartphone but admitted being active users, using parents' or siblings' smartphones on a regular basis, spending 1–5 hours a day on smartphones, with 7 of them (6.6%) reporting use of more than 5 hours per day. From a social behavior aspect, 107 or 49.5% of students considered themselves heavy users addicted to spending a lot of time (more than 5 hours per day) using a smartphone. One hundred seven students or 49.5% had been cautioned by family members and friends for spending too much time on smartphones, and 71 or 33% of them reported experiencing emotional changes such as growing moody, nervous, or depressed while away from a smartphone. From an academic performance aspect, more than half of the respondents (114) supported the use of a smartphone in school projects, although a higher number of them (59.7%) thought it can also be a source of distraction and can cause decline in academic results. From a health aspect, 123 students experienced neck pain after using smartphones for a long time; 44% thought that smartphone use can cause wrist pain leading to carpal tunnel syndrome. Less than half of the respondents (58) agreed that smartphone use causes obesity. Use of smartphones was high among primary school children of all races and economic strata. It has benefits in terms of communication and providing tools to acquire knowledge, but it can also lead to social misbehavior and academic performance disturbances. However, the students had a moderate perception on smartphones from the aspect of health; maybe they were too young and not well informed of the dangers yet. Further study needs to be done using better sampling and methodology involving a bigger number of students from a larger overage area or maybe from several states in order to ascertain the true effects of smartphone use among primary school children in Malaysia. Some

101

form of cybersecurity can also be imposed to improve safety, and time restrictions on smartphone use can be implemented. Digital knowledge platforms and their use is very desirable to be taught as a topic in schools to promote responsible use of smartphones among schoolchildren.

1 INTRODUCTION

1.1 Overview

The earliest mobile phone was released during the 1990s, but, of course, the technologies were such that people could do very limited data transfer with it. As technology improved with the invention of the touch screen, android, and better IOS with high-speed internet, data could be easily shared and with that, more people owned and resorted to using smartphones to complete daily chores, including finding information, banking, shopping, participating in online meetings, and many other online applications. With this rapid development, data traffic among users increased drastically, which also caused an explosion in mobile phone use. It has been predicted that by 2020, data transfer will comprise 80% of mobile traffic (Lunden 2015). Furthermore, the advent of 4G and soon 5G technology will bring changes to how the human population lives, plays, and trades, and people will become more dependent on smartphones in their daily life. This can also then lead to overreliance on, dependency on, and addiction to mobile technology.

Many criteria can indicate mobile phone overuse. In the twenty-first century, the age groups between 18 and 29 have the highest percentage of smartphone use (Center 2014). Recently, according to data from Common Sense Media, more than one-third of children under age 2 are also found to be starting to use smartphones, which covers around 38% of all children's use of smartphones (Krumboltz 2013).

1.2 Advantages of smartphones for schoolchildren

Smartphones have invariably provided advantages in our everyday life. With smartphones, children can interact with other people at the tip of their fingers. They can communicate easily with other people outside their house. This allows children to keep in touch with their parents. This is especially true for schoolchildren when there is a change in their school schedule or an emergency. A growth in smartphone use can be seen in the United States when there is a rise in mass school shootings. A research firm, Youth Beat, has found that 1 in 10 kids between the ages of 6 and 10 owns a mobile phone (Meredith 2012).

Other benefits of smartphones include a new and interesting dimension of learning. According to the *PBS Parents' Guide to Children & Media*, smartphones allow children to explore the virtual reality of media, which includes augmented reality, dictionaries, and encyclopedias. It is separate from the traditional way of teaching (Jones 2013).

Children can also download health applications (apps) to their smartphones such as an app that can keep track of their diet. This helps them to be fit and healthy. The inclusion of a pedometer can also help them to track the number of steps they take per day.

1.3 Disadvantages of smartphones for schoolchildren

Smartphone use also has many harmful effects on schoolchildren. The most important matter is that most smartphone content, applications, and social media interactions are difficult to control or regulate. Children are therefore more easily exposed to pornography, cyberbullying, anorexia and suicidal pact sites, and many more. Their real-life social interaction can be impaired in the future.

On the other hand, many consider smartphones a distraction for schoolchildren. Some schoolchildren abuse the smartphones meant as their study tools. This makes them focus more on their smartphones as entertainment and gaming tools rather than listening to their lecture of the day. Their academic performance will decline due to these smartphones.

Distractions not only affect individuals but also irritate teachers or others in a group. The sound or ringtone from a smartphone can affect the flow or pattern of a class, leading to class disruptions.

Smartphone use also can make schoolchildren feel sleepless and tired. They will spend more time gaming and watching videos, playing and using the smartphone. They will have delayed and inadequate sleep. Because of this, they cannot concentrate during class the next day. This will cause poor focus and deterioration in performance, jeopardizing academic skills. They will not feel fresh and confident the next morning.

1.4 *Significance of this study*

Schoolchildren using smartphones of course has pros and cons. We conducted research into how smartphones affect the lifestyle of schoolchildren. This is to understand how smartphones affect social performance, academic performance, and health. This study also aimed to examine how children perceive smartphones.

2 LITERATURE REVIEW

2.1 *Introduction*

The invention of the smartphone has influenced different aspects of lifestyle. Three important aspects – social behavior, health, and academic performance – have felt the biggest impact from the use of smartphones (Archer 2013).

2.2 *Literature review*

A good number of research studies have suggested various good or bad influences of smartphone use. Horstmanshof and Power (2004) proposed that smartphone technology has given young people opportunities to create and maintain relationships. Aoki and Downes (2003) stated that smartphones provide youngsters social fulfillment, convenience, and mobility. Frissen (2000) and Jones (2013) suggested that with this technology, busy parents can better keep in touch with their children, thus lessening their stress. It is a necessity for college students to own a smartphone in order to maintain contact with their families. Cova (1997) contended that youngsters seek peer group acceptance by using their mobile phones. In contrast with these statements, other researchers have highlighted the bad influences that smartphones now have on the social lives of young people. A study conducted by Australian students concluded that there are adverse consequences of excessive use of smartphones such as financial costs, emotional stress, damaged relationships, and falling literacy. A link has emerged between criminal activities like alcohol abuse, fighting, theft, and narcotics use and the use of smartphones (Jacobsen & Forste 2011). Srivastava (2005) claimed that students use their mobile phones while attending lectures. Griffiths and colleagues (2003) argued that debt and other financial worries among adolescents is one of the consequences of the higher use of mobile phones. Thompson and Ray (2007) emphasized the security of children using mobile phones by pinpointing the risks of uncontrolled expenditures, exposure and access to prohibited, damaging, or adult material, and bullying via mobile phones.

Sánchez-Martínez and Otero (2009) pointed out that intensive cell phone use has been related to school failure as well as to other negative behaviors such as smoking and excessive alcohol use. Several studies have revealed that students frequently report using a variety of electronic media, including cell phones, while in class, studying, and doing homework (Jacobsen & Forste 2011; Junco 2012; Wood et al. 2012). They measured the influence of multitasking with an array of electronic media on students' ability to learn from typical, university classroom lectures. Results showed that multitasking with any of the technologies was associated with lower scores on follow-up tests compared with students who did not multitask.

Immense research has been conducted on the effects of smartphone use on health. Aoki and Downes (2003) learned that most students from the United States are inclined to make calls at night, which can lead to adverse outcomes such as sleep loss. Niaz (2008) proposed that awareness of the dangers associated with excessive and addictive mobile phone use must be created in society because a public health problem has appeared due to unwise use of smartphones.

In conclusion, a study ought to be done on this matter in order to create awareness of the influence of smartphones on the lifestyle of people, including college students and schoolchildren.

2.3 *Objectives*

The general objective of this study was to determine the extent of smartphones' influence on schoolchildren and to examine schoolchildren's perceptions of smartphones. In order to achieve the objective, three ideas were generated:

1. To determine the demography of smartphone users among primary school children (aged 10–12 years old) in three different schools representing the three biggest ethnic groups in Muar, Johor.
2. To determine the influence of smartphones on the social behavior and academic perform-ance of primary school children.
3. To determine the influence of smartphone use on the health of primary school children.

2.4 *Hypotheses*

Schoolchildren are easily influenced by smartphones and have a bad perception of the effect of smartphones on their social behavior.

Schoolchildren are easily influenced by smartphones and have a bad perception of the effect of smartphones on their academic performance.

Schoolchildren are easily influenced by smartphones and have a bad perception of the effect of smartphones on their health.

3 METHODOLOGY

3.1 *Study design*

The approach for this research consisted of a cross-sectional quantitative study. Data were collected to be analyzed descriptively. The target populations of this study included students from Standards 4–6 in:

– SK Parit Setongkat (Malay dominance)
– SJK (C) Chung Hwa (Chinese dominance)
– SJK (T) Jalan Khalidi (Tamil dominance)

The data were collected through convenient sampling.

3.2 *Sample population*

All students who consented from Standards 4–6 (10–12 years old) were encouraged to partici-pate provided that they were from:

– SK Parit Setongkat
– SJK (C) Chung Hwa
– SJK (T) Jalan Khalidi

3.3 Sampling method

The research was conducted using nonprobability sampling. The sampling method was convenient sampling through choosing three primary schools, each representing one of three ethnic groups – Malay, Chinese, and Tamil. Students from Standards 4–6 were chosen from each school. The schools were selected when the headmasters consented to the survey. The data were then collected and analyzed.

3.4 Survey instrument

A questionnaire was distributed in three different primary schools – a Malay primary school, a Chinese primary school, and a Tamil primary school – for the purpose of data collection. The questionnaires targeted Standard 4–6 students. The answered questionnaires were collected after completion of the questionnaires on the spot itself.

3.5 Questionnaire design

The questionnaire was validated via a pretest. The questionnaire was divided into four parts – namely a sociodemographic section with questions associated with general perceptions of smartphones, a social behavior portion, and sections on academic performance and health.

3.6 Data analysis

The primary data analysis was coded and analyzed by PASW Statistics Student Version 18. Descriptive statistics were used to describe the demographic features of the participants, and tables with frequencies and percentages were used to interpret the results.

3.7 Ethical considerations

The protocols were reviewed by the Medical Research Ethical Committee (MREC) of Asia Metropolitan University to ensure full protection of the rights of study subjects. National Malaysian Research Register (NMRR) registration was carried out after the approval by the MREC. Following the acquisition of the NMRR registration number, the questionnaires were distributed to the study subjects. To assure the confidentiality of all the information provided, a cover letter was attached to the questionnaire.

4 RESULTS AND DISCUSSION

In this research, a total of 216 primary school children participated in the survey. The survey comprised 110 schoolchildren (62 males, 48 females) (Table 1) who owned a smartphone while 106 participants admitted to using smartphones owned by family members and regularly spent 1–5 hours daily on a smartphone. According to these data, male students used smartphones more than females. Supporting our finding, Chen and Katz (2009) have stated that boys spend more time on smartphones (mostly for gaming and entertainment) compared to girls. On the contrary, researchers such as Hakoama and Hakoyama (2011) have suggested that females are more likely to depend heavily on their phones to maintain social relationships, surf websites, and shop.

We concluded that more students are using smartphones at the age of 10, compared to children at the ages of 11 and 12. But, if we compare the percentages, most of the 11-year-old students (60.86%) owned a smartphone.

Among the races, Chinese had the highest number of students using smartphones followed by Malays (40%) and Indians (21%).

With regard to family income, more students from lower-income families used smartphones compared to students from middle-income (34%) and higher-income (28%) families. This

Table 1.   Sociodemographic data.

Variables		Frequency (n), Percentage (%)	
		With smartphone	Without smartphone
Gender	Male	62 (57.40)	46 (42.60)
	Female	48 (44.40)	60 (55.60)
Age	10	43 (44.32)	54 (55.67)
	11	42 (60.86)	27 (39.14)
	12	25 (50.00)	25 (50.00)
Type of primary school	Malay primary school	35 (40.70)	51 (59.30)
	Chinese primary school	54 (67.50)	26 (32.50)
	Tamil primary school	21 (42.00)	29 (58.00)
Race	Malay	40 (44.00)	51 (56.00)
	Chinese	48 (66.70)	24 (33.30)
	Indian	21 (40.40)	31 (59.60)
	Others	1 (100.00)	0 (00.00)
Age of having first smartphone	< 7 years old	20 (90.90)	2 (9.10)
	8–12 years old	82 (91.10)	8 (8.90)
	Do not own smartphone	8 (7.70)	96 (92.30)
Parents' income	< RM2500	48 (51.10)	46 (48.90)
	RM2500–RM4000	34 (55.70)	27 (44.30)
	> RM4000	28 (45.90)	33 (54.10)
Perception of time spend on smartphone	<2 hours	58 (46.00)	68 (54.00)
	2–3 hours	23 (50.00)	23 (50.00)
	3–5 hours	15 (62.50)	8 (34.80)
	>5 hours	14 (66.70)	7 (33.30)
Perception of the types of apps preferred	Social media	22 (61.10)	14 (38.90)
	Education	15 (38.50)	24 (61.50)
	Entertainment	63 (54.80)	52 (45.20)
	Others	10 (38.50)	16 (61.50)
Perception of the usefulness of smartphone	Yes	101 (59.10)	70 (40.90)
	No	9 (20.00)	36 (80.00)

statement is supported by Brown and colleagues (2011) who found that lower-income students used their mobile phones more.

When examining the apps students use on their smartphones, we found that most students (75%) prefer to use smartphones as entertainment devices. However, educational apps were a popular choice for students (39%); they have a higher percentage compared to students using smartphones for social media. This is very different from what Kibona and Mgaya (2015) found in their study, which was that most of the students in higher learning institutions tend to use smartphones for social rather than academic purposes.

Some 32.9% (71) (Table 2) of the children using smartphones admit to be nervous, moody and depressed while they are away from their smartphones but most of the schoolchildren (67.1%) disagreed on emotional changes while they are away from them. This statement is in contrary with the opinion of (Archer 2013) where Archer suggested smartphone users will easily get annoyed when they are unable to access to smartphone. Besides that, 76.9% of the schoolchildren disagree to spend more time on smartphone when they are attending an occasion. On top of that, 68.5% of schoolchildren do not support the act of extending the time period spend on smartphones.

Table 2. Social behavior data.

Aspects	Items	Frequency (n), Percentage (%)	
		Yes	No
Influences	Did you ever use a smartphone in the presence of your family or friends?	129 (59.7)	87 (40.3)
	Do your family or friends tell you that you spend a lot of time on your smartphone?	107 (49.5)	109 (50.5)
Perception	Do you think you will get nervous, moody and depressed while away from cell phone?	71 (32.9)	145 (67.1)
	Will you spend more time on Smartphone when you are attending an event such as wedding or at a beach?	50 (23.1)	166 (76.9)
	Do you think Smartphone is a better communication tools than meeting people face to face?	109 (50.5)	107 (49.5)
	Do you think there is a need to increase the time spent in your smartphones even when you used for long time?	68 (31.5)	148 (68.5)

As Table 3 shows, 79.2% of schoolchildren admitted to getting useful and important information from smartphones, 65.7% admitted to using educational applications, and 52.8% admitted to using smartphones as means of communication and discussion for school projects.

However, from the academic performance perspective, the respondents understood that improper use of smartphones can cause a decline in their academic results. The findings revealed that 69.9% agreed that it can cause a decline in their academic results, although 30.1% disagreed. More than half of the respondents (64.4%) thought that the smartphone should be banned in school to prevent misuse, but 35.6% disagreed with this statement, When

Table 3. Academic performance data.

Aspects	Items	Frequency (n) Percentage (%)	
		Yes	No
Influences	Did you ever find useful information using your smartphone?	171 (79.2)	45 (20.8)
	Did you ever download useful and educational application (apps) into your smartphone?	142 (65.7)	74 (34.3)
	Did you ever discuss school projects with your friends through chat messages in your smartphone?	114 (52.8)	102 (47.2)
Perception	Do you think smartphones are a distraction when doing homework?	129 (59.7)	87 (40.3)
	Do you think a smartphone is a better teaching tool than teachers?	70 (32.4)	146 (67.6)
	Do you think smartphones can cause your academic results to decline?	151 (69.9)	65 (30.1)
	Do you think your school should ban smartphone use in your school compound?	139 (64.4)	77 (35.6)

Table 4. Health data.

Aspects	Items	Frequency (n) Percentage (%)	
		Yes	No
Influences	Do you constantly have pain in your thumb after you use a smartphone?	70 (32.4)	146 (67.6)
	Do you have a headache after using a smartphone for a long period of time?	110 (50.9)	106 (49.1)
	Do you have blurry vision after a long period of using a smartphone?	82 (38)	134 (62)
	Do you have neck pain after you use a smartphone while you lie down?	123 (56.9)	93 (43.1)
Perception	Do you think the smartphone is one of the causes of obesity?	58 (26.9)	158 (73.1)
	Do you think smartphones can emit radiation?	145 (67.1)	71 (32.9)
	Do you think smartphones can cause wrist pain (carpal tunnel syndrome)?	95 (44)	121 (56)

comparing the results, we found that the students were aware of the side effects of smartphone addiction on academic results. This finding is supported by the research of Kibona and Mgaya (2015). They found that the misuse of smartphones has a direct negative impact on the academic performance of higher-level students.

Many of the students (56.9%) (Table 4) reported neck pain after they used smartphones while lying down, although 43.1% disagreed. Their perception of smartphones was moderate when compared to the other two previous aspects. Of the student population, 73.1% disagreed with smartphones being a cause of obesity as they can promote a sedentary lifestyle. Meanwhile, 67.1% of the students agreed that smartphones can emit radiation, but 32.9% disagreed. Although long periods of smartphone use can cause wrist pain, 56% of the schoolchildren disagreed and 44% agreed that smartphones can cause wrist pain (carpal tunnel syndrome).

This study shows that the students are lacking in the knowledge of health issues regarding smartphone addiction as half of the participants did not acknowledge the risks of obesity and carpal tunnel syndrome that smartphones present. This is contrary to the claims of professional orthopedic surgeon Jonathan Dearing, spokesman for the Royal College of Surgeons of Edinburgh, who stated that the technology revolution has led to reduced physical activity and increased obesity (Goldhill 2015). Dallas (2014) wrote that carpal tunnel syndrome is one of the health problems stemming from smartphone addiction.

5 CONCLUSION

Schoolchildren are not easily influenced by smartphones, and they have a good perception of smartphones and their impact on social behavior and academic performance.

Schoolchildren are not easily influenced by smartphones, and they have a moderate perception of smartphones and their impact on health.

6 RECOMMENDATIONS

The data showed a very positive outcome in this research, but the students lacked knowledge of the impact of smartphones on health. We can overcome this through seminars or talks in primary schools on the harms of smartphones.

Nevertheless, every coin has two sides; smartphones can bring certain advantages to this digital generation. For instance, students can learn better with the aid of smartphones. In addition, there is a lot of information on the Internet that we cannot learn from a textbook, hence smartphones can expose children to the wonderful things in this world. Last but not least, having a smartphone can shape children into more responsible and disciplined individual as they have to control their time and desires.

REFERENCES

Aoki, K., & Downes, E. J. 2003. An analysis of young people's use of and attitudes toward cell phones. *Telematics and informatics* 20(4): 349–364.

Archer, D. 2013. Smartphone addiction. *Psychology today*.

Brown, K., Campbell, S. W., & Ling, R. 2011. Mobile phones bridging the digital divide for teens in the US? *Future Internet* 3(2): 144–158.

Center, P. 2014. Mobile technology fact sheet. *Pew Research Internet Project, Janeiro*.

Chen, Y. F., & Katz, J. E. 2009. Extending family to school life: College students' use of the mobile phone. *International journal of human-computer studies* 67(2): 179–191.

Cova, B. 1997. Community and consumption. *European journal of marketing*,31(3/4): 297–316.

Dallas, K. 2014. 4 Unexpected health risks of smartphone use. *Deseret News*. [Online] Retrieved from: www.deseretnews.com/article/865616128/4-unexpected-health-risks-of-smartphone-use.html.

Frissen, V. A. 2000. ICTs in the rush hour of life. *Information society* 16(1): 65–75.

Goldhill, O. 2015. Why smartphones are making you ill. *The Telegraph*.

Griffiths, M., Renwick, B., & Griffiths, M. 2003. *Misfortune or mismanagement: A study of consumer debt issues*. Central Coast School of Business.

Hakoama, M. & Hakoyama, S. 2011. The impact of cell phone use on social networking and development among college students. *American Association of Behavioral and Social Sciences Journal* 15(1):1–20.

Horstmanshof, L & Power, MR 2004, 'YYSSW (Yeah, yeah, sure sure whatever)', paper presented to 90th National Communications Association Annual Convention, Chicago, Ill., 11–14 November.

Jacobsen, W. C. & Forste, R. 2011. The wired generation: Academic and social outcomes of electronic media use among university students. *Cyberpsychology, behavior, and social networking* 14(5): 275–280.

Jones. R. 2013. 5 benefits of giving your kids a smartphone. Edupad. [Online]. Retrieved from: www.edupad.com/blog/2013/11/5-benefits-giving-kids-smartphone/.

Junco, R. 2012. Too much face and not enough books: The relationship between multiple indices of Facebook use and academic performance. *Computers in human behavior* 28(1): 187–198.

Kibona, L. & Mgaya, G. 2015. Smartphones' effects on academic performance of higher learning students. *Journal of multidisciplinary engineering science and technology* 2(4): 777–784.

Krumboltz, M. 2013. *Study 38% of kids under 2 use smartphones or tablets*. Yahoo News. October 28, 2013.

Kwon, M., Lee, J. Y., Won, W. Y., Park, J. W., Min, J. A., Hahn, C., ... & Kim, D. J. 2013. Development and validation of a smartphone addiction scale (SAS). *PloS one* 8(2): e56936.

Lunden, I. 2015. 6.1 b smartphone users globally by 2020, overtaking basic fixed phone subscriptions. TechCrunch.

Meredith, L. 2012. How cellphones are changing school emergency plans. *Tech News Daily*. [Online]. Retrieved from: https://mashable.com/2012/12/15/cellphones-school-emergency/.

Niaz, U. 2008. Addiction with internet and mobile: An overview. *Journal of Pakistan Psychiatric Society* 5(2): 72.-75.

Sánchez-Martínez, M. & Otero, A. 2009. Factors associated with cell phone use in adolescents in the community of Madrid (Spain). *Cyberpsychology & behavior* 12(2): 131–137.

Srivastava, L. 2005. Mobile phones and the evolution of social behaviour. *Behaviour & information technology* 24(2): 111–129.

Thompson, R. & Ray, G. 2007. More safety for children with mobiles. *Card technology today* 19(9): 10.

Wood, E., Zivcakova, L., Gentile, P., Archer, K., De Pasquale, D., & Nosko, A. 2012. Examining the impact of off-task multi-tasking with technology on real-time classroom learning. *Computers & education* 58(1): 365–374.

Biomedical and health technology

Medical Technology and Environmental Health – Abdullah, Widiaty & Abdullah (eds)
© 2020 Taylor & Francis Group, London, ISBN 978-0-367-86053-0

Significance of preoperative measurement of tibial reference point in knee replacement with tibial valgus deformity

J.C.P. Butarbutar, T. Mandagi, R. Aditya & L. Siahaan
Department of Orthopedic and Traumatology, Siloam Hospital, Lippo Village, Tangerang, Indonesia
University of Pelita Harapan, Tangerang, Indonesia

ABSTRACT: A tibial cut, referring to the center of the intercondylar eminence in knee replacement, often leads to malalignment due to unnoticed preexisting tibial deformities. This research preoperatively measured the proximal tibial reference point in full-length, weight-bearing lower limb radiographs, and the same reference point was replicated intraoperatively. Postoperative measurements were taken to evaluate the lower limb alignment. The result shows that optimal lower limb alignment and tibial component angle were achieved. The proximal tibial reference point was placed medial from the center of the intracondylar eminence through preoperative measurements, increasing the survival of knee replacements with tibial valgus deformities.

1 INTRODUCTION

One of the important factors that affect the long-term survival of a knee replacement (KR) is postoperative limb alignment (Thippana & Kumar 2017), where it defines the dynamic load distribution across the medial and lateral knee articular surfaces against the weight-bearing line, with an ideal angle of ± 3° (Alghamdi et al. 2014; Khanna et al. 2017; Massin 2017; Palanisami et al. 2019).

Previous studies have found a relatively high incidence of extra-articular deformities (EAD) in Asian patients with osteoarthritis of the knee. Lateral tibia vara and varus orientation of the joint line are not infrequent in Asian patients undergoing KR (Fang et al. 2009; Yau et al. 2007).

An unaccounted EAD in KR could deviate the optimal weight-bearing line and distribute the load unevenly; thus it is liable to polyethylene wear and can compromise the survival of the KR (Fang et al. 2009; Khanna et al. 2017; Palanisami et al. 2019).

This could be prevented by an accurate placement of the tibial component. Traditional methods of proximal tibial cuts perpendicular to the line between the center of intercondylar eminence (IE) and the center of talus often lead to malalignment due to unaccounted preexisting tibial deformities (Thippana & Kumar 2017). In the case of tibial valgus deformity, shifting the reference point (RP) medial from the center of IE will result in a better lower limb and tibial component alignment (Alghamdi et al. 2014; Khanna et al. 2017).

The purpose of this research was to show that preoperative RP measurement contributes to postoperative lower limb alignment in KR with tibial valgus deformity. An acceptable degree of accuracy can be achieved with the use of an extramedullary jig for tibial component alignment by medially shifting the RP.

2 METHODS

This was a prospective study involving pre- and postoperative evaluation of the coronal plane profile of the lower limbs in a patient with EAD who underwent KR.

Pre- and postoperative full-length, weight-bearing lower limb radiographs were taken in true anteroposterior (AP) position with the patella facing forward. This standard position ensured that the tibia faced forward with minimal rotation. Radiograph measurements were then made using the PACS software.

A line was drawn joining the center of the femoral head to the center of the distal femoral intercondylar notch and another line was drawn joining the center of the talus to the center IE (Figure 1). The hip-knee-ankle (HKA) angle is defined as the angle between both lines, with negative and positive angles indicating varus and valgus deviation subsequently.

The tibial valgus angle (x) was measured between line (b) proximal metaphysis axis and line (c) distal third of tibial shaft axis (Figure 2). Proximal metaphysis axis is defined as the line perpendicular (a) to the proximal tibia articular surface. Proximal RP was determined using the distance from center of IE and the point of intersection between lines (a) and (b), producing a lateral or medial shift.

Preoperative RP was then replicated intraoperatively by placing the extra-articular jig medial to the center of IE and the center of the talus (Figure 3).

The same full-length radiograph was obtained at follow-up and the postoperative HKA was measured with 180 ± 3° as acceptable angle. The medial proximal tibial component angle (MPTA) between the mechanical axis of the tibia and the tangent to the tibial component drawn at the component–cement interface of the base plate was measured on the medial side of the knee. A range of 90 ± 2° was considered acceptable.

Figure 1. HKA is measured between two mechanical axes of the lower limb: the femoral (a) and tibial (b) mechanical axes. (HKA: Hip-knee-ankle angle)

Figure 2. Modified Thippana method: (a) Line drawn along the tibial articular surface using the healthy articular surface as a guide. (b) Line drawn along the anatomical axis of the tibia; the line is drawn from the center of the talus, the middle of the distal third tibial diaphysis, and to the tibial articular surface. (c) Line drawn perpendicular to line (a) at the center of the tibial IE. (x) Angle between line (b) and (c). (y) Distance in mm the RP shift from center of intercondylar eminence, lateralization or medialization of the RP. (IE: Intercondylar eminence, RP: Reference point).

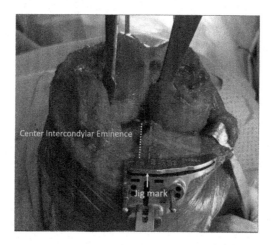

Figure 3. Intraoperative replication of the preoperative RP measurements by placing the jig mark medial from the center of IE.

3 CASE PRESENTATION

A 62-year-old male presented with bilateral knee pain worsened by walking and squatting. Physical examination showed a knee varus deformity, mild effusions, and tenderness on the medial side of the right knee. A plain radiograph showed advanced knee osteoarthritis (OA) right and left with significant "S-shaped" tibial valgus. The patient underwent KR in both knees. Preoperative measurements (Figure 4) and postoperative evaluations (Figure 5) were applied to the patient.

By shifting the proximal tibial reference point medial from the center of IE in the patient with an extra-articular tibial valgus deformity, an optimal angle of HKA and MPTA were achieved.

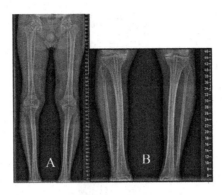

Figure 4. Preoperative measurements: (a) HKA of right and left lower limbs is 7.47° and 5.03°, varus limb alignment is noted. (b) Tibial deformity of right and left is 5.11° and 4.37°; a bilateral extra-articular tibial valgus deformity is noted. Medial shift of reference point from the center of IE in both knees is noted, 1.15 cm on the right knee and 1.04 cm on the left knee.

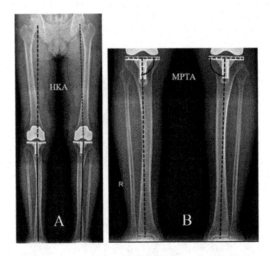

Figure 5. Postoperative measurements: (a) Hip-knee-ankle angle of right and left lower limbs is 0.75° and 2.89° respectively. (b) Medial proximal tibial angle of right and left is 88.62° and 89.69° respectively.

4 DISCUSSION

In KR, it is common knowledge that optimal lower limb and component alignment is critical for long-term success (Dimitrios et al. 2018; Tucker et al. 2019). Malalignment due to tibial deformities and PE wear causing malalignment has been a common cause of knee arthroplasty revisions. Neutral alignment was traditionally recommended to maximize the survival rate in TKA, with navigation to avoid error exceeding 3° (Bottros et al. 2008). Retrieval studies showed that > 5° malalignment increased implant thickness loss by 0.11 mm per year in the concavity of the deformation (Massin 2017).

An EAD can mislead the tibial cuts if not detected or if an intramedullary rod is used without accounting for the tibial deformity (Thippana & Kumar 2017). Alghamdi and colleagues (2014) found only 1% tibial valgus deformity in the varus knee group, and 53% in the valgus knee group. Such deformity should be considered during KR, as this changes the way surgeons approach the procedure in each case.

When evaluating component alignment post arthroplasty, the long-leg films are recommended, especially with tibial valgus where the tibial component may appear varus in relation to the proximal tibia. Standing long-leg films have been shown to be more accurate than short-leg films in determining lower limb alignment (Bottros et al. 2008).

After performing a standing long-leg film, a preoperative measurement shows tibial valgus deformity and the center of the tibial plateau is lateral to the anatomical axis center of the tibial shaft. A tibial cut centered to the IE could cause postoperative misalignment in such cases. Earlier studies have recommended the use of the lateral or medial edge of the intercondylar eminence instead of the center as the RP (Palanisami et al. 2019; Thippana & Kumar 2017).

Achieving optimal HKA and MPTA postoperative measurements means distributing the load across the articular surface of the PE tibial component, which is a crucial factor in increasing the TKA survival rate. Furthermore, an intramedullary guide for tibia cut should not be used in cases of EAD, which could result in excessive bone resection on the lateral tibial plateau (incorrect tibial cut) (Alghamdi et al. 2014; Khanna et al. 2017).

In our study, we used an extramedullary jig and did not use the center IE as a fixed RP. The tibial RP should be individualized based on the degree of an EAD, and it can shift even further medial from the center IE in the case of tibial valgus deformity. Careful preoperative planning gives good postoperative component placement and better correction of the mechanical axis, therefore increasing the survival of the KR.

5 CONCLUSION

In the case of KR in patients with tibia valgus deformity, an optimal alignment can be achieved with the use of an extramedullary jig for tibial component alignment by a medial shift of the proximal tibial RP from the center of IE. We recommend individualized RP based on KR with tibial deformity through careful preoperative planning.

REFERENCES

Alghamdi, A., Rahmé, M., Lavigne, M., Massé, V., & Vendittoli, P. A. 2014. Tibia valga morphology in osteoarthritic knees: Importance of preoperative full limb radiographs in total knee arthroplasty. *Journal of arthroplasty* 29(8): 1671–1676.
Bottros, J., Klika, A. K., Lee, H. H., Polousky, J., & Barsoum, W. K. 2008. The use of navigation in total knee arthroplasty for patients with extra-articular deformity. *Journal of arthroplasty* 23(1): 74–78.
Dimitrios, N., George, S., & Ioannis, M. 2018. Primary total knee arthroplasty in valgus deformity. In *Primary Total Knee Arthroplasty*. IntechOpen.
Fang, D. M., Ritter, M. A., & Davis, K. E. 2009. Coronal alignment in total knee arthroplasty. *Journal of arthroplasty* 24(6): 39–43.
Khanna, V., Sambandam, S. N., Ashraf, M., & Mounasamy, V. 2017. Extra-articular deformities in arthritic knees: A grueling challenge for arthroplasty surgeons: An evidence-based update. *Orthopedic reviews* 9(4).
Massin, P. 2017. How does total knee replacement technique influence polyethylene wear? *Orthopaedics & traumatology: Surgery & research* 103(1): S21–S27.
Palanisami, D., George, M. J., Hussain, A. M., MD, C., Natesan, R., & Shanmuganathan, R. 2019. Tibial bowing and tibial component placement in primary total knee arthroplasty in valgus knees: Are we overlooking? *Journal of orthopaedic surgery* 27(3).
Thippana, R. K. & Kumar, M. N. 2017. Lateralization of tibial plateau reference point improves accuracy of tibial resection in total knee arthroplasty in patients with proximal tibia vara. *Clinics in orthopedic surgery* 9(4): 458–464.
Tucker, A., O'Brien, S., Doran, E., Gallagher, N., & Beverland, D. E. 2019. Total knee arthroplasty in severe valgus deformity using a modified technique: A 10-year follow-up study. *Journal of arthroplasty*, 34(1): 40–46.
Yau, W. P., Chiu, K. Y., Tang, W. M., & Ng, T. P. 2007. Coronal bowing of the femur and tibia in Chinese: Its incidence and effects on total knee arthroplasty planning. *Journal of orthopaedic surgery* 15(1): 32–36.

Medical Technology and Environmental Health – Abdullah, Widiaty & Abdullah (eds)
© 2020 Taylor & Francis Group, London, ISBN 978-0-367-86053-0

Reliability of a common digital body scale to determine body composition

F.A. Yulianto, H.S. Rathomi, E. Nurhayati, R.G. Ibnusantosa & E.R. Indrasari
Universitas Islam Bandung, Bandung, Jawa Barat, Indonesia

ABSTRACT: The escalating prevalence of metabolic diseases needs effective screening as a preventive strategy. Many tools are used to measure various metabolic parameters, both for research and clinical purposes. However, there is still doubt about the tools' reliability especially the digital type. This research aims to asses a Body Index Analyzer reliability available in the market for calculating body composition consisting of body fat, visceral fat and skeletal muscle. There were 89 respondents that were measured twice by two same brands and easy to find, digital body scales. The Bland-Altman plot was used to assess the body composition measurement reliability. Mean difference of body fat was 0.33 (95% CI -0.34 to 0.99), visceral fat was 0.53 (95% CI -0.37 to 1.45) and skeletal muscle was 0.01 (95% CI -0.11 to 0.13). Among those three parameters, the instrument has the most precise agreement in skeletal muscle measurement, although it still can be used to quantify body fat and visceral fat as well.

1 INTRODUCTION

Obesity is the new epidemic of the millennium. Excessed food intake and lower energy expenditure are the culprit of the problem. Consequently, the frequency of metabolic diseases sufferers soaring high resulting in death by non-communicable diseases as the highest cause of death (Engin 2017). Fortunately, people nowadays tend to be more aware of their health and many of them are monitoring their body weight at home.

Body weight surveillance is very easy; people will easily find body scales in the nearest stores. However, due to limited usefulness of body weight that is usually used to measure body mass index (BMI), body composition analyzer (BCA) is the method to analyze more detail components of body weight. In scientific area, the most valid and reliable BCA is dual-energy X-ray absorptiometry (DEXA) scan but it requires radiologist, very expensive, and unfamiliar procedures for the laymen (Shepherd et al. 2017). In other words, it is barely possible for people to have it at home.

Alongside conventional methods to quantify body composition such as BMI, waist circumference, waist-hip ratio, and skinfold measurements, bioelectrical impedance analyzer (BIA) arises as the most applicable BCA method (Kuriyan 2018). When we type the keywords in a web browser, we will have 12.900.000 results with vast choices from the cheapest to the most expensive ones, and most of them were digitals. Despite of their simplicity, there are rooms for error in a non-golden standard method especially in a mass-produced BIA that is common and easy to find.

2 METHODS

2.1 *Study design and rationale*

This cross-sectional study was designed to evaluate the reliability of a common digital BIA. There were 2 examination rooms, each was equipped by 2 digital BIAs and 2 examiners in the day of collecting data. The whole respondents were separated into 2 groups and each subject entered only one room. Although those BIAs were located in the same room, the set up was designed to be clean-contaminated information between two assessors in the same room. Results from devices were transferred to a form in cloud and gathered as a spreadsheet file to be analyzed. The body composition parameters were consisted of body fat, visceral fat, and skeletal muscle percentage, factors that weighing the most in individual body weight.

2.2 *Participants*

There were 197 subjects registered as the first year students of UNISBA medical faculty. 108 of them were randomized into this study representing their group, but only 89 subjects entered the study after they agreed to participate in the study. 73.03% of the respondents were female, with age median in 19 years old. The youngest were 17, while the oldest were 21 years old.

2.3 *Statistical methods*

The distribution of numeric variables from three body composition were evaluated in terms of their mean-median difference, standard deviation, skewness, kurtosis and Shapiro-Wilk's test followed by Bland-Altman plot, a method to assess the reliability of numeric variables, which interpret the mean difference between two measurements. The mean difference was assessed to determine if the difference between two measurements wide or narrow (Giavarina 2015). Stata version 11 was used as a statistical tool in this study.

3 RESULTS

Tables 1 and 2 show insignificant differences between the first and second examination. Table 1 describes the distribution frequency in fat that located in body and visceral. The visceral fat is normally below the body (total) fat, as illustrated in the Table 1. The mean difference in fat measurement is small, 0.33% and 0.55% in body and visceral fat. The most visible difference only happened between minimum value of the first and second body fat examination. The data distribution in body fat is more normal compared to visceral fat distribution according to given standard deviation.

Table 1. Distribution frequency of body and visceral fat percentage.

	body fat 1 (%)	body fat 2 (%)	visceral fat 1 (%)	visceral fat 2 (%)
min	9.6	3.1	0.5	0.5
median	29.7	29.3	4.5	4.5
max	40.7	40.7	45	20
mean	28.92	28.59	6.57	6.02
SD	6.74	7.37	6.26	4.61

Table 2. Distribution frequency of skeletal muscle percentage.

	skeletal muscle 1	skeletal muscle 2
min	20.8	20.8
median	26.1	26
max	37.5	37.6
mean	27.34	27.33
SD	4.2	4.2

The skeletal muscle measurement in Table 2 describes slightest difference between the first and second measurement compared to previous fat measurements. It has 0.01% mean difference.

Related to previous statement that the smallest mean difference was found in skeletal muscle measurement, while difference in body and visceral fat measurements were still acceptable according to Table 3. 95% CI range included mean difference zero, therefore there was no significant difference between first and second assessment in 3 composition measurements.

Lower limit of difference in body fat was -5.93 (-1.96 SD) while the higher limit was 6.59 (+1.96 SD). Hence, the limit of agreement for body fat was 12.52. In visceral fat measurement reliability, the lower limit of difference was -8.09 and the higher limit was 9.17. Therefore, the limit of agreement for visceral fat was 17.26. Different from previous body composition measurements, the skeletal muscle had the smallest limits of agreement, 2.2 as a result from -1.09 and 1.11 as the lower and higher limit consecutively.

Three next figures tell distribution of the mean difference on each body composition measurement consecutively. The red line indicates mean difference while the blue lines indicate -1.96 SD as the lower line and +1.96 SD as the higher line. The following figures confirm the finding in analysis above. Figure 1 depicts difference and average plot of body fat that looked similar with plot in visceral fat and skeletal muscle, where the mean differences located near zero value. Despite of observed skeletal muscle data are scattered more compared to other variables, the distance between lower and higher line is the shortest. Therefore, the difference between first and second examination is the smallest in skeletal muscle measurement.

4   DISCUSSION

The digital BCA that is used in this research is based on bioelectrical impedance (BIA). This method is a lot more common nowadays due to its portability, usability, safety, and low cost with minimal participant burden. This method is based on electrical conductive properties of the body and involves measuring the impedance (Khalil et al. 2014). Water and electrolytes are known as a good electrical conductor while fat, which is hydrophobic, is known as a poor conductor.

Table 3.   Mean of difference in first and second measurement.

	mean of difference	95% CI	
between body fat	0.33	-0.34	0.99
between visceral fat	0.54	-0.37	1.45
between skeletal muscle	0.01	-0.11	0.13

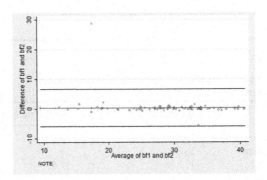

Figure 1.   Bland Altman plot of body fat.

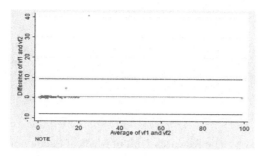

Figure 2.   Bland Altman plot of visceral fat.

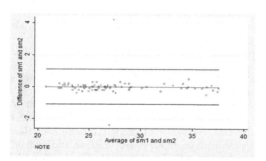

Figure 3.   Bland Altman plot of skeletal muscle.

Position of the subjects in examination was standing upright with their feet above feet electrode and holding the palm electrode in 90° arms extension. There were slight difference feet position due to various size of respondents feet, different holding quality and arm extension between subjects that hard to excluded in this research. There were few subjects who had mean-average difference plot far beyond ± 1.96 SD according to Figure 1 to 3. According to reference, BIA has possible source of error due to differences in limb length, physical activity, nutrition status, hydration level, blood chemistry, ovulation and placement of electrodes. (National institutes of health technology assessment conference statement 1994).

According to majority samples consisting of women, the mean of body fat is normal compared to standard. It is the optimum fat amount that is needed to run in daily activities (Lohman 1986). The result is related to age median ranging around teenage and healthy criteria to be ruled in the study. Possible cause may be due to imbalance between energy demand and supply. The first year students are students forbidden to take elevator, they have to climb the stairs to get their wanted floors every day. There are 9 floors in UNISBA medical faculty building. The skeletal measurement data show more uniform data related to their nearly uniform activities as the student, regarding minimal muscular involvement. They have tasks that much different than previous education level, no physical education topics and some obligations that must be fulfilled that come from the seniors.

How the BIA measures skeletal muscle more reliable than body and visceral fat is …. (must find reference why skeletal muscle. Pietilainen et al described how BIA measurement superior in determining the skeletal muscle better than fat.

5 CONCLUSION

The measurement of body composition is important to minimize the limitations of BMI interpretation; therefore, simple method is required to get the result consistently at home. This common digital BIA can be used as a reliable device to measure muscle and fat mass as well.

ACKNOWLEDGEMENT

This study was sponsored by The Research Unit of Medical Faculty of Universitas Islam Bandung (UNISBA), Indonesia.

REFERENCES

Engin, A. 2017. The definition and prevalence of obesity and metabolic syndrome. In *Obesity and Lipotoxicity* (pp. 1–17). Springer, Cham.
Giavarina, D. 2015. Understanding bland altman analysis. *Biochemia medica: Biochemia medica, 25*(2), 141–151.
Khalil, S. F., Mohktar, M. S., & Ibrahim, F. 2014. The theory and fundamentals of bioimpedance analysis in clinical status monitoring and diagnosis of diseases. *Sensors, 14*(6), 10895–10928.
Kuriyan, R. 2018. Body composition techniques. *The Indian journal of medical research, 148*(5), 648.
Lohman, T. G. 1986. Body composition; a round table. *Physician Sportsmed, 14*(3), 144–162.
National Institutes of Health (US). Office of Medical Applications of Research. (1994). *Bioelectrical impedance analysis in body composition measurement: National Institutes of Health Technology Assessment Conference Statement, December 12–14, 1994.* NIH Office of Medical Applications of Research.
Shepherd, J. A., Ng, B. K., Sommer, M. J., & Heymsfield, S. B. 2017. *Body composition by DXA. Bone, 104, 101–105.*

Medical Technology and Environmental Health – Abdullah, Widiaty & Abdullah (eds)
© 2020 Taylor & Francis Group, London, ISBN 978-0-367-86053-0

A literature review: Biomarker CD 31+ as a sign of endothelial dysfunction in children and adults

I. Rahmawaty & L.A. Garina
Universitas Islam Bandung, Bandung, Indonesia

ABSTRACT: Endothelial Dysfunction is defined as an imbalance between coagulation and thrombosis factors, loss of vasodilatation response ability in endothelium-dependent stimuli, loss of bioavailability of NO, increase in vasoconstrictive agents, interference with inflammatory regulation, and stimulate the release of micro particles. This could be a risk on cardiovascular disease in the future. The purpose of this study is to build the knowledge about micro particles as a potential biomarker for endothelial dysfunction that in the future could be a cardiovascular risk for children and adults. The method used in this study is literature review of 20 journal articles by Boolean searching using micro particles as keyword to detect endothelial dysfunction. Many studies from the articles found that CD 31+ as one of the microparticle biomarkers increased in endothelial dysfunction. Microparticles could be released from endothel when the endothel got activation or apoptosis, especially by chronic inflammatory that could be found in children such as obstructive sleep apnea or in adults such as obesity and many hypoxia condition. This hypoxia causes endothelial stress which in turn causes apoptosis can be a danger signal of acute ischemia and the incidence of hypoxia and triggers tissue repair mechanism.

1 INTRODUCTION

Endothelium is a simple monolayer, the healthy endothelium is optimally placed and able to respond to physical and chemical signals by production of a wide range of factors that regulate vascular tone, cellular adhesion, thromboresistance, smooth muscle cell proliferation, and vessel wall inflammation (Deanfield et al. 2007). Vascular endothelium is not only as a passive barrier between flowing blood and the vascular wall but also uses this strategic location to maintain vascular homeostasis. This plays an important role in modulating vascular tone, calibre, and blood flow in response to humoral, neural, and mechanical stimuli (Behrendt & Ganz 2002).

A variety of insults may damage endothelial structure and function, which include physical injuries, biochemical injury, and immune mediated damage. These insults cause alteration in endothelial physiology resulting in impairment or loss of its normal functions (Tousoulis et al. 2008). Endothelial Dysfunction is defined as an imbalance between coagulation and thrombosis factors, loss of vasodilatation response ability in endothelium-dependent stimuli, loss of bioavailability of NO, increase in vasoconstrictive agents, interference with inflammatory regulation, and stimulate the release of microparticles (Berezin et al. 2015, Shantsila 2009). Circulating markers of such endothelial cell damage include endothelial microparticles derived from activated or apoptotic cells, and whole endothelial cells (Deanfield et al. 2007, Mallat 2000).

Microparticles are plasma membrane-derived vesicles shed from stimulated cells, in the broad sense of the term. Their presence is interpreted by proximal or remote cells in fundamental physiological processes including intercellular communication, hemostasis, and immunity. On the other hand, variations of their number or characteristics are frequently observed in pathophysiological situations (Hugel et al 2005), that circulate in the peripheral blood and play active roles in thrombosis, inflammation and vascular reactivity. Microparticles can be released from nearly every cell type, such as platelet, leucocyte and endothelial cell origin. Cells can release microparticles during activation or death, while microparticles are present in peripheral blood of healthy individuals, marked elevations occur in many disease states. These conditions include inflammatory and autoimmune diseases, atherosclerosis, malignancy and infection among others (Ardoin et al. 2007).

Interestingly, hypoxia, oxidative stress, inflammation, and coagulation can cause different degrees of vascular endothelial injury and stimulate release of endothelial microparticles (EMPs). The several conditions can induce the release of EMP both *in vivo* and *in vitro*, such as acute coronary syndromes (ACS), myocardial infarction, and stroke (Deng et al. 2017). Furthermore, endothelial microparticles that harbour markers of cellular activation have shown to predict poor cardiovascular outcomes in at-risk patients are also discussed (Trzepizur et al. 2014).

The purpose of this study is to review method for analyzing endothelial dysfunction earlier by releasing microparticles as s potential biomarker cardiovascular risk in children and adult.

2 METHODS

This study is a literature review discussing 20 journal articles by Boolean searched using microparticles as the keyword. The method to find this microparticles is technically challenging because of their small size (0.1–1μm), heterogeneous densities, and overlap with other particulate structures and platelets. Microparticles carry antigenic marker characteristics of their parent cell, which is exploited for identifying their cellular origin, usually by fluorescently labelled monoclonal antibodies using flow cytometry or an ELISA (Trzepizur et al. 2014). Methods for microparticles detection include flow cytometry, ELISA, fluorescence-based antibody array system, and functional coagulative assays. In particular, ELISA and flow cytometry are widely used in clinical settings to measure microparticles in plasma samples (Saleh & Kabeer 2015).

The majority of endothelial microparticles studies to date have employed flow cytometry as the primary method. Classification of the methods of some representative clinical studies is described in Table 1 (Horstman, 2004).

Analysis of microparticles includes four essential steps: isolation, detection, typing, and counting (Saleh & Kabeer 2015, Gradziuk & Radziwon 2017). Flow cytometry is the most common method of identification, quantification, and size assessment of the microparticles. In flow cytometry, microparticles are detected in two stages: based on the intensity of light scattering and using fluorescently-labeled antibodies against specific surface antigens. Flow cytometry allows simultaneous analysis of morphology and subtype determinations of thousands of microparticles per second (Gradziuk & Radziwon 2017).

Endothelial microparticles (EMPs) can be defined according to membrane glycoproteins, as they possess antigens constitutively expressed on endothelial cells such as CD31+ (Deng et al. 2017).

Table 1. Representative methods of endothelial microparticles analysis in clinical studies.

Analysis	Application	Authors	Year
Flow Cytometric			
CD 31+, CD 51+	Lupus anticoagulant	Combes et al	1999
CD 31+, CD 42-	TTP	Jiminez et al	2001
CD 31+, CD 42-	Multiple sclerosis	Minagar et al	2001
CD 31+, CD 42-	Hypertension	Preston et al	2002
CD 31+, CD 42-	Acute coronary syndrome	Bernal-Mizrachi et al	2003
CD 31+, CD 42-	Preeclampsia	Gonzales-Quintero et al	2003
CD 51+	Diabetes	Sabatier et al	2002
CD 62E+	TTP	Jiminez et al	2003
AnnexV+/CD 146+	Sicle cell disease	Shet et al	2002, 2003
Immuno-capture/ELISA	Acute coronary syndrome	Mallat et al	2000

3 RESULTS

PECAM-1 endothelial microparticles, defined as CD 31+ endothelial microparticles, are concentrated at endothelial junction. CD31+ endothelial microparticles may play diverse roles in vascular biology, regulating the platelet function, angiogenesis, T cell and B cell activation, endothelial cells permeability and transmigration across the endothelium. Therefore, the released endothelial microparticles likely reflect the apoptosis of injured endothelial cells (Deng et al. 2017). However, CD31+ is also expressed on endothelial cells and other myeloid cells such as monocytes, granulocytes, and B cells but is most abundant in endothelial cells (Saleh & Kabeer 2015).

Circulatory hypoxia, including heart failure, shock, embolism and arterial stenosis, represents a decrease in tissue blood flow resulting in tissue oxygen deficiency. Endothelial microparticles have been investigated in several human circulatory hypoxia diseases. Endothelial microparticles CD31+ subtype may play role in circulatory hypoxia-related diseases (Table 2) (Deng et al. 2017), including obstructive sleep apnea and chronic obstructive pulmonary disease (Deng et al. 2018).

Many studies from the articles searched found that CD 31+ as one of microparticles biomarker increased in endothelial dysfunction. Microparticles could be released from endothel when the endothel got activation or apoptosis, this was especially chronic inflammatory that could be found in children such as obstructive sleep apnea (OSA) or in adults such as obesity and many hypoxia condition (Figure 1) (Nieri et al 2016).

4 DISCUSSION

Based on the various stages of cell life, cells are stimulated to produce submicron fragments from the plasma membrane, known as microparticles or microvesicles (Mallat 2000). In 1955, O'Brien described "platelet-like activity" in the serum of normal people. Twelve years later, Wolf made the same conclusion, introducing "platelet dust" in plasma. The two discoveries are the first statement of what is now known as microparticles. The formation of microparticles comes from the pathway that has been identified, namely cell activation and apoptosis. Besides originating from platelets, microparticles can be produced by endothelial cells, leukocytes, and erythrocytes (Sierko 2015).

Microparticle composition depends on both their cellular origin and the stimuli involved during their generation. Microparticles may content membrane and cytosolic protein, transcription factors, and genetic material, as well as lipids or organelles from their cells of origin (Deng et al. 2017). However, microparticles are now ascribed prominent roles in cardiovascular diseases and contribute to the regulation of pathophysiological processes including endothelial function, inflammation, coagulation, angiogenesis, and cellular remodeling (Trzepizur et al. 2014).

Table 2. CD31+ in circulatory hypoxia-related disease.

Endothelial microparticles	Diseases	Aspect related to diseases
CD31+	Coronary arterial disease	Coronary endothelial function, cardiovascular events, risk stratification, left ventricle dysfunction and endothelium-dependent vasodilatation
	Acute coronary syndrome	Acute endothelial injury, risk stratification
	Myocardial infarction	Recurrence, the size of myocardial at risk in DTEMI patients, thrombus occlusion
	Cerebrovascular atherosclerosis	The endothelial microparticles level significantly discriminates extracranial and intracranial arterial stenosis

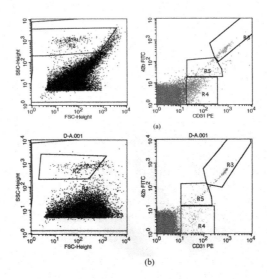

Figure 1.   CD31+ in Suspect Paediatric OSA (a), and in Adult with Night Shift Worker (b) (from our case study, 2019).

Endothelial microparticles or microvesicles are 0.1–1µm nucleated vesicles formed from cytoskeletal and reorganized cell membranes and released to extracellular due to apoptosis or activation of endothelial cells, reflecting the degree of damage to endothelial cells, and increased endothelial microvesicles occur in diseases of the extracellular due to apoptosis or activation of endothelial cells, reflecting the degree of damage to endothelial cells, and increasing endothelial microvesicles occur in diseases of the extracellular due to apoptosis or activation of endothelial cells, reflecting the degree of damage to endothelial cells, and increasing endothelial microvesicles occurring in diseases of the endothelial hypoxic hypoxia (Gradziuk & Radziwon 2017).

It is now widely recognized that the vascular endothelium plays a pivotal role in the pathogenesis of numerous thrombotic and inflammatory disorders. Accordingly, much effort has been devoted to identify plasma markers of endothelial disturbance. PECAM-1 (CD31), ICAM-1 (CD105), E selectin (CD62E) and P selectin (CD62P), these are routinely used to identify endothelial microparticles (Saleh & Kabeer 2015), that this difference is attributable to CD31+ endothelial microparticles (primarily released from apoptotic endothelial cells and reflecting hypoxia-induced acute injury, while CD62E+ endothelial microparticles are primarily released from activated endothelial cells) (Gradziuk & Radziwon 2017).

Obesity is a state of chronic oxidative stress and inflammation (Nieri et al 2016, Furukawa 2017). Endothelial microparticles are also elevated in obese women and independently involved in the pathogenesis of endothelial dysfunction. The results of the present study demonstrate that circulating endothelial and platelet-derived microparticles are higher in obese women as compared with age-matched lean women, and also negatively correlated with endothelium-dependent (Furukawa 2017).

Therefore, the released CD31+ endothelial microparticles are likely to reflect the apoptosis of injured endothelial cells. As markers of endothelial cells apoptosis, CD31+/CD42b– endothelial microparticles levels were strongly correlated with obstructive sleep apnea severity, endothelial dysfunction and carotid intima media thickness (Gradziuk & Radziwon 2017). Based on the research, an increase in endothel microparticles levels (CD31 +/CD42b-, CD62E +/CD42b-) in OSA children with apnea hypopnea index/AHI ≥1 (Horstman 2004, Esposito 2006).

Endothelial microparticles CD31+ subtype may play role in circulatory hypoxia-related diseases (Ardoin et al. 2007). PECAM-1 (CD31+) has predictably been implicated in a number of clinically-relevant disorders, ranging from thrombosis and cardiovascular disease to inflammation and cancer (Kim et al 2011).

5 CONCLUSION

Biomarker CD 31+ is a sign of endothelial dysfunction in both children and adults. Hypoxia causing endothelial stress which in turn causes apoptosis can be a danger signal of acute ischemia and the incidence of hypoxia and triggers tissue repair mechanism.

ACKNOWLEDGEMENTS

This study was supported by Medical Faculty of Universitas Indonesia and Medical Faculty of Universitas Islam Bandung.

REFERENCES

Ardoin, S. P., Shanahan, J. C., & Pisetsky, D. S. 2007. The role of microparticles in inflammation and thrombosis. *Scandinavian journal of immunology, 66*(2-3), 159–165.

Behrendt, D., & Ganz, P. 2002. Endothelial function: from vascular biology to clinical applications. *The American journal of cardiology, 90*(10), L40–L48.

Berezin, A., Zulli, A., Kerrigan, S., Petrovic, D., & Kruzliak, P. 2015. Predictive role of circulating endothelial-derived microparticles in cardiovascular diseases. *Clinical biochemistry, 48*(9), 562–568.

Deanfield, J. E., Halcox, J. P., & Rabelink, T. J. 2007. Endothelial function and dysfunction: testing and clinical relevance. *Circulation, 115*(10), 1285–1295.

Deng, F., Wang, S., & Zhang, L. 2017. Endothelial microparticles act as novel diagnostic and therapeutic biomarkers of circulatory hypoxia-related diseases: a literature review. *Journal of cellular and molecular medicine, 21*(9), 1698–1710.

Deng, F., Wang, S., Xu, R., Yu, W., Wang, X., & Zhang, L. 2018. Endothelial microvesicles in hypoxic hypoxia diseases. *Journal of cellular and molecular medicine, 22*(8), 3708–3718.

Esposito, K., Ciotola, M., Schisano, B., Gualdiero, R., Sardelli, L., Misso, L., ... & Giugliano, D. 2006. Endothelial microparticles correlate with endothelial dysfunction in obese women. *The Journal of Clinical Endocrinology & Metabolism, 91*(9), 3676–3679.

Furukawa, S., Fujita, T., Shimabukuro, M., Iwaki, M., Yamada, Y., Nakajima, Y., ... & Shimomura, I. 2017. Increased oxidative stress in obesity and its impact on metabolic syndrome. *The Journal of clinical investigation, 114*(12), 1752–1761.

Gradziuk, M., & Radziwon, P. 2017. Methods for detection of microparticles derived from blood and endothelial cells. *Acta Haematologica Polonica, 48*(4), 316–329.

Horstman, L. L., Jy, W., Jimenez, J. J., & Ahn, Y. S. 2004. Endothelial microparticles as markers of endothelial dysfunction. *Front biosci, 9*(1), 118–135.

Hugel, B., Martínez, M. C., Kunzelmann, C., & Freyssinet, J. M. 2005. Membrane microparticles: two sides of the coin. *Physiology, 20*(1), 22–27.

Kim, J., Bhattacharjee, R., Kheirandish-Gozal, L., Spruyt, K., & Gozal, D. (2011). Circulating microparticles in children with sleep disordered breathing. *Chest, 140*(2), 408–417.

Mallat, Z., Benamer, H., Hugel, B., Benessiano, J., Steg, P. G., Freyssinet, J. M., & Tedgui, A. 2000. Elevated levels of shed membrane microparticles with procoagulant potential in the peripheral circulating blood of patients with acute coronary syndromes. *Circulation, 101*(8), 841–843.

Nieri, D., Neri, T., Petrini, S., Vagaggini, B., Paggiaro, P., & Celi, A. 2016. Cell-derived microparticles and the lung. *European Respiratory Review, 25*(141), 266–277.

Saleh, H. A., & Kabeer, B. S. 2015. Microparticles: Biomarkers and effectors in the cardiovascular system. *Global Cardiology Science and Practice, 2015*(3), 38.

Shantsila, E. 2009. Endothelial microparticles: a universal marker of vascular health? *Journal of human hypertension, 23*(5), 359–361.

Sierko, E., Sokół, M., & Wojtukiewicz, M. Z. 2015. Endothelial microparticles (EMP) in physiology and pathology. *Postepy higieny i medycyny doswiadczalnej (Online), 69*, 925–932.

Tousoulis, D., Charakida, M., & Stefanadis, C. 2008. Endothelial function and inflammation in coronary artery disease. *Postgraduate medical journal, 84*(993), 368–371.

Trzepizur, W., Martinez, M. C., Priou, P., Andriantsitohaina, R., & Gagnadoux, F. 2014. Microparticles and vascular dysfunction in obstructive sleep apnoea. *European Respiratory Journal, 44*(1), 207–216.

Biopharmaceutical product and engineering

Skin penetration enhancement of polyphenolic compounds from cocoa pod husk topical serum using a phytosomal system

A.S.E. Priani, B.S. Aprilia, C.R. Aryani & D.D. Mulyanti
Universitas Islam Bandung, Bandung, Jawa Barat, Indonesia

ABSTRACT: Cocoa pod husks containing polyphenol compounds have many beneficial topical applications. Phytosome is a complex between a water-soluble phytoconstituent like polyphenol and a phospholipid known to enhance oral and topical penetrations. The objective of this research was to determine the effect of a phytosomal system on polyphenol skin penetration from a topical serum preparation containing cocoa pod husk extract and to determine its kinetic release profile. The phytosome complex was made using a thin layer hydration method at a 1:1 comparison of extract and phospholipid. The phytosome complex was then developed into a topical serum preparation. Skin penetrations of polyphenol compounds were determined using Franz diffusion cells. The results showed that the phytosome system could enhance polyphenol skin penetration from the preparation that was significantly different with a non-phytosomal system (P < 0.05). The released kinetic of the compounds from the topical serum preparation followed the Higuchi model.

1 INTRODUCTION

Cocoa is one of the important commodities in Indonesia and across the world. Cocoa pod husk is a waste of cocoa processing that can reach 75% of the whole cocoa fruit (Daud et al. 2013; Irwanto et al. 2018). Cocoa pod husk is known to contain many active compounds. Our previous study showed that cocoa pod husk contains high levels of polyphenol. The polyphenol content of cocoa pod ethanolic extract was 15.5%. Polyphenols can be used as an active agent for cosmetics and pharmacy due to their beneficial biological properties. Polyphenols are secondary plant metabolites with antioxidant, antimicrobial, anti-inflammatory, photoprotective, and antiaging activities (Priani et al. 2019; Zillich et al. 2015).

Many phytochemical constituents show excellent biological in vitro activities, but demonstrate fewer in vivo activities due to the properties of the phytoconstituents like poor lipid solubility. Some studies have revealed that polyphenol and other phytoconstituent are hydrophilic in nature, which causes poor absorption through the lipid membrane (Das et al. 2013). Characteristic modification of phytoconstituents must be applied for enhancing oral absorption and skin penetration. One of the formulation strategies to enhance biological absorption of phytoconstituent is through the use of phytosomes. Phytosomes are complex yet natural active ingredients like polyphenol and phospholipids that could increase absorption of herbal extracts or isolated active ingredients when applied topically or orally (Singh et al. 2011). Phytosomes are cell-like structures that result from the reaction of the phospho-lipids with the standardized extract or polyphenolic. A previous study has demonstrated the enhancing of polyphenol penetration through skin with phytosomal formulation compared to non-phytosomal formulation. For cosmetic application, phytosome complex could be developed for various topical applications such as cream, gel, or serum (Surini et al. 2018).

Our previous study has developed the topical serum containing cocoa pod husk phytosomes (Priani et al. 2019). This study was conducted to determine the effect of the phytosomal system on polyphenol skin penetration from a topical serum containing cocoa pod husk extract, and also to determine its kinetic release model. A skin penetration test was conducted by a Franz diffusion cell using HT-Tuffryn as a diffusion membrane.

2 METHOD

2.1 *Preparation of cocoa pod husk phytosomes*

Closed-loop supply chain management (CLSCM) can be defined as a system of design, control, and operation to maximize the creation of value throughout a product's life cycle by recovering the value of the product dynamically, because the type and volume of the returned product are not the same at different time (Schultmann, 2003). This activity is a combination of forward and reverse supply chain activities.

According to Grassi et al. (2005), CLSCM activity took place as follows. Cocoa pod extract was prepared by a maceration process using ethanol 70% as a solvent. The phytosome complex was prepared by reacting cocoa pod extract with phosphatidylcholine in ethanol at a weight comparison of 1:1. The mixture sonicated for 30 minutes and then evaporated via a rotary evaporator (40°C) until resulting in a thin layer and stored overnight at room temperature. A thin layer was then hydrated using double distilled water then submitted to ultrasonication to decrease the particle size of the phytosome complex (Priani et al. 2019).

2.2 *Preparation of topical serum containing cocoa pod husk phytosomes*

The gel-based topical serum was prepared using viscolam Mac 10 as a gelling agent at a concentration of 3%. The formula is shown in Table 1.

The topical serum was prepared with disperse Viscolam Mac 10 on distilled water by adding triethanolamine (TEA) to reach a pH of 6–7. Methyl and propyl paraben in propylene glycol and ascorbic acid in distilled water were added to the Viscolam gel. The phytosome complex was then added to the gel and homogenized at 300 rpm for 10 minutes using s magnetic stirrer (Priani et al. 2019; Surini et al. 2018).

2.3 *In vitro skin diffusion test of the cocoa pod serum*

The study was conducted using HT-Tuffryn (3.14 cm^2) in Franz diffusion cells with a diffusional area. The receiver compartment was filled with phosphate buffer at a pH of 7.4 (37 ± 0.5 °C, 600 rpm). A half gram of gel was applied on the membrane and 2 mL samples of the receiving solution were collected at 30, 45, 60, 90, 120, and 150 minutes. The receptor phase was replaced at each sampling time by 2 mL of phosphate buffer. The polyphenol

Table 1. Formulation of cocoa pod topical serum.

Component (%)	F1	F2
Cocoa pod phytosome	1	-
Cocoa pod extract	-	1
Propyl paraben	0.02	0.02
Methyl paraben	0.18	0.18
Ascorbic acid	0.05	0.05
Propyleneglycol	15	15
Viscolam mac 10	3	3
Trietanolamine	Qs	qs
Aquadest ad	100	100

F1 : Phytosomal system
F2: Non-phytosomal system

content of the samples was determined using the Folin-Ciocalteu method using gallic acid as a reference (Priani et al. 2019). The total amount of penetrated drug per diffusion area (Q), percentage of diffusion (%), and flux (J) of polyphenol from the preparation were calculated (Iskandarsyah et al. 2017).

2.4 *Kinetic release profile determination*

The kinetic release profile was evaluated by fitting the experimental data to kinetic order equations. Linier regression analyses were made to zero order (Q vs t), first order (Ln Q vs t), Higuchi (Q and $t^{1/2}$), and Korsmeyer-Peppas (Ln Q vs Ln t), and the best coefficient correlation of the equation was chosen (Cojocaru et al. 2015).

3 RESULTS AND DISCUSSION

We have described the physical evaluation results of the cocoa pod phytosome and topical serum in a previous publication (Priani et al. 2019). Therefore, the focus of this publication is on skin diffusion profile. A skin diffusion test was conducted using a Franz diffusion cell, which is the most common in vitro release testing method for semisolid formulations. The advantages of using the Franz diffusion cell is less handling of tissues, no continuous sample collecting, and the low amount of preparation required for analysis (Bartosova & Bajgar 2012). The results of the skin penetration test of preparation are shown in Table 2, Table 3, and Figure 1.

The results show that the phytosome system could enhance the amount of polyphenol penetrated through the membrane compared to the non-phytosomal system that was significantly different (P < 0.05). The phytosome could enhance drug skin penetration associated with phosphatidylcholine as a carrier. The cell-like structure of the phytosome facilitates the penetration process of active compounds through skin. Besides the phytosomal system, propylene glycol in the formulation also acts as a penetrant enhancer that increases the skin penetration of polyphenol in the phytosomal and non-phytosomal systems.

Another study aimed to determine the kinetic release model of the cocoa pod topical serum. The results are presented in Table 4.

Table 2. In vitro skin diffusion result of F1.

Time	Q (μg/cm^2)	% diffusion	J (μg/cm^2.h)
15	159.02	27.13	636.10
30	260.56	44.44	521.12
60	375.87	64.11	375.87
90	446.37	76.13	297.58
120	485.46	82.80	242.73
150	551.09	93.99	220.44

Table 3. In vitro skin diffusion result of F2.

Time	Q (μg/cm^2)	% diffusion	J (μg/cm^2.h)
15	122.84	20.94	491.36
30	220.21	37.56	440.42
60	331.76	56.58	331.76
90	384.89	65.64	256.59
120	437.90	74.68	218.95
150	460.49	78.56	184.25

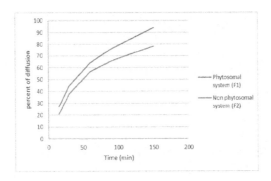

Figure 1. Percent diffusion of topical serum.

Table 4. Kinetic release model.

Model	X value	Y value	Regression equation	R^2 (Coefficient correlation)
Orde 1	Q	t	y = 2.72x + 168.69	0.9405
Orde 2	Ln Q	t	y = 0.008x + 5.22	0.8423
Higuchi	Q	$t^{1/2}$	y = 45.30x + 3.81	0.9873
Korsmeyer-Peppas	Ln Q	Ln t	y = 0.52x + 3.72	0.9851

Based on the result, we can conclude that the most appropriate kinetic release model of polyphenol from the preparation is Higuchi (Figure 2). The polyphenol release was found to be proportional to the square root of time, suggesting that all formulations followed a diffusion-controlled release mechanism associated with the phytosomal system (Sareen et al. 2015). The Higuchi model suggests a pure diffusion release mechanism of the drug from a vehicle, with no occurring erosion or swelling of the matrix therefore (Panapisal et al. 2012).

4   CONCLUSION

The phytosome system could enhance polyphenol skin penetration from the preparation that is significantly different from the non-phytosomal system ($P < 0.05$). The release kinetic of the compounds from topical serum preparation follows the Higuchi model.

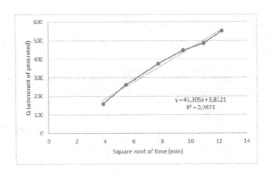

Figure 2. Higuchi kinetic release profile.

ACKNOWLEDGMENT

The author would like to acknowledge UNISBA (LPPM) for the financial support for this research.

REFERENCES

Bartosova, L. & Bajgar, J. 2012. Transdermal drug delivery in vitro using diffusion cells. *Current medicinal chemistry* 19(27): 4671–4677.

Cojocaru, V., Ranetti, A. E., Hinescu, L. G., Ionescu, M., Cosmescu, C., Poştoarcă, A. G., & Cinteză, L. O. 2015. Formulation and evaluation of in vitro release kinetics of Na3CaDTPA decorporation agent embedded in microemulsion-based gel formulation for topical delivery. *Farmacia* 63(5): 656–664.

Das, M. & Kalita, B. 2013. Phytosome: An overview. *Journal of pharmaceutical and scientific innovation* 2: 7–11.

Daud, Z., Kassim, M., Sari, A., Mohd Aripin, A., Awang, H., Hatta, M., & Zainuri, M. 2013. Chemical composition and morphological of cocoa pod husks and cassava peels for pulp and paper production. *Australian journal of basic and applied sciences* 7(9): 406–411.

Grassi, D., Lippi, C., Necozione, S., Desideri, G., & Ferri, C. 2005. Short-term administration of dark chocolate is followed by a significant increase in insulin sensitivity and a decrease in blood pressure in healthy persons. The American journal of clinical nutrition, 81(3), 611–614.

Irwanto, D., Wiratni, W., Rochmadi, R., & Syamsiah, S. 2018. Lactic acid production from cocoa pod husk by studying further the influence of alkaloids on fermentation process using *Lactobacillus Plantarum* bacteria. *Reaktor* 18(1): 51–56.

Iskandarsyah, I., Puteri, A. W., & Ernysagita, E. 2017. Penetration test of caffeine in ethosome and desmosome gel using an in vitro method. *International journal of applied pharmaceutics* 9: 120–123.

Panapisal, V., Charoensri, S., & Tantituvanont, A. 2012. Formulation of microemulsion systems for dermal delivery of silymarin. *Aaps Pharmscitech* 13(2): 389–399.

Priani, S. E., Aprilia, S., & Purwanti, L. 2019. Antioxidant and tyrosinase inhibitory activity of face serum containing cocoa pod husk phytosome *(Theobroma Cacao L) Journal of Applied Pharmaceutical Science*, 9(10), 110–115.

Sareen, R., Bhardwaj, V., Mehta, V., & Sharma, A. 2015. Topical gel incorporated with non-ionic surfactant based solid lipid microspheres of ketoprofen: Physicochemical analysis and anti-inflammatory evaluation. *International journal of pharmacy and pharmaceutical sciences* 7(10), 200–206.

Schultmann, F., Engels, B., & Rentz, O. 2003. Closed-loop supply chains for spent batteries. Interfaces, 33(6), 57–71.

Singh, A., Saharan, V. A., Singh, M., & Bhandari, A. 2011. Phytosome: Drug delivery system for polyphenolic phytoconstituents. *Iranian journal of pharmaceutical sciences* 7(4): 209–219.

Surini, S., Mubarak, H., & Ramadon, D. 2018. Cosmetic serum containing grape (*Vitis vinifera* L.) seed extract phytosome: Formulation and in vitro penetration study. *Journal of young pharmacists* 10(2): S51–S55.

Zillich, O. V., Schweiggert-Weisz, U., Eisner, P., & Kerscher, M. 2015. Polyphenols as active ingredients for cosmetic products. *International journal of cosmetic science* 37(5): 455–464.

Medical Technology and Environmental Health – Abdullah, Widiaty & Abdullah (eds)
© 2020 Taylor & Francis Group, London, ISBN 978-0-367-86053-0

In silico analysis of multi-target antimelasma aloe vera compound

D. Hikmawati, T. Respati, Y. Yuniarti & L. Yuniarti
Universitas Islam Bandung, Bandung, Indonesia

ABSTRACT: Melasma is a predominant hyper-melanosis disorder on the face often experienced by middle-aged women. Melasma occurs due to the excessive melanin formation. The target of melasma therapy is to inhibit melanocyte proliferation and melanosome formation and to increase melanosome degradation. De-creasing the inflammatory response is a way to inhibit melanin formation. Conventional melasma therapy has side effects like erythema and burning, ochronosis, and depigmentation such as confetti. One of the natural ingredients with the potential to be developed as an anti-melasma treatment is aloe vera. This study aimed to analyze the potential of aloe vera protein as an anti-melasma treatment. This study employed the in silico method via the pathway analysis method with PubChem software, Swiss Target Prediction, STRING, and Cystoscope. The results show that of the many phytochemicals present in aloe, several phytochemical compounds are predicted to act and inhibit the formation of melanin Tyrosinase-related protein 1 (TYRP1), Estrogen Receptor 1 (ESR1), Interleukin 10 (IL10), and inducible NO synthase INOS. The conclusion is that after silico analysis, the active substance can inhibit the enzyme TYRP1, increase IL10, and suppress ESR1 and INOS.

1 INTRODUCTION

Melasma is an increase in skin pigmentation or symmetrical hyper-melanosis characterized by irregular light to gray macules. Melasma is the most common chronic hyper-melanosis of the skin and the most commonly encountered hyperpigmentation disorder. Patients with melasma constitute 40% of patients who come to dermatology clinics (Achar & Rathi 2011; Handel et al. 2014; Sarkar et al. 2014). Melasma features are usually symmetrical, irregular, jagged hyperpigmentation in certain areas. The most common locations for melasma on the face are the cheeks, upper lip, chin, and forehead, but other areas exposed to the sun may also be occasionally involved (Hexsel et al. 2009).

Melasma often causes a decrease in the quality of life, especially in social relations, in the form of shame, low self-esteem, and dissatisfaction with appearance. This disorder is challenging to treat because it also causes anhedonia or discomfort and stress in carrying out daily activities. The lack of confidence in appearance, in some cases with high melasma area severity index (MASI) scores, causes people to commit suicide. Besides significantly disrupting the quality of life, the cost of treatment is very expensive, but usually does not provide satisfactory results (Ali et al. 2016; Guarneri 2014).

The present treatment for melasma still has side effects such as those seen with the use of hydroquinone, which causes burning, redness, and hypopigmentation on the face. In animal testing, hydroquinone causes kidney damage (Sahu et al. 2013). Tretinoin and retinoids (RA) can cause birth defects. Azelaic acid (AA) is also used. However, this compound is known to be less effective with unsatisfactory results (Nobakht et al. 2006).

Aloe vera, containing many antioxidants and other active ingredients, is often used to make the skin brighter. It also shows potential to cure melasma, but scientific evidence and research into the mechanism of action of these natural ingredients are still lacking (Tamega et al. 2013).

Previous research has shown that aloe vera has many phenolic fractions that can act as anti-oxidants and reduce the effects of ultraviolet radiation (Ray et al. 2013).

Computation-based methodology has become an essential component in drug discovery, ranging from identifying targets, optimization, and optimization with a structured ligand approach based on virtual screening techniques. The key to this method is docking a small molecule on a protein binding site (Kitchen et al. 2004).

2 RESEARCH METHODS

The research method used in this study was descriptive and qualitative, aiming to process and interpret data obtained from the database. Steps taken during the study included, first, a literature search on previous research into aloe vera based on active ingredient, and, second, testing in silico using the pathway analysis method with PubChem software. Compound structure searches were performed through the PubChem database, then the 4-D structure of the compound and the SMILE canonical data were selected for use in the next step.

The third step involved tracking target proteins of aloe vera-based active compounds in the SMILE canonical structure using Swiss Target Prediction (www.SwissTargetPrediction.ch/).

The SMILE can predict the target protein of interaction of a compound based on structural similarity between the structure we want to predict (query structure) with the structure of drugs that have been approved by the FDA. It also can predict the interaction on non-drug compounds that have been analyzed in vitro and in vivo (Dunkel et al. 2008; Gfeller et al. 2014). Based on these findings, the target proteins selected were the ones above 70%. The next step was a pathway analysis of these proteins on protein targets that play a role in the melanogenesis process using STRING and a cystoscope.

STRING is a database that can be used to predict the interaction of a protein, its expression in a network, and trace its relationship to other proteins in the cell mechanism.

3 RESULTS AND DISCUSSION

More than 200 compounds are found in *Aloe barbadensis*, 75 of which conduct biological activity. Aloe vera leaves contain a variety of compounds. They are anthraquinone (e.g., aloe emodin), antrones and their glycosides (e.g., 10- (5-anhydroglucosyl), aloemodin-9-anthrone (also known as aloin A and B), chromon, carbohydrate, protein, glycoprotein, amino acids, organic acids, lipids, sugars, vitamins, and minerals. For centuries, it has been used medically for diseases such as fever, minor injuries, wounds and burns, digestive disorders, and diabetes. It is also used for sexual vitality and fertility problems, cancer, immune modulation, AIDS, and various skin diseases (Ali et al. 2015). The results of tracing the content of the active compounds of aloe vera and its chemical structure (canonical smile) based on the PubChem database and the target protein according to the target prediction of the active substance can be seen in Table 1.

Table 1 shows 13 active substances in aloe vera that have a target protein.

Pathway analysis of the target protein from aloe vera's active compound in the inflammatory and tyrosinase pathways is shown in Figure 1.

Aloe (Xanthorrhoeaceae), which consists of about 400 species, is a perennial plant that has been used as traditional medicine for around 3,000 years. The most famous species is *Aloe vera* L., a plant without stems that grows up to 100 cm. The green leaves contain clear, colorless, and unpleasant ingredients as a gel. This gel is used in cosmetics, health drinks and beverages, and medicines. Substances containing aloe vera consist of 99.5% water and 0.5% active ingredients, including vitamins, polysaccharides, phenolic compounds, and organic acids. This gel has many biological activities, including anti-inflammatory, antiviral, antibacterial, anticancer, anti-diabetes, and anti-allergen activities. It also protects against radiation. The main phytochemicals in the gel are derivatives of anthraquinone and C-glucosilantrone: aloin A and B, emodin, desoksialoin, aloinosida B and C, and elgonica dimer A1. Aloesin, aloe emodin, and

Table 1. Active compounds, canonical smile, and target protein of aloe vera.

No	Active compound	Canonical smile	Target protein
1	α- tocopherol	CC1=C(C(=C(C2=C1OC(CC2)(C)CCCC(C)CCCC(C)CCCC(C)C)C)OC(=O)C)C	SLC5A7
2	choline	C[N+](C)(C)CCO	SLC5A7
3	folic acid	C1=CC(=CC=C1C(=O)NC(CCC(=O)O)C(=O)O)NCC2=CN=C3C(=N2)C(=O)NC(=N3)N	TYMS, DHFR
4	niasin (nico-tinamide)	C1=CC(=CN=C1)C(=O)N	SIRT3, SIRT2
6	anthranol	C1=CC=C2C(=C1)C=C3C=CC=CC3=C2O	EPHX2,
7	aloin A and B	C1=CC2=C(C(=C1)O)C(=O)C3=C(C2C4C(C(C(C(O4)CO)O)O)O)C=C(C=C3O)CO	EPHX2,
8	emodin	CC1=CC2=C(C(=C1)O)C(=O)C3=C(C2=O)C=C(C=C3O)O	ESR1, PIM1, ESR2, CSNK2A1PTP4A3
9	Campesterol	CC(C)C(C)CCC(C)C1CCC2C1(CCC3C2CC=C4C3(CCC(C4)O)C)C	AR, NR1H3, HMGCR, CYP51A1NPC1L1, CYP17A1
10	β-sitosterol	CCC(CCC(C)C1CCC2C1(CCC3C2CC=C4C3(CCC(C4)O)C)C)C(C)C	AR, HMGCR, CYP51A1NPC1L1 NR1H3, CYP19A1
11	arachidonic acid	CCCCCC=CCC=CCC=CCC=CCCCC(=O)O	FABP4 PPARG PPARA PPARD
12	linolenic acid	CCC=CCC=CCC=CCCCCCCCC(=O)O	PPARG PPARA PPARD FABP4 FFAR1 PTGS1 FABP3
13	phenylalanine	C1=CC=C(C=C1)CC(C(=O)O)N	CACNA2D1

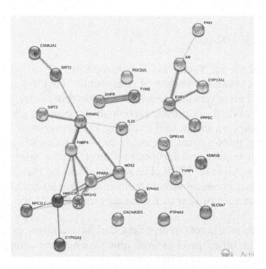

Figure 1. Pathway analysis of apoptotic target protein from soursop leaf active compounds.

aloin A are reported to have antioxidant activity, whereas aloinoside B and C inhibit the activity of soluble epoxide hydrolase and phosphodiesterase-4D (Kim et al. 2017).

Figure 1 shows that the content of aloe vera can target TYRP1, as an enzyme that plays a role in melanin formation. Melasma occurs because of the formation of excess melanin. The target of melasma therapy is inhibition of melanocyte proliferation and melanosome formation and the increase of melanosome degradation. Depigmentation materials can be classified according to their work methods – namely (1) inhibiting tyrosinase or tyrosinase transcription, tyrosinase-related protein (TYRP) 1, TYRP-2, and peroxidase; (2) inhibiting melanosome transfer; (3) cytotoxic melanocytes transcription of tyrosinase-related genes (TYRP) 1, TYRP-2, and peroxidase; (4) inhibiting melanosome transfer; (5) cytotoxic melanocytes arranged by microphthalmia-associated transcription factor (MITF). Picardo and Carrera's studies found that aloesin from aloe vera works as an uncompetitive tyrosinase inhibitor, influencing the complex action of tyrosinase in the substratum and reducing the conversion of DOPA to melanin. Tan and colleagues and Cheng and colleagues have reported that the aloin isolated from aloe vera leaf extract acts as a natural skin lightener. It binds not only to the tyrosinase enzyme but also to the enzyme-substrate complex, which causes the inactivation of enzymes that result in skin lightening (Gupta et al. 2018).

One of the causes of increased melanin production (melanogenesis) is exposure to ultraviolet A rays through the induction of oxidative stress caused by the production of free radicals (oxidants) that involve the NOS/NO system, and interference with oxidative defenses (Panich et al. 2011). Thus, the presence of aloe vera that can regulate INOS can be anti-melanogenic, form oxidants, and produce nitric oxide by eNOS and iNOS.

4   CONCLUSION

This research concluded that in silico, the content of active compounds of aloe vera can modulate the expression of proteins that play a role in melanogenesis such as enzymes tyrosinase, INOS, and IL10 so that aloe vera has potential as an anti-melasma treatment.

ACKNOWLEDGMENT

This work was funded by the Institute for Research and Community Service of the Islamic University of Bandung, through Hibah Dosen Utama.

REFERENCES

Achar, A. & Rathi, S. K. 2011. Melasma: A clinico-epidemiological study of 312 cases. *Indian journal of dermatology* 56(4): 380.

Ali, R., Aman, S., Nadeem, M., & Kazmi, A. H. 2016. Quality of life in patients of melasma. *Journal of Pakistan Association of Dermatology* 23(2): 143–148.

Ali, S. A., Choudhary, R. K., Naaz, I., & Ali, A. S. 2015. Understanding the challenges of melanogenesis: Key role of bioactive compounds in the treatment of hyperpigmentory disorders. *Journal of pigment disorders* 2(11) 21–29.

Dunkel, M., Günther, S., Ahmed, J., Wittig, B., & Preissner, R. 2008. SuperPred: Drug classification and target prediction. *Nucleic acids research* 36(2): W55–W59.

Gfeller, D., Grosdidier, A., Wirth, M., Daina, A., Michielin, O., & Zoete, V. 2014. Swiss Target Prediction: A web server for target prediction of bioactive small molecules. *Nucleic acids research* 42(W1): W32–W38.

Guarneri, F. 2014. Etiopathogenesis of melasma. *Pigmentary disorders* S4: 2376–0427.

Gupta, B., Ahmed, K. M., Dhawan, S., & Gupta, R. 2018. Aloe vera (medicinal plant) research: A scientometric assessment of global publications output during 2007–16. *Pharmacognosy Journal* 10(1): 01–08.

Handel, A. C., Miot, L. D. B., & Miot, H. A. 2014. Melasma: A clinical and epidemiological review. *Anais brasileiros de dermatologia* 89(5): 771–782.

Hexsel, D., Rodrigues, T., Dal'Forno, T., Zechmeister-Prado, D., & Lima, M. 2009. Melasma and pregnancy in southern Brazil. *Journal of the European Academy of Dermatology and Venereology* 23(3): 367–368.

Kim, J. H., Yoon, J.-Y., Yang, S. Y., Choi, S.-K., Kwon, S. J., Cho, I. S., Jeong, M. H., Ho Kim, Y., & Choi, G. S. 2017. Tyrosinase inhibitory components from aloe vera and their antiviral activity. *Journal of enzyme inhibition and medicinal chemistry* 32(1): 78–83.

Kitchen, D. B., Decornez, H., Furr, J. R., & Bajorath, J. 2004. Docking and scoring in virtual screening for drug discovery: Methods and applications. *Nature reviews drug discovery* 3(11): 935.

Nobakht, M., Zirak, A., Mehdizadeh, M., & Tabatabaeei, P. 2006. Teratogenic effects of retinoic acid on neurulation in mice embryos. *Pathophysiology* 13(1): 57–61.

Panich, U., Onkoksoong, T., Kongtaphan, K., Kasetsinsombat, K., Akarasereenont, P., & Wongkajornsilp, A. 2011. Inhibition of UVA-mediated melanogenesis by ascorbic acid through modulation of antioxidant defense and nitric oxide system. *Archives of pharmacal research* 34(5): 811–820.

Ray, A., Gupta, S. D., & Ghosh, S. 2013. Evaluation of anti-oxidative activity and UV absorption potential of the extracts of *Aloe vera* L. gel from different growth periods of plants. *Industrial crops and products* 49: 712–719.

Sahu, P. K., Giri, D. D., Singh, R., Pandey, P., Gupta, S., Shrivastava, A. K., Kumar, A., & Pandey, K. D. 2013. Therapeutic and medicinal uses of aloe vera: A review. *Pharmacology & pharmacy* 4(08): 599.

Sarkar, R., Arora, P., Garg, V. K., Sonthalia, S., & Gokhale, N. 2014. Melasma update. *Indian dermatology online journal* 5(4): 426.

Tamega, A. D. A., Miot, L., Bonfietti, C., Gige, T., Marques, M. E. A., & Miot, H. A. 2013. Clinical patterns and epidemiological characteristics of facial melasma in Brazilian women. *Journal of the European Academy of Dermatology and Venereology* 27(2): 151–156.

Clove leaf oil compound in combination with standard drugs for effective liver cancer therapy

S.A.D. Trusda, T. Respati, E. Hendryanny, L. Yuniarti & M. Tejasari
Universitas Islam Bandung, Bandung, West Java, Indonesia

ABSTRACT: Liver cancer is the sixth most common cancer and the third highest cause of deaths due to cancer. Failure in default therapy, side effects, and high costs of cancer therapy have lead researchers to look for effective and selective new cancer therapies. The aim of this study was to explore the effect of flavonoid active compound from clove leaf oil on liver cancer cells and its combination with standard drugs for cancer therapy. The cytotoxicity of active compounds isolated from clove leaf oil and standard drugs was analyzed on liver cancer cell culture HepG$_2$ and examined by (3-(4,5-dimethylthiazol-2-yl)-2,5-diphenyltetrazolium bromide) tetrazolium (MTT) assay. The drug combination synergisms were indicated by the combination index (CI) using CompuSyn 1.4. The results showed that flavonoid active compound from clove leaf oil alone, or in combination with standard drugs, has a CI index below 1. It was concluded that flavonoid active compounds from clove leaf oil were shown to have anticancer activity on liver cell cultures and also to have synergistic effect in combination with standard drugs for cancer therapy.

1 INTRODUCTION

Hepatocellular carcinoma (HCC), often called hepatoma, is the most commonly found cancer in the liver. This cancer is one of the most common types of cancer found in the world and causes many deaths (Cardoso et al. 2010; Ho et al. 2009; Huynh et al. 2010). This liver cancer is reported to cause half a million deaths annually. The American Cancer Society in 2010 estimated that HCC incidence and mortality will continue to rise until 2020 (Ho et al. 2009; Jemal et al 2011).

The palliative approach – i.e. administering systemic chemotherapy to the advanced HCC stage – is still the main choice based on the Barcelona Clinic Liver Cancer (BCLC) classification. Curative approaches such as resections and transplants can only be done on early-stage HCC. Chemotherapy is still the best choice for patients with advanced HCC, but so far the effectiveness of chemotherapy in patients with HCC remains debated. The limited choice of therapies for patients with HCC is also the cause of its poor prognosis (Abou-Alfa 2004; Park et al. 2006; Wirth et al. 2005).

In addition to the lack of successful standard therapies for cancer, the side effects of cancer treatment in general are not yet insurmountable. Side effects of cisplatin and other cancer medications are the result of the death of normal cells, despite the death of the cancer cell itself. The unprecedented treatments of cancer and the sheer number of side effects, as well as the price of costly drugs, encourage researchers to seek out effective and selective cancer medications (Fajarningsih et al. 2008).

One of the attempts to find anticancer treatments is to explore natural compounds. The potential active compounds that are often isolated from medicinal plants are flavonoids. Studies in vitro demonstrate anticancer activity of genistein, a flavonoid, caused by antioxidant effects and anti-proliferative effects through modulation of gene expression involved in cell cycles and apoptosis (Hussain et al. 2012; Li et al. 2005).

Isolation of genistein from soybeans requires large amounts of soybeans and is expensive, therefore some researchers try to synthesize genistein and its derivatives to obtain large quantities of compounds at a lower cost. One of the active compounds derived from flavonoids that are being investigated as anticancer therapy is a compound derived from clove leaf oil (eugenol), which is abundant in Indonesia (Herwiyanti 2015).

The aim of this study was to explore the effect of flavonoid active compounds from clove leaf oil on liver cancer cells and its combination with a standard drug for cancer therapy (doxorubicin).

2 METHODS

The cytotoxicity of active compounds isolated from clove leaf oil (epoxy-EPI) and doxorubicin (DOX) as a standard drug was analyzed on liver cancer cell culture $HepG_2$ and examined by MTT assay. The drug combination synergisms were indicated by the combination index (CI), which was obtained using CompuSyn 1.4 software.

2.1 *Subject and research material*

The subjects in this study were $HepG_2$ liver cancer cell cultures obtained from the parasitology laboratory, Faculty of Medicine, Gadjah Mada University, Yogyakarta.

The test material was in the form of compound 1,2-epoxy-3(3-(3,4-dimethoxyphenyl)-4H-1benzopyran-4on. The propane was synthesized at the Organic Chemistry Laboratory MIPA Faculty, Universitas Gajah Mada, Yogyakarta. Doxorubicin was used as a positive control (Sigma Chem. Co., St. Louis, USA). The concentrations used were obtained from preliminary research in the form of IC50 Epoxy 50.62 µg/mL and IC50 Dox 20.25 µg/mL.

Materials used were DMEM (Sigma Chem. Co., St. Louis, USA), fetal bovine serum (Sigma Chem. Co., St. Louis, USA), penicillin, streptomycin (Gibco), fungizone (Gipco), sodium bicarbonate (E Merck), and filter paper 0.2 µm (Whatman), phosphate buffer saline (PBS), MTT reagent and SDS 10% stopper.

2.2 *Procedure*

2.2.1 *Cell culture*
$HepG_2$ cells were suspended in a media culture containing 1.5×10^4 cells/100 µL; the suspension was inserted into each micro well plate 96 and incubated in an incubator at 37°C and 5% CO_2 for 24 hours. DMEM was used as growth medium, with an addition of 10% FBS (fetal bovine serum), 0.5% fungison, and 1% penicillin-streptomycin.

2.2.2 *Combination test*
The method used to evaluate drug combinations in this study was isobologram and the combination index (CI). CI analysis produces a quantitative parameter value that describes the combination efficacy using the equation: $CI = (D)_1/(Dx)_1 + (D)_2/(Dx)_2$.

Dx is the concentration of one single compound needed to give the effect of the effect of the combination – i.e., IC50 on cell growth $HepG_2$, and $(D)_1$, $(D)_2$ is the concentration of the two compounds needed to give the same effect. The CI value is used to determine the effect of the two compound combinations, whether in the form of synergistic, additive, or antagonistic effects.

3 RESULTS

The combination test aims to obtain a combination of the drug with a lower dose but effective results so as to reduce the likelihood of severe side effects.

Table 1. Combination index (CI) value of EPI and DOX compounds against HepG$_2$ cell culture.

EPI concentration μg/mL	DOX concentration μg/mL	Viability %	CI (Combination index)
25.31	12.13	5	0.28
18.98	12.13	3	0.11
12.65	12.13	6	0.18
6.33	12.13	13	0.24
25.31	9.09	7	0.43
18.98	9.09	5	0.21
12.65	9.09	8	0.25
6.33	9.09	13	0.24
25.31	6.06	6	0.35
18.98	6.06	4	0.16
12.65	6.06	14	0.53
6.33	6.06	16	0.32
25.31	3.03	4	0.21
18.98	3.03	4	0.16
12.65	3.03	13	0.48
6.33	3.03	15	0.29

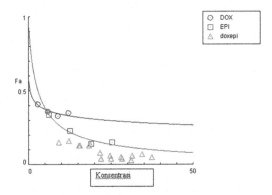

Figure 1. Cytotoxic curve combination EPI and DOX compound.

Based on the IC$_{50}$ values of each test compound, the concentrations of EPI used in combination tests were: 25.31 μg/mL, 18.98 μg/mL, 12.65 μg/mL, and 6.33 μg/mL, while DOX: 12.13 μg/mL, 9.09 μg/mL, 6.06 μg/mL, and 3.03 μg/mL. The drug synergy effect was measured using isobologram with a combination index (CI) with CompuSyn software. The cytotoxic test results of the combination of EPI and DOX compounds can be seen in Table 1.

From Table 1 we found that the lowest viability was in concentrations of 18.98 μg/mL EPI and 12.13 μg/mL DOX, and this concentration had the lowest combination index (Figure 1).

4 DISCUSSION

In the process of HCC carcinogenesis, apoptosis dysregulation causes inhibition of apoptosis in the initiation phase, both extrinsic and intrinsic pathways, so that the cell is not subjected to death. HCC's incidence relates to the low expression of Fas, which is a cell death receptor, thereby inhibiting the initiation of apoptosis of extrinsic pathways. In HCC, there is also an increase in molecular inhibitors of apoptosis (IAP) and an increase in BCl$_2$ expression, which is an anti-apoptosis protein that inhibits the initiation of apoptosis of intrinsic pathways (He et al. 2011; Hsieh et al. 2009; Liu et al. 2008; Lu et al. 2005; Yildiz et al. 2008).

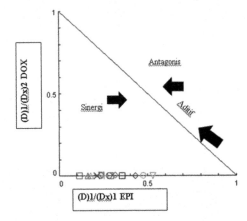

○ · Epoxy administered HepG₂ group
□ · Doxorubicin administered HepG₂ group
△ Epoxy and Doxorubicin HepG₂ group

Figure 2.   Normalization curve isobologram EPI and DOX combination.

The results of compound combination tests of EPI with DOX are presented in Table 1 and Figure 2. The combination index of EPI and DOX compounds in HepG$_2$ cells at concentrations of IC$_{50}$ EPI and DOX is below 1. This value indicates that the compound EPI in that concentration synergizes with DOX. The lowest viability was in concentrations of 18.98 µg/mL EPI and 12.13 µg/mL DOX, and this concentration had the lowest combination index.

These results are in line with those in previous studies that demonstrate that isoflavone compounds such as genistein and daidzein are synergies with various standard medications in some cancer cells such as pancreatic cancer and MCF-7 cells (Banerjee et al. 2007). Overexpression and activation of NF-κB may inhibit apoptosis response thereby reducing the efficacy of chemotherapy. The structure of polyphenols on genistein was instrumental in the anticancer mechanisms through the protein suppression of BCL-2 and BCL-XL and down-regulation of NF-κB in pancreatic cancer cells (Lewandowska et al. 2014). Another study has shown that genistein is synergistic with some standard medicines on some cancer cell cultures such as: prostate cancer cell cultures (PC-3), pancreatic cancer (Bx PC-3), lung cancer cells (H460), and breast cancer cells (MDA-MB-231) (Li et al. 2005).

5   CONCLUSION

The flavonoid compounds of the leaf oil (epoxy) have moderate levels of cytotoxic activity against liver cancer cells, and have a strong synergistic effect with doxorubicin.

ACKNOWLEDGMENT

We wish to thank Kementerian Riset, Teknologi, dan Pendidikan Tinggi Republik Indonesia (Ministry of Research, Technology and Higher Education, Republic of Indonesia) for the PDUPT Grant (contract number: 156/B.04/Rek/III/2018).

REFERENCES

Abou-Alfa, G. K. 2004. Current and novel therapeutics for hepatocellular carcinoma. *Am Soc Clin Oncol Educational Book*. Alexandria, VA, 192–197.

Banerjee, B., Vadiraj, H. S., Ram, A., Rao, R., Jayapal, M., Gopinath, K. S., ... & Hegde, S. 2007. Effects of an integrated yoga program in modulating psychological stress and radiation-induced genotoxic stress in breast cancer patients undergoing radiotherapy. *Integrative cancer therapies* 6(3): 242–250.

Cardoso, A. C., Moucari, R., Figueiredo-Mendes, C., Ripault, M. P., Giuily, N., Castelnau, C., ... & Carvalho-Filho, R. J. 2010. Impact of peginterferon and ribavirin therapy on hepatocellular carcinoma: Incidence and survival in hepatitis C patients with advanced fibrosis. *Journal of hepatology* 52(5): 652–657.

El Bassiouny, A. E., El-Bassiouni, N. E., Nosseir, M. M., Zoheiry, M. M., El-Ahwany, E. G., Salah, F., ... & Ibrahim, R. A. 2008. Circulating and hepatic Fas expression in HCV-induced chronic liver disease and hepatocellular carcinoma. *Medscape journal of medicine* 10(6): 130.

Fajarningsih, N. D., Nursid, M., Wikanta, T., & Marraskuranto, E. 2008. Bioaktivitas Ekstrak Turbinaria decurrens sebagai Antitumor (HeLa dan T47D) serta efeknya terhadap Proliferasi limfosit. *Jurnal Pascapanen dan Bioteknologi Kelautan dan Perikanan* 3(1): 21–28.

He, H. B., Wu, X. L., Yu, B., Liu, K. L., Zhou, G. X., Qian, G. Q., ... & Chen, X. Y. 2011. The effect of desacetyluvaricin on the expression of TLR4 and P53 protein in Hepg 2.2. 15. *Hepatitis monthly* 11(5): 364.

Herwiyanti, S. 2015. POTENSI SENYAWA 1, 2-Epoksi-3 [3-(3, 4-dimetoksifenil)-4H-1-benzopiran-4on) propana SEBAGAI ANTIKANKER Kajian in vitro Aktivitas Molekuler Sel Kanker Payudara T47D dan MCF-7 serta Aktivitas Molekuler in vivo pada Hewan Model yang Diinduksi 7, 12-Dimetilbenz (a) Anthracene (Doctoral dissertation, Universitas Gadjah Mada).

Ho, H. K., Pok, S., Streit, S., Ruhe, J. E., Hart, S., Lim, K. S., ... & Ullrich, A. 2009. Fibroblast growth factor receptor 4 regulates proliferation, anti-apoptosis and alpha-fetoprotein secretion during hepatocellular carcinoma progression and represents a potential target for therapeutic intervention. *Journal of hepatology* 50(1): 118–127.

Hsieh, S. Y., Hsu, C. Y., He, J. R., Liu, C. L., Lo, S. J., Chen, Y. C., & Huang, H. Y. 2009. Identifying apoptosis-evasion proteins/pathways in human hepatoma cells via induction of cellular hormesis by UV irradiation. *Journal of proteome research* 8(8): 3977–3986.

Hussain, A., Harish, G., Prabhu, S. A., Mohsin, J., Khan, M. A., Rizvi, T. A., & Sharma, C. 2012. Inhibitory effect of genistein on the invasive potential of human cervical cancer cells via modulation of matrix metalloproteinase-9 and tissue inhibitors of matrix metalloproteinase-1 expression. *Cancer epidemiology* 36(6): e387–e393.

Huynh, H., Koong, H. N., Poon, D., Choo, S. P., Toh, H. C., Thng, C. H., ... & Smith, P. D. 2010. AZD6244 enhances the anti-tumor activity of sorafenib in ectopic and orthotropic models of human hepatocellular carcinoma (HCC). *Journal of hepatology* 52(1): 79–87.

Jemal, A., Bray, F., Center, M. M., Ferlay, J., Ward, E., & Forman, D. 2011. Global cancer statistics. *CA: A cancer journal for clinicians* 61(2): 69–90.

Lewandowska, U., Gorlach, S., Owczarek, K., Hrabec, E., & Szewczyk, K. 2014. Synergistic interactions between anticancer chemotherapeutics and phenolic compounds and anticancer synergy between polyphenols. *Postepy Hig Med Dosw*, 68(4468): 528–540.

Li, Y., Ahmed, F., Ali, S., Philip, P. A., Kucuk, O., & Sarkar, F. H. 2005. Inactivation of nuclear factor κB by soy isoflavone genistein contributes to increased apoptosis induced by chemotherapeutic agents in human cancer cells. *Cancer research* 65(15): 6934–6942.

Liu, Q., Fu, H., Sun, F., Zhang, H., Tie, Y., Zhu, J., ... & Zheng, X. 2008. miR-16 family induces cell cycle arrest by regulating multiple cell cycle genes. *Nucleic acids research* 36(16): 5391.

Lu, X., M ML, Tran, T., & Block, T. 2005. High level expression of apoptosis inhibitor in hepatoma cell line expressing hepatitis B virus. *International journal of medical science* 2005 2(1): 30–35.

Park, S. H., Lee, Y., Han, S. H., Kwon, S. Y., Kwon, O. S., Kim, S. S., ... & Cho, E. K. 2006. Systemic chemotherapy with doxorubicin, cisplatin and capecitabine for metastatic hepatocellular carcinoma. *BMC cancer* 6(1): 3.

Wirth, T., Kühnel, F., Fleischmann-Mundt, B., Woller, N., Djojosubroto, M., Rudolph, K. L., ... & Kubicka, S. 2005. Telomerase-dependent virotherapy overcomes resistance of hepatocellular carcinomas against chemotherapy and tumor necrosis factor–related apoptosis-inducing ligand by elimination of Mcl-1. *Cancer research* 65(16): 7393–7402.

Yildiz, L., Baris, S., Aydin, O., Kefeli, M., & Kandemir, B. 2008. Bcl-2 positivity in B and C hepatitis and hepatocellular carcinomas. *Hepato-gastroenterology* 55(88): 2207–2210.

Community and occupational health

Scabies and the development of clean and healthy living behavior tools for Islamic boarding schools (*pesantren*)

Y. Triyani, E. Hendryanny, R.A. Indriyanti, W. Purbaningsih & T. Respati
Universitas Islam Bandung, Bandung, Jawa Barat, Indonesia

ABSTRACT: The incidence of scabies, especially in Islamic boarding schools (*pesantren*), is still a problem in Bandung City, Indonesia. Previous studies found multifactorial causes for scabies that are difficult to eliminate. Various efforts have been made to treat and eliminate scabies, including the introduction of healthy and clean-living behaviors (PHBS). This study aimed to describe the use of a PHBS pocketbook to promote healthy and clean-living behaviors among students in *pesantren* in Bandung. Screening for scabies was carried out on 18 *pesantren* selected from around Bandung using a purposive sampling method. A PHBS pocketbook, especially on scabies prevention, was introduced in the *pesantren* with the highest incidence of scabies. An assessment of PHBS knowledge and scabies using pretest and posttest methods was performed using the Wilcoxon test. The research was conducted from February to August 2019 in Bandung. Results showed the students' level of knowledge about PHBS and prevention of scabies was good. However, the knowledge did not contribute to PHBS. In conclusion, the level of knowledge of PHBS is not related to a teaching material.

1 INTRODUCTION

Based on data from the World Health Organization (WHO) in 2015, scabies, a skin disease caused by *Sarcoptes scabiei* var. *ectoparasites hominins*, is 1 of 20 neglected tropical diseases (WHO 2015). Difficulties in eradicating scabies are due to factors that cause scabies, originating from the host, the agent, and the environment (Middleton et al. 2018; Triyani et al. 2019).

The host factors include low knowledge of healthy and clean-living behavior (PHBS), dense population, poverty, and poor nutrition (Heukelbach et al. 2003). The main environmental factors comprise density in one-bedroom public health facilities, late diagnosis and treatment, temperature, and many others (Abou-El-Naga 2015; Micali et al. 2018). In Indonesia, scabies remains a problem difficult to overcome, especially in a *pesantren* environment (Nuraini & Wijayanti 2016; Ratnasari & Sungkar 2014). In West Java, the *Santri Budug* term is already known and it has become a myth that scabies is a disease that is inevitable in the *pesantren* environment (Tresnasari et al. 2019). The problem of scabies in Islamic boarding schools is very closely related to the knowledge of PHBS (Triyani et al. 2019). The results are very alarming because Islam presents hygiene teachings called *thaharah* with the strongest message of "cleanliness is half of faith" (Abdat 2019).

PHBS is also related to the availability of teaching materials or modules as guidelines that the Ministry of Education has already provided (Menteri Kesehatan Republik Indonesia 2011). Based on a report from the Ministry of Religion in Bandung in 2019, no teaching materials or manuals are available to students regarding PHBS.

Based on these issues, a different approach is needed to provide tools for use as practical guidelines for PHBS in *pesantren*. This study aimed to describe the incidence of scabies and to analyze the level of student's knowledge before and after being provided with practical PHBS guidelines.

2 METHOD

This was a pilot project to develop tools to educate Islamic students (*santri*) about PHBS and scabies. This study was preceded by a survey at 18 *pesantren* chosen from the 169 *pesantren* throughout Bandung and conducted through a purposive sampling method. At each *pesantren*, 30 *santri* were selected using a simple random sampling method. Each santri underwent a physical examination to define his health status, especially scabies. The examination was conducted by Islamic University of Bandung medical students supervised by a lecturer. The *santri* were also assessed on their knowledge of PHBS and scabies. The tools used to assess scabies knowledge included a validated questionnaire consisting of 20 questions about the causes, symptoms, prevention, transmission, treatment, and risk of scabies. Other validated questionnaires measured students' level of knowledge in a PHBS program.

The research team developed a *santri* health pocketbook using results from the survey. A focus group discussion was also conducted as one of the development processes. The information provided in the pocketbook was modified from the information in a healthy student guideline from the Ministry of Education. After the *santri* pocketbook content had been developed, it was illustrated by a professional illustrator and printed for the intervention stage.

A *pesantren* with the highest incidence of scabies was selected for intervention regardless of the *santri* knowledge level. The intervention was a quasi-experiment using pretest and posttest methods. *Santri* were introduced to the health pocketbook as a practical guideline for daily living. A pretest was given before the santri were allowed to use the pocketbook. The posttest was conducted after 1 week. A physical examination was also conducted for the 33 *santri* in the selected *pesantren*. Statistical analysis was conducted using the Wilcoxon test in SPSS ver. 24.

3 RESULTS AND DISCUSSION

Results from the survey showed that scabies cases found in the 18 representative *pesantren* varied greatly from 10% to 82%, as can be seen in Table 1.

Table 2 shows the results during the intervention phase.

The survey found that, in all *pesantren*, no PHBS handbooks were available. The introduction of the health pocketbook showed that a good level of knowledge can be obtained. The provision of the pocketbook made it easier for students to teach themselves. The observation also showed that students teach each other about PHBS and scabies.

Pesantren G was the *pesantren* with the highest incidence of scabies. However, the pretest and posttest analyses showed good results for the level of knowledge of the students in terms of PHBS and scabies prevention. This can be seen from the results of the pretest and posttest with values of 86.67 and 94.8. This attracts attention; it turns out that good knowledge about PHBS has not made the *pesantren* free of scabies. Heukelbalch and colleagues have highlighted multifactorial causes of scabies including poverty, availability of healthcare facilities, dwelling in slums, access to medical services, failure of treatment, delays and misdiagnosis, limitations of laboratory services, temperature and humidity, assistance from health workers, behavior and habits, etc. (Heukelbach et al. 2003). The Islamic Boarding School is located in the center of Bandung; this is a very alarming condition, considering that Bandung is a large city with easy access to information and healthcare services. The incidence of scabies in the *pesantren* should be easily reduced (Tresnasari et al. 2019). However, based on research, the incidence of scabies depends on variables that are significantly related to the respondents' gender, the respondents' knowledge, the respondents' personal hygiene, the provision of socialization or information about personal hygiene, the support of boarding school caretakers, the support of friends, the support of health workers, and the support of the Ministry of Religion (Triyani et al. 2019).

A good level of knowledge is not enough to make the students develop the necessary habits for a clean and healthy life, because they need examples and assistance from the teachers (Ratnasari & Sungkar 2014). A high level of knowledge still requires assistance and examples of PHBS, especially since Islam already teaches that cleanliness is half of faith (Abdat 2019).

Table 1. Scabies cases in *pesantren*.

Schools	Scabies (+)		Scabies (−)	
	N	%	n	%
A	19	63	11	37
B	18	60	12	40
C	20	67	10	33
D	11	37	19	63
E	8	27	22	73
F	23	77	7	23
G	27	82	6	18
H	5	17	25	83
I	3	10	27	90
J	6	20	24	80
K	19	63	11	37
L	8	27	22	73
M	5	17	25	83
N	6	20	24	80
O	23	77	7	23
P	13	43	17	57
Q	8	27	22	73
R	11	37	19	63
Total	271	49	272	51

Table 2. Pretest and posttest result at intervention.

	Mean	N	P-value
Pretest	86.67	33	0.008*
Posttest	94.80	33	

* Wilcoxon test

4 CONCLUSIONS

Health pocketbooks can significantly increase the level of knowledge about PHBS. A good level of knowledge was not related to the incidence of scabies.

ACKNOWLEDGMENT

This research was funded by LPPM UNISBA through the Hibah Penelitian Dosen Utama.

REFERENCES

Abdat, A.-U. A. H. bin A. *Kitab Thaharah Bab Air. 2019*. [Online]. Retrieved from https://almanhaj.or.id/2899-kitab-thaharah-bab-air-1-4.html.

Abou-El-Naga, I. F. 2015. Demographic, socioeconomic & environmental changes affecting circulation of neglected tropical diseases in Egypt. *Asian Pacific journal of tropical medicine* 8(11): 881–888.

Heukelbach, J., Van Haeff, E., Rump, B., Wilcke, T., Sabóia Moura, R. C., & Feldmeier, H. 2003. Parasitic skin diseases: Health care-seeking in a slum in north-east Brazil. *Tropical medicine & international health* 8(4): 368–373.

Menteri Kesehatan Republik Indonesia. 2011. *Pedoman Pembinaan Perilaku Hidup Bersih Sehat (PHBS). Kementrian Kesehatan.* Jakarta: Menteri Kesehatan Republik Indonesia.

Micali, G., Giuffrida, G., & Lacarrubba, F. 2018. Scabies. *Diagnostics to Pathogenomics of Sexually Transmitted Infections* 367: 357–371.

Middleton, J., Cassell. J. A., Jones, C. I., Lanza, S., Head, M. G., & Walker, S. L. 2018. Scabies control: The forgotten role of personal hygiene – Authors' reply. *Lancet infectious diseases* 18(10): 1068–1069.

Nuraini, N. & Wijayanti, R. A. 2016. Faktor Risiko Kejadian Scabies Di Pondok Pesantren Nurul Islam Jember (Scabies risk factors in Pondok Pesantren Nurul Islam Jember). *Jurnal Ilmiah Inovasi* 16(2): 137–141.

Ratnasari, A. F. & Sungkar, S. 2014. Prevalensi Skabies dan Faktor-Faktor Yang Berhubungan Di Pesantren X, Jakarta Timur. *EJournal Kedokteran Indonesia* 2(1): 7–12.

Tresnasari, C., Respati, T., Maulida, M., Triyani, Y., Tejasari, M., Kharisma, Y., & Ismawati, I. 2019. Understanding scabies in religious boarding school. *Pesantren* 307(2018): 520–522.

Triyani, Y., Yuniarti, L., Tejasari, M., Purbaningsih, W., Ismawati, I., & Respati, T. 2019. A journey to a better community service in religious boarding school. *Pesantren* 307(2018): 497–499.

World Health Organization (WHO). 2015. *Neglected tropical diseases Scabies.* Geneva: World Health Organization, 4–6.

Medical Technology and Environmental Health – Abdullah, Widiaty & Abdullah (eds)
© 2020 Taylor & Francis Group, London, ISBN 978-0-367-86053-0

Empowering local women to promote community health in Indonesia

N. Yuliati, P. Pawito, M. Wijaya & P. Utari
Universitas Sebelas Maret, Surakarta, Indonesia

ABSTRACT: AIDS has become a serious threat not only for health but also for humans and humanity. One reason is because AIDS is transmitted through sexual contact. People cannot talk about sexuality openly and honestly, so HIV/AIDS is difficult to discuss and thus difficult to control as well. AIDS prevention in Indonesia is included in the development strategy through community participation. Local Residents Concerned about AIDS (Warga Peduli AIDS) (WPA) is the first form of community action and participation to respond to AIDS in Indonesia and has the readiness, ability, and willingness to participate in preventing and overcoming HIV/AIDS. The majority of WPA members are women who have become agents in promoting community health and education on HIV/AIDS. This study was conducted at WPAs in four districts in the city of Bandung, Indonesia. The results of this study showed that some of the important factors in overcoming HIV/AIDS are local women serving as health agents, critical exploration of women cadres concerning HIV/AIDS among community members, and participatory communication about resolving the low health literacy related to HIV/AIDS. Using a qualitative descriptive paradigm with a case study method, data were collected through in-depth interviews and documentation.

1 INTRODUCTION

One of the most critical health problems in Indonesia is HIV/AIDS. As a developing country, Indonesia is also one of the nations in Asia experiencing the relatively rapid growth of HIV/AIDS. In 2004, only 16 provinces reported AIDS cases, but in 2007, such reports came from 32 provinces (Aripurnami 2013; UNAIDS 2008). The cumulative HIV/AIDS epidemic in the city of Bandung has spread throughout the city. From 1991 to November 2011, the city reported 2,570 people with AIDS, the highest incidence of AIDS reported in West Java (Komisi Penanggulangan AIDS Kota Bandung 2011), as reported by human resources in Bandung. The spread of HIV/AIDS is related not only to health problems but also to social and economic problems, which is worrying. Its influence is also widespread at the national, regional, and global levels.

Women have become important in the context of HIV/AIDS. They are among the vulnerable groups seeing increases in AIDS cases today (Hidayana et al. 2010; Nasution et al. 2011). Women have not yet reached their full potential either individually or collectively; likewise in developing institutional approaches to overcoming women's problems. Discriminatory prejudices and practices marginalize women from resources, markets, and politics. Women's talents, skills, and energy remain untapped. The Global Program on AIDS and the WHO Global Action on Women and AIDS framework document emphasize the importance of reducing women's social helplessness in the face of HIV infection by improving the prospects and level of women's health, education, legal status, and economy. One such strategy is empowering women. Empowerment means expanding community options and opportunities that allow individuals to make choices that improve their lives.

Women's empowerment in efforts to promote the prevention and control of HIV/AIDS are an integral part of various HIV/AIDS prevention activities across sectors. Diverse studies show the strengthening of women in combating HIV/AIDS (Gupta & Srivastava 2012; McCreary et al. 2013; Ramírez-Johnson et al. 2013; Romero et al. 2006). Many cases prove the importance of women in tackling HIV/AIDS (Nicholas et al. 2009; Nyamathi 2014). These studies justify the contention that women and HIV/AIDS are interrelated. Women always find a way to dispel and overcome obstacles and harsh realities that confront them.

Women are one of the segments in Indonesian development, and the Ministry of Women's Empowerment has proclaimed women's independence in the prevention of HIV/AIDS. To provide protection against HIV/AIDS for themselves and their families, women must be actively involved or included, both as individuals and as members of women's organizations. HIV/AIDS prevention is carried out in an integrated manner with the existing Community Empowerment Program under the principles of transparency, participation, and accountability, and reflects the religious and cultural values of Indonesia (Peraturan Menteri Dalam Negeri 2007).

Local Residents Concerned about AIDS or *Warga Peduli* AIDS (WPA) is the first form of community action and participation to respond to HIV and AIDS in Indonesia and is officially recognized by the National AIDS Commission (KPA). In order to encourage increased knowledge and awareness of the community toward HIV/AIDS, various efforts are needed to disseminate information so that there is a growing concern in the community to participate actively in the prevention and control of HIV/AIDS. The WPA operates at the district level so that community activities are more coordinated and targeted (Hidayat 2012).

In the context of HIV/AIDS, community participation is important, especially in severely affected countries. The response from community members is the most effective because they understand their local context, they are cheaper, they grasp local needs and the available resources, and they gain the understanding and support of community members. Community participation in response to HIV/AIDS has increasingly been identified as an "important supporter" for effective measures against HIV/AIDS (Kabore et al. 2010; Rodriguez-García et al. 2011; Schwartländer et al. 2011).

The WPA is a community-based group of citizens who have the readiness, ability, and willingness to participate in preventing and overcoming problems caused by medical diseases transmitted through behavior – in this case, HIV/AIDS along with other comorbidities – in order to create a healthy and prosperous society. Through the WPA, women play an active role in meetings and activities. They participate in the dissemination of information about HIV/AIDS and provide assistance to groups affected by HIV/AIDS.

Based on this background, the objectives of this research are to (1) reveal participating local women as health agents, (2) identify critical exploration of women cadres concerning HIV/AIDS among community members, and (3) organize communication approaches for resolving the low health literacy related to HIV/AIDS.

2 RESEARCH METHOD

This research used a qualitative approach with a case study. Informants were women who are members of the WPA in Bandung City. The criteria were WPA women with companion and non-companion roles. The selection of the informants was done purposively based on their willingness to reveal their experiences. In this study, the informants were women who were members of AIDS Care Citizens in Bandung.

The research locations comprised two regions that represent the urban Bandung area – Sumur Bandung, Tamansari, and Cibeunying Kidul – and the industrial Bandung area – Kecamatan Bojongloa Kidul. Four WPA cadres were interviewed in the field, three of whom are WPA companions and one of whom is a WPA non-companion. A WPA companion is a cadre who in addition to participating in promotive and preventive efforts also provides assistance to patients with HIV/AIDS in obtaining treatment, care, and consultation. The data in this research consisted of primary and secondary data. Because this study used qualitative methods, the data collection techniques included conducting in-depth interviews (deep interviews), observation, and analysis of documents.

3 RESULTS AND DISCUSSION

3.1 *Participation of local women as agents in community health*

The AIDS Prevention Commission (KPA) of Bandung City and several nongovernmental organizations (NGOs) confirm the contributions of women to WPA activities. The majority of WPA members are housewives involved in Family Welfare Empowerment (Pemberdayaan Kesejahteraan Keluarga) (PKK) and Integrated Health Post (*Posyandu Pos Pelayanan Terpadu*). These women are cadres or activists promoting HIV/AIDS awareness and prevention in the smallest community groups at the *kelurahan* level. They are initiators or communicators who mobilize to spread information about and promote action against HIV/AIDS. The informants in this study represent the village. They work in the territorial area. However, this does not mean that they are isolated from the government, because these cadres interact with one another through interdistrict WPA forums. In general, WPA cadres (1) participate in various activities concerning HIV/AIDS, (2) promote preventative and rehabilitative activities in the form of information dissemination and mobilization for HIV testing, and (3) mentor people living with HIV/AIDS.

The WPA provides a positive site for informants to actively organize; the real benefit of this community is to increase knowledge and skills, broaden horizons, improve self-image, self-confidence, and tolerance, and serve as an intellectual training ground for informants. On the other hand, becoming a WPA cadre requires demanding an active role, initiative, and great responsibility to participate in reducing the stigmatization people with HIV/AIDS often experience. As initiators of communication, WPA cadres have received training directly related to HIV/AIDS, health, organization, gender, and human rights.

Becoming a WPA cadre is an active process in which women interact with their social environment, where they reconstruct their understanding of HIV/AIDS. Women cadres of the WPA are housewives who initially engaged in PKK activities. The role of wife and mother is very important and prominent in Indonesian society. This identity as wives and mothers seems to be attached to the informants when they take part in PKK activities outside the home during their free time. The PKK and WPA give a new "identity" to these women. Although currently the public sphere is already open for Indonesian women, the PKK, which is in the neighborhood and the community that is close and well known by these women, makes their involvement easier.

For informants, involvement as a WPA cadre builds a new identity as a social agent where they can actualize themselves. The WPA community has empowered women while emphasizing awareness of gender rights, placing women as equal partners with men and not just as men's companions. Thus, as informants, women are transformed from housewives into WPA activists. Their work not only dwells on domestic activity but extends to social activities outside of the home. The transformation of an informant's identity into an activist against HIV/AIDS in her environment is of course based on her views, inner attitudes, and life principles and thus brings out her personal characteristics.

3.2 *Critical exploration of WPA women in building community awareness of HIV/AIDS*

The involvement of all WPA informants begins with their involvement in the PKK. Each informant has her own story, motivation, and argument. Informants explained the diverse challenges they have confronted in their efforts to deal with HIV/AIDS. They face various difficulties as happened in the following cases:

Informant 1:

In 2014, there was [an] ODHA [person living with HIV/AIDS or PLWHA] who was abandoned by his family. According to his family, he will die soon. He was dumped in the market, and the people there were looking for me. I checked, then I talked to him. When he had nothing, we coordinated with the kelurahan [the village]. Because of an emergency, I finally helped him. I continued to coordinate with the puskesmas [local hospital] so that we received a free ambulance to take [him] to RSHS [a public hospital].

Informant 2:

I had to accompany [an] ODHA; he did not seek treatment because he did not have money. After coordinating with Aisyiyah [an NGO], I finally got him to Hasan Sadikin [a public hospital]. When he was treated, we were confused because we had to tell his sister. Nobody in his family knew yet. Finally, he told his biological sister. I then called her. I explained that HIV transmission is not like watching TV [not easily transmitted].

Informant 3:

One family all died. First the father, second the mother, and finally the child. When the child died, I took care of it, because other people did not want to. I became a companion even though his family was still a lot. I helped. If it's really bad, no one cares, so me and the other WPA [members] intervene.

Informant 4 explained his approach to community members:

The approach is through talking, visiting homes. Some are interested; some are angry. If they are interested, I continue while for those who [are] angry we postpone it first. For VCT [voluntary consulting and testing] is not easy.

The informants mentioned that there remains a high level of public rejection of AIDS patients. Various forms of discriminatory treatment are accepted by persons with AIDS, while the issue of HIV/AIDS is still difficult to address. Residents feel afraid to even hear the words HIV/AIDS. However, they do not feel the need to know and check themselves.

Based on these interviews with informants, their involvement in the WPA can be described as divided into two categories – namely the initial pattern and the management pattern.

3.2.1 *Initial pattern*

The initial pattern is the *perception* of informants before they become involved in the WPA. Their attitude regarding HIV/AIDS is based on *emotion*, fear, and suspicion, but there is also compassion and curiosity. These emotions in turn form a desire to care, to help. This form of motivation is the basic *framing* and is manifested through the willingness to be involved in overcoming HIV/AIDS.

3.2.2 *Management pattern*

The management pattern is the result of the interaction of informants after joining the WPA, where they learn *reinterpretation* so they have a new understanding of HIV/AIDS. They learn *rationalization*, which is managing understanding based on factual (true) knowledge about HIV/AIDS. They have knowledge that is not based on prejudice and emotions. Informants become rational in understanding HIV/AIDS after receiving training. This rational understanding makes informants critical in the fight against HIV/AIDS because they can make accurate observations about their social and environmental conditions. They are also able to design various strategies to cope with HIV/AIDS. It means that they learn *reframing* in the end and can build new values – namely individual and collective critical awareness of HIV/AIDS and how to treat and prevent it.

Figure 1 depicts a model of the initial and management patterns.

3.3 *Participatory communication approaches for resolving low health literacy related to HIV/AIDS.*

The PKK already has an organized structure and cadres scattered throughout Indonesia. These PKK activists become WPA cadres and play an active role in tackling HIV/AIDS. The PKK has existed since the end of the 1960s, and the WPA was formed around 2010 in response to the outbreak of HIV/AIDS in Indonesia, whose transmission is increasing, while the level of public knowledge of HIV/AIDS remains low. The marginalization of AIDS patients remains high. The WPA certainly aims to create social change. Social movements, of course, require a variety of strategies, tactics, and forms of organization to achieve their goals.

As WPA cadres, the informants who participated in this study have become social agents. Basically everyone is a social agent because social expectations are communicated in many ways and in every interaction whether intentionally or not. These informants are initiators

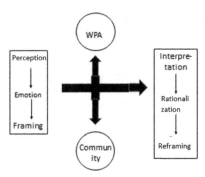

Figure 1.   Model of the initial and management patterns.

and, in the context of communication, they are communicators. For this reason, the informants must have the ability to communicate and to remember that the information they provide
addresses HIV/AIDS, which is still a frightening specter for the community. Participatory
communication by the WPA consists of both interpersonal and group communication. (1)
Group communication is used to disseminate information about preventive efforts, whereas
(2) interpersonal communication is carried out in counseling or assistance.

Human interaction at any level (between persons, groups, and the masses), including health
communication on HIV/AIDS, has a similar goal – namely to influence the audience or recipi-
ent. In developing literacy on HIV/AIDS, participatory communication is used.
A participatory communication approach has emerged to replace the emphasis on moderniza-
tion, diffusion of knowledge, and technology transfer. Participatory communicators place
their highest value on local knowledge and abilities. Instead of encouraging communities to
participate in development initiatives planned by outsiders, they encourage citizens to make
their own plans.

To be effective, communication must occur between all parties affected, ensuring all have
the same opportunity to influence the results of the initiative. Ideally, participatory communication should be part of the whole process whereby (a) two-way communication must be
adopted from the beginning and applied consistently, and (b) full participation by all stakeholders in each step of the process is not possible and in some cases may not be desirable.
Thus broad consensus might be enough. (c) Inclusiveness must be balanced with consideration
of the time, resources, interests, and knowledge of stakeholders. After their input is taken into
account, stakeholders may not need to be involved in detailed decisions outside the scope of
their interests.

Free and open dialogue remains the core of participatory communication where dialogue is defined as "a human meeting in understanding the world (life)." Free and open
dialogue where people can "name and interpret the world" is the principle of horizontal
communication. Thus, participatory communication is an approach based on dialogue,
which enables the exchange of information, perceptions, and opinions among various
stakeholders and thus facilitates empowerment, especially for the most vulnerable and
marginalized.

Information, dissemination of information, and education on HIV/AIDS in the city of Bandung was carried out by informants serving as WPA cadres through diverse activities. The
form of communication used was group communication featuring lectures, dialogues, and dis-
cussions. Interpersonal communication was carried out in prevention and assistance programs
for people living with HIV/AIDS. Communication level took the form of:

3.3.1 *Group communication*
Group communication was conducted through socialization programs, information dissemination, and education of the community. The activities WPA and PKK cadres carried out
included (1) counseling youth groups (2) giving special attention to pregnant women and

infants, (3) mobilizing voluntary consulting and testing, (4) and initiating various activities to increase knowledge about, awareness of, and concern for HIV/AIDS.

3.3.2 *Interpersonal communication*

Interpersonal communication was demonstrated in rehabilitative acts in the form of assistance for people living with HIV/AIDS. Informants also used two-way communication as a form of *dialogue* and *advocacy* in order to give assistance in obtaining defense or support for HIV/AIDS health programs.

The approach taken in developing health literacy concerning HIV/AIDS is shown in Figure 2:

Figure 2.   The approach taken in developing health literacy concerning HIV/AIDS.

4   CONCLUSION

The participation of WPA female cadres begins with their involvement as PKK cadres who care about their environment. The WPA is a community-based program in which, in this case, the members were women. Becoming a WPA cadre is an active process of interaction with the social environment where women reconstruct their understanding of HIV/AIDS. The volunteerism of WPA women stems from their concern for the surrounding environment. The negative stigma attached to people with HIV/AIDS excludes them from the environment. The desire to recover turns into fear to confront the reality of society. The role of the WPA in changing the way people think about HIV/AIDS sufferers is needed.

ACKNOWLEDGMENT

Gratitude is conveyed to all of my dissertation supervisors in the communications postgraduate program at Sebelas Maret University in Indonesia who have guided this research.

REFERENCES

Aripurnami, S. 2013. Defisit Keadilan Gender dan Agenda Pasca-2015 di Indonesia 2013. *Jurnal analisis sosial* 18 (1)Agustus. Yayasan Akatiga. Bandung.

Gupta, M. D. & Srivastava, R. 2012. Indexing women's empowerment: A methodological study. *Pakistan journal of women's studies* 19(1): 59–74.

Hidayana, I. M., Noor, I. R., & Pakasi, D. 2010. *Hak Seksual perempuan dan HIV/AIDS: Studi pada Perempuan Muda (15–24 th) di Tiga Kota di Jawa Barat*. Jakarta: Pusat Kajian Gender dan Seksualitas FISIP UI.

Hidayat, R. 2012. *Buku Pedoman Pembentukan Forum Warga Peduli AIDS Tingkat Kecamatan dan Kelurahan di Kota Bandung*. Bandung: Komisi Penanggulangan Aids Kota Bandung.

Kabore, I., Bloem, J., Etheredge, G., Obiero, W., Wanless, S., Doykos, P., . . . & Bostwelelo, J. 2010. The effect of community-based support services on clinical efficacy and health-related quality of life in

HIV/AIDS patients in resource-limited settings in sub-Saharan Africa. *AIDS patient care and STDs* 24(9): 581–594.

Komisi Penanggulangan AIDS Kota Bandung. 2011. *Rencana Strategis Penanggulangan HIV-AIDS Kota Bandung Tahun 2012–2016*. Bandung: Komisi Penanggulangan AIDS Kota Bandung.

McCreary, L. L., Kaponda, C. P., Davis, K., Kalengamaliro, M. & Norr, K. F. (2013). Empowering peer group leaders for HIV prevention in Malawi. *Journal of nursing scholarship* 45(3): 288–297.

Nasution, B. R., Mukuan, O. S., Sebayang, P. M. & Aznur, S. D. 2011. *Kualitas dan Rekomendasi Perbaikan Layanan PMTCT Bagi Perempuan Terinfeksi HIV di Empat Kota di Indonesia*. Jakarta: Ikatan Perempuan Positif Indonesia.

Nicholas, P. K., Adejumo, O., Nokes, K. M., Ncama, B. P., Bhengu, B. R., Elston, E. & Nicholas, T. P. 2009. Fulbright Scholar opportunities for global health and women's health care in HIV/AIDS in sub-Saharan Africa. *Applied nursing research* 22(1): 73–77.

Nyamathi, A. 2014. Engaging community health workers in HIV/AIDS care: A case exemplar among rural Indian women living with AIDS. *Journal of HIV/AIDS & social services* 13(4): 330–336.

Peraturan Menteri Dalam Negeri No 20 tahun 2007 tentang Pedoman Umum Pembentukan Komisi Penanggulangan AIDS dan Pemberdayaan Masyarakat alam Rangka Penanggulangan HIV dan AIDS di Daerah. 2007. Departemen Dalam Negeri Republik Indonesia.

Ramírez-Johnson, J., Díaz, H. L., Feldman, J. B., & Ramírez-Jorge, J. 2013. Empowering Latino church leaders to deal with the HIV-AIDS crisis: S strengths-oriented service model. *Journal of religion and health* 52(2): 570–588.

Rodriguez-García, R., Bonnel, R., N'Jie, N. D., Olivier, J., Pascual, F. B., & Wodon, Q. 2011. *Analyzing community responses to HIV and AIDS: Operational framework and typology*. Washington, DC: The World Bank.

Romero, L., Wallerstein, N., Lucero, J., Fredine, H. G., Keefe, J. & O'Connell, J. 2006. Woman to woman: coming together for positive change—using empowerment and popular education to prevent HIV in women. *AIDS Education & Prevention* 18(5): 390–405.

Schwartländer, B., Stover, J., Hallett, T., Atun, R., Avila, C., Gouws, E., … & Alsallaq, R. 2011. Towards an improved investment approach for an effective response to HIV/AIDS. *Lancet* 377(9782): 2031–2041.

Joint United Nations Programme on HIV/AIDS. 2008. *Report on the Global Acquired Immunodeficiency Syndrome Epidemic*. Unaids.

Medical Technology and Environmental Health – Abdullah, Widiaty & Abdullah (eds)
© 2020 Taylor & Francis Group, London, ISBN 978-0-367-86053-0

Clinical profile in adult typhoid fever in patients at hospital X, East Jakarta, Indonesia, January–March 2018

R.S. Kuddah & S.W.M. Husnah
Universitas Muhammadiyah Jakarta, Jakarta, Indonesia

ABSTRACT: Typhoid fever is an acute disease caused by the *Salmonella typhi* bacterium. The spread of the bacteria is mainly oral-fecal with the intermediary of contaminated water and cooking. Many people in Indonesia do not care about environmental cleanliness, especially in densely populated cities such as Jakarta. Varied clinical symptoms and laboratory results make diagnosis difficult. Therefore, research needs to be done to detect clinical symptoms commonly found in typhoid fever patients. To examine the clinical symptoms, laboratory results for adults with typhoid fever in January–March 2018 were acquired. This study was a retrospective descriptive study using a cross-sectional study design with a total sample of 84 medical records. The study was conducted from October to December 2018 at Hospital X, East Jakarta. Data were obtained from patients' medical records. Adult typhoid fever patients were hospitalized in Hospital X: 32.24% were aged 19–24 years, 58.33% were female, and 40.48% stayed for 3 days with an average visit length of 3.06 days. Patients who had a fever were as much as 92.86%. Widal O titer was 1/320 positive at 34.52% and H titer was 1/160 positive at 21.43%, normal Hb at 54.76%, normal leukocytes at 50%, and normal platelets at 77.39%. Typhoid fever often occurs in young adults who have anemia.

1 INTRODUCTION

1.1 *Background*

Typhoid fever is still an important problem in countries with tropical climates, including Indonesia (Widoyono 2008).

Typhoid fever is an endemic disease in Indonesia, driven by poor sanitation conditions in big cities, especially Jakarta (Purnama 2016). The managing director of Bappenas Irrigation, Abdul Malik Sadat Idris, has stated that, according to the results of Bappenas's research, 96% of urban water is highly or heavily polluted. For example, 13 rivers in Jakarta have poor water content so that they are not good for consumption (David 2018).

Typhoid fever is a systemic infection caused by the *Salmonella typhi* bacterium, which can be spread through contaminated food or water. This disease is highly contagious and can cause an outbreak (Setiati et al. 2014). According to the latest estimates, 11–21 million cases and 128,000–161,000 deaths related to typhoid fever occur annually throughout the world (World Health Organization, 2018). The estimated incidence rate of typhoid fever is 150/100,000 per year in South America and 900/100,000 per year in Asia.

Based on a report of the Directorate General of Medical Services of the Ministry of Health of the Republic of Indonesia in 2009, typhoid and paratyphoid fever ranked third out of the 10 most common diseases of hospitalized patients in Indonesia out of 80,850 cases, with a proportion of 1.25% (Dep Kes RI, 2009). Based on data at Hospital X, East Jakarta has an incidence rate of typhoid fever of 554/12,904 annually.

Salmonella is part of the Enterobacteriaceae family of bacteria that are pathogenic to humans. *Salmonella* is a gram-negative, facultative anaerobic rod-shaped bacterium. The morphology of *Salmonella* resembles that of other enteric bacteria (see Figure 1). Most of

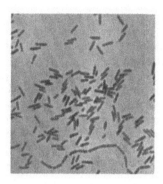

Figure 1. Morphology of *Salmonella typhi* (Todar 2008).

these bacteria are motile isolates with peritric flagella (having more than one flagella found throughout the surface of their body) and capsules (Apriyadi & Sarwili 2018).

Salmonella typhi, one of the subspecies of the genus *Salmonella* that causes typhoid fever, causes prolonged fever, headaches, nausea, vomiting, loss of appetite (anorexia), and constipation or not infrequently diarrhea. Severe cases can result in serious complications or even death. The average dose that can cause symptoms of clinical or subclinical infection in humans is around 10^5–10^8 *Salmonella* bacteria (*Salmonella typhi* is enough with the number of 10^3 bacteria that may already be causing clinical symptoms). Infection is transmitted through food or water contaminated by feces or urine from the patient or carrier. Raw fruits and vegetables are the main cause of the spread of *S. typhi* bacteria. This happens in some countries where human feces are used as fertilizer for plants (fruits and vegetables) and contaminated water is used to make fruits look more attractive in the market (Brooks et al. 2010; Setiati et al. 2014).

Typhoid fever remains an endemic disease in Indonesia; the World Health Organization (WHO) states that the number of typhoid fever sufferers in Indonesia is 81% per 100,000 cases per year with a mortality rate of 3.1–10.4% (Apriyadi and Sarwili 2018; Seran et al. 2015).

Typhoid fever is classified as an infectious disease, in accordance with Indonesian Law No. 6 of 1962 (Indang et al. 2013).

Diagnosis of the disease must be established early and be supported by knowledge of its clinical features and symptoms; sometimes additional investigations are needed. Typhoid fever has an incubation period of 10–14 days. Symptoms vary widely, from asymptomatic to very characteristic symptoms, to the emergence of complications that can cause death if not immediately treated (Setiati et al. 2014).

Clinical symptoms of typhoid fever include fever, dirty tongue (coated tongue), disorders of the digestive tract, disturbance of consciousness, hepatosplenomegaly, and relative bradycardia.

Fever is stated when the body temperature is above the normal value (> 36.7°C). Fever is usually the main complaint for people with typhoid fever. At first the fever is not too high, but at the second week the fever's intensity grows higher. Usually in the morning the fever feels lower than in the afternoon or evening because the body's metabolism decreases in the afternoon. Note that the typical type of fever in typhoid fever is not always there. This can be caused by treatment interventions or complications that occur earlier than the disease (RI Minister of Health 2006).

Disturbances in the digestive system very often occur due to the entrance of *S. typhi* bacteria through the digestive tract. The most typical disorder is a dirty tongue or the presence of a white membrane on the tongue (coated tongue). This can also cause bad breath. Not infrequently, the lip mucosa will feel dry until cracking occurs. Abdominal pain (usually in the epigastric region), nausea, and vomiting can also occur. Constipation and bloating (meteorismus) usually occur early after exposure to the source of infection and subsequently diarrhea can occur (RI Minister of Health 2006).

In general, disturbance of consciousness experienced by patients is mild. However, if the clinical symptoms are severe, a decrease in patient awareness can range from somnolence to coma and can cause delirium in patients with toxic symptoms accompanied by psychosis (RI Minister of Health 2006).

Hepatosplenomegaly often occurs. This is because these organs, including reticuloendothelial organs, are good breeding sites for *S. typhi* bacteria (RI Minister of Health 2006).

Bradycardia is relatively rare because carrying out the examination is difficult. Bradycardia is said to be relative if an increase in body temperature is not followed by an increase in pulse frequency. The standard used is when any increase in temperature of 1°C is not followed by an increase in the pulse frequency of 8 beats in 1 minute (RI Minister of Health 2006).

During routine blood tests, leukopenia is usually found, but leukocyte levels within normal limits or leukocytosis are possible. Mild anemia and thrombocytopenia can also occur. The blood sludge rate can increase, and examination of the type of leukocytes may find no eosinophils and lymphocyte counts that are below normal. SGOT and SGPT are often found in a state of increase, but if the patient has recovered, they will return to normal conditions. Another routine check is a bacterial culture. Culture testing remains the standard for diagnosing typhoid fever. In addition to cultures or isolation of germs, researchers can also employ serological methods such as widal tests. But the widal test has many weaknesses. Another serological examination – namely the Salmonella IgM or IgG test – is faster and its level of sensitivity and specificity is high, but the price is still quite costly (Setiati et al. 2014).

2 METHODS

This research comprised a nonanalytic retrospective descriptive study with a cross-sectional approach. Data processing was carried out using Excel Windows 2010. Using secondary data from medical records of clinical symptoms and laboratory results, this study included as many as 84 adult patients with a diagnosis of typhoid fever in total sampling of patients over the age of 18 who suffer from typhoid fever at Hospital X, East Jakarta, in the period of January–March 2018, while patients with incomplete data and concomitant diseases (other than typhoid fever) were excluded from the sample. Laboratory results were adjusted to the reference value from the Hospital X laboratory in tabular form.

3 RESULTS

Hospital X, East Jakarta, handled as many as 621 inpatients with typhoid fever in 2016, 557 in 2017, and 483 in 2018. This figure is still relatively high as it comprises 0.4% of the total number of inpatients in Hospital X.

Distribution by age of 84 adult patients with typhoid fever obtained the following patient classification data (WHO) (Table 1):

Table 1 and Figure 2 show 17 patients (20.24%) were aged 19–24, 11 patients (13.10%) were aged 25–30, 16 patients (19.05%) were aged 31–36, 11 patients (13.10%) were aged 37–42, 8

Table 1. Distribution of adult patients with typhoid fever by age.

Category	Number	Percent (%)
19–24	17	20.24
25–30	11	13.10
31–36	16	19.05
37–42	11	13.10
43–48	8	9.52
49–54	9	10.71
55–60	5	5.95
61–65	7	8.33
Total	84	100.00

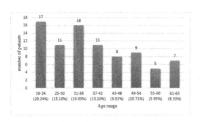

Figure 2. Distribution of adult patients with typhoid fever based on age.

patients (9.52%) were aged 43–48, 9 patients (10.71%) were aged 49–54, 5 patients (5.95%) were aged 55–60, and 7 patients (8.33%) were aged 61–65. The table shows that most adult patients with typhoid fever were aged 19–24 years, or 20.24%.

Distribution of 84 adult patients with typhoid fever based on gender obtained the following data classification:

Table 2 and Figure 3 show 35 patients (41.67%) were male and 49 patients (58.33%) were female. These data indicate that the population of women with typhoid fever was greater than that of men.

Distribution of 84 adult patients with typhoid fever based on length of hospital stay obtained the following data:

Table 3 and Figure 4 show that the length of stay of adult typhoid fever patients was 1 day at the shortest and 6 days at the longest. Out of these 84 patients, the average hospitalization was 3.06 days.

Research on adult patients with typhoid fever treated at Islamic Hospital Jakarta Pondok Kopi East Jakarta obtained the following overview of clinical symptoms and physical examination results:

Table 4 and Figure 5 show that out of the adult typhoid fever patients included in the sample, 78 (92.86%) had fever, 59 (70.24%) suffered nausea, 28 (33.34%) experienced vomiting, 30 (35.71%) had headache, 16 (19.05%) developed a limp, 12 (14.29%) endured diarrhea and coughing, 5 (5.95%) had stomachache, 4 (4.76%) complained of bitter tongue/mouth, 2 (2.38%) had constipation, and 1 (1.19%) became bloated. The results of the physical examination indicated that 10 patients (11.90%) experienced epigastric pain, 2 (2.38%) had a coated tongue, and 1 (1.19%) was pale.

Table 2. Distribution of adult patients with typhoid fever based on gender.

Category	Number	Percent (%)
Male	35	41.67
Female	49	58.33
Total	84	100.00

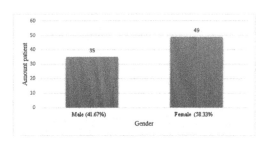

Figure 3. Distribution of adult patients with typhoid fever based on gender.

163

Table 3. Distribution of adult patients with typhoid fever based on length of hospitalization.

Days	Number	Percent (%)
1	4	4.76
2	24	28.57
3	34	40.48
4	12	14.29
5	5	5.95
6	5	5.95
Total	84	100.00

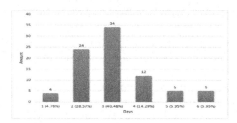

Figure 4. Distribution of adult patients with typhoid fever based on length of hospitalization.

Table 4. Distribution of adult patients with typhoid fever based on clinical symptoms and physical examination results.

Symptoms and results of physical examination	Frequency	Percent (%)
Fever	78	92.86
Nausea	59	70.24
Headache	30	35.71
Vomiting	28	33.34
Fatigue	16	19.05
Diarrhea	12	14.29
Cough	12	14.29
Epigastric pain*	10	11.90
Decreased appetite	5	5.95
Stomach ache	4	4.76
Bitter tongue/mouth	2	2.38
Constipation	2	2.38
Coated tongue*	2	2.38
Bloating	1	1.19
Pale*	1	1.19

* Physical examination results

3.1 *Supporting investigation*

A widal examination was performed on 84 of the 78 patients; the results were as follows:

Table 5 and Figure 6 show that of the 78 patients who underwent widal examination, 29 (34.52%) had a titer O of 1/320 and 18 (21.43%) had a titer H of 1/160.

A tubex examination was performed on 6 of the 84 patients in the sample; the results were as follows:

164

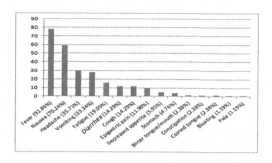

Figure 5.   Distribution of adult patients with typhoid fever based on clinical symptoms and physical examination results.

Table 5.   Distribution of adult patients with typhoid fever based on widal test results.

Titer	Category	Frequency	Percent (%)
Titer O	1/80	15	17.86
	1/160	13	15.48
	1/320	29	34.52
	1/640	1	1.19
Titer AO	1/80	4	4.76
	1/160	4	4.76
	1/320	5	5.95
Titer BO	1/80	9	10.71
	1/160	16	19.05
	1/320	5	5.95
Titer CO	1/80	1	1.19
	1/160	5	5.95
	1/320	6	7.14
Titer H	1/80	15	17.86
	1/160	18	21.43
	1/320	11	13.10
Titer AH	1/80	5	5.95
	1/160	8	9.52
	1/320	5	5.95
Titer BH	1/80	6	7.14
	1/160	14	16.67
	1/320	8	9.52
Titer CH	1/80	8	9.52
	1/160	10	11.90

Information
O: *S. typhi* (body)
AO: *S. paratyphi* type A (body)
BO: *S. paratyphi* type B (body)
CO: *S. paratyphi* type C (body)
H: *S. typhi* (flagellum)
AH: *S. paratyphi* type A (flagellum)
BH: *S. paratyphi* type B (flagellum)
CH: *S. paratyphi* type C (flagellum)

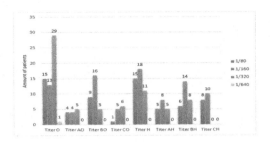

Figure 6. Distribution of adult patients with typhoid fever based on widal test results.

Table 6 and Figure 7 show that of the six patients who underwent a tubex examination, three were reactive.

Of the 84 patients who underwent peripheral blood tests, the hemoglobin examination results were as follows:

Table 7 and Figure 8 show that of the 84 patients who underwent hemoglobin examination, 36 (42.86%) had anemia, 46 (54.76%) had normal results, and 2 (2.38%) had high hemoglobin.

From the 84 patients who had their leukocytes examined, the following results were obtained:

Table 8 and Figure 9, show that of the 84 patients whose leukocytes were examined 17 (20.24%) had leukopenia (20.24%), 25 (29.76%) had leukocytosis, and 42 (50%) had normal test results.

From the 84 patients who underwent peripheral blood tests, the following hematocrit results were obtained:

Table 9 and Figure 10 show that of the 84 patients who underwent hematocrit examination, 34 (40.48%) had low hematocrit levels, 48 (57.14%) had normal levels, and 2 (2.38%) had high levels.

Table 6. Distribution of adult patients with typhoid fever based on the results of the tubex examination.

Category	Frequency
≥4	3
<4	3

Information:
Tubex ≥4: Positive indicates active typhoid fever infection
Tubex <4: Does not show active typhoid fever infection

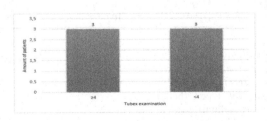

Figure 7. Distribution of adult patients with typhoid fever based on the results of the tubex examination.

Table 7.   Distribution of adult patients with typhoid fever based on hemoglobin value.

Category (mg/dL)	Number	Percent (%)
<13.5	36	42.86
13.5–17.5	46	54.76
>17.5	2	2.38
Total	84	100.00

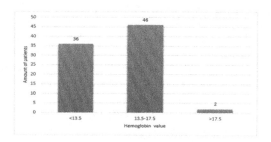

Figure 8.   Distribution of adult patients with typhoid fever based on hemoglobin value.

Table 8.   Distribution of adult patients with typhoid fever based on leukocyte values.

Category	Number	Percent (%)
Leucopenia (<5x10^3/uL)	17	20.24
Normal (5x10^3–10x10^3/uL)	42	50.00
Leukocytosis (>10x10^3/uL)	25	29.76
Total	84	100.00

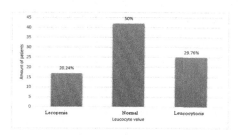

Figure 9.   Distribution of adult patients with typhoid fever based on leukocyte values.

From the 84 patients who underwent peripheral blood tests, the platelet examination results were as follows:

Table 10 and Figure 11 show that of 84 patients undergoing platelet examination, 12 (14.27%) had thrombocytopenia, 65 (77.39%) had normal results, and 7 (8.34%) had thrombocytosis.

Table 9.   Distribution of adult patients with typhoid fever based on hematocrit values.

Category (%)	Number	Percent (%)
<40	34	40.48
40–50	48	57.14
>50	2	2.38
Total	84	100.00

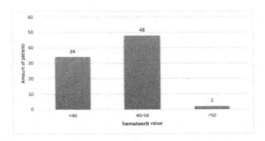

Figure 10.   Distribution of adult patients with typhoid fever based on hematocrit values.

Table 10.   Distribution of adult patients with typhoid fever based on platelet values.

Category	Number	Percent (%)
Thrombocytopenia (<150x10^3/uL)	12	14.27
Normal (<150x10^3-400x10^3/uL)	65	77.39
Thrombocytosis (>400x10^3/uL)	7	8.34
Total	84	100.00

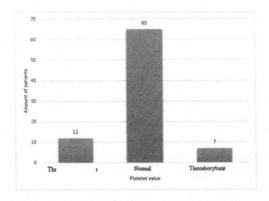

Figure 11.   Distribution of adult patients with typhoid fever based on platelet values.

4 DISCUSSION

Results of clinical profiles of inpatients with a diagnosis of typhoid fever in the January–March 2018 period at Hospital X, East Jakarta, showed that most were aged 19–24 and 31–36 years.

The study was conducted in 2018 at the Dinoyo Community Health Center, Malang, East Java, Indonesia. The study has similar results to those of previous research conducted by the author – namely the highest number of typhoid fever patients were 18–24 years and 31–35 years of age (Awa et al. 2019). This is in line with research conducted at Sanglah Hospital Denpasar in 2016–2017, which indicated that most typhoid fever patients were aged 18–34 years with a total of 31 out of 55 patients (Melarosa et al. 2019). Similarly, research conducted at the Shashemene referral hospital in southern Ethiopia in 2016 found that most typhoid fever patients are aged 21–40 years, or 267 patients out of 421 patients (Habte et al. 2018).

The results of the sex distribution of adult typhoid fever patients in this study fall in line with those of research conducted at Jakarta Cempaka Putih Islamic Hospital in Central Jakarta in 2018. The population of adult typhoid fever patients was dominated by women, with 40 patients out of a total of 60 being women. This is also in line with research conducted at the Shashemene referral hospital in 2016, which noted 235 female patients and 186 male patients (Habte et al. 2018).

Adult typhoid fever patients had a mean hospital stay of 3.06 days in this study. This is similar to a study conducted in 2014 at the Bancak Public Health Center in Semarang, where typhoid fever patients had a mean treatment stay of 3.45–4 days. The mean short duration of hospitalization can be related to the complication status of typhoid fever patients; the sample used in this study included typhoid fever patients without complications or other comorbidities (Fithria et al. 2017).

According to this study, the most common clinical symptom is fever. Fever in typhoid fever tends to increase at night because the body's metabolism will decrease at night (RI Minister of Health 2006). The results of this study are commensurate with research conducted at Cempaka Putih Hospital, Central Jakarta in 2018, which indicated that three clinical symptoms most often occur – namely fever, nausea, and vomiting (Daramuna 2018). Similarly, a study conducted at the Shashemene referral hospital in 2016 found that the most frequent clinical symptoms were fever, with 358 patients out of a total of 421 experiencing fever (Habte et al. 2018).

Widal examination carried out on typhoid fever patients in this study had the most results – namely 1/320 O titer with 34.52% and 1/160 H titer with 21.43%. In 2014, a study was conducted at one hospital in the city of Padang during which 46 patients underwent a widal test. A result of 1/160 O titer was seen in 34 patients, and 9 patients had 1/320 O titer. Likewise with the highest H titers – namely 1/160 (22 patients) and 1/320 (21 patients). O or H titers 1/160 and 1/320 were the most common (Velina et al. 2016). Widal examination is still frequently used as a supporting examination for typhoid fever patients. The interpretation of the widal test is assessed based on the increase in agglutinin titer four times, especially agglutinin O. The application of agglutinin O titer ranges from > 1/160 to > 1/320. This interpretation works if the patient has not been vaccinated or recovered from typhoid fever within the previous 8 months (Garna 2012). This statement is in line with the results of the aforementioned study, which shows that 29 patients had 1/320 O titer.

Tubex testing is often done because it has a sensitivity of 75–80% and specificity of 75–90% (Garna 2012). However, in this study, only six patients underwent the tubex test and three were reactive. In a study conducted on 1,266 patients in the clinical laboratory of Nikki Medika Denpasar in 2014, 1,021 patients were negative and 242 were positive for the tubex test. This can happen because patients who have symptoms of typhoid fever may not actually have typhoid fever; the symptoms are almost the same as those seen with influenza, dengue fever, or gastroenteritis and other diseases. It could also be because the patient is in a state of chronic typhoid fever or in the healing period. The tubex test can only detect IgM and not IgG, whereas in chronic typhoid fever, IgG becomes an immunoglobulin that circulates in the patient's blood. Based on the foregoing explanation, the tubex test does have better accuracy compared to widal examination in its sensitivity and specificity. However, many other factors can influence the results of the tubex test. In this study, due to the small number of patients who underwent a tubex examination, the results cannot be used as a reference.

Typhoid fever patients often have routine blood tests, especially to determine the levels of hemoglobin, leukocytes, hematocrit, and platelets (Garna 2012). In this study, data obtained from a total of 84 patients showed that 46 patients had normal Hb levels. However, 36 patients experienced decreased Hb levels.

Routine blood laboratory results of typhoid fever patients can indicate anemia. For leukocyte levels, out of the 84 patients examined in this study, 42 had normal results, 25 had leukocytes, and 17 had leukopenia. Most patients show normal results for routine laboratory blood tests on leukocyte levels.

Based on the literature, leukopenia is usually found, but leukocyte levels may fall within normal limits or the patient may have leukocytosis. Hematocrit levels can also be used as a reference because hematocrit levels always follow hemoglobin levels (Garna 2012). The results of this study showed that as many as 34 patients out of the sample of 84 had decreased hematocrit levels.

Finally, the results of platelet examination on the sample of 84 patients examined here found that 65 patients had normal results, 17 had thrombocytopenia, and 7 had thrombocytosis.

The results of these routine blood tests can be compared with those in studies first conducted in 2018 at Cempaka Putih Hospital, which obtained results for hemoglobin examinations showing that out of a total of 65 patients, 26 had normal results and 22 had anemia. The results of examination of leukocyte levels showed that 42 patients had normal levels, 17 had leukocytosis, and 6 had leukopenia. The results of the hematocrit examination found 26 patients had decreased hematocrit levels. Finally, a platelet examination revealed 53 normal patients, 7 with thrombocytosis, and 5 with thrombocytopenia (Daramuna 2018). The results of the two studies were not much different from each other. Hematologic changes in typhoid fever patients can be due to bone marrow suppression or intestinal bleeding. This can cause anemia. Changes in leukocyte and platelet levels can also be due to bone marrow depression by endotoxins.

However, if a patient's immune system is good enough, then hematological examinations also show good or normal results (Handayani & Mutiarasari 2017).

5   CONCLUSION

This research was conducted in 2018 at Hospital X, East Jakarta. The research subjects were adult patients diagnosed with typhoid fever and hospitalized. Eighty-four patients met the inclusion requirements as research subjects. The results of this study showed that 20.24% of typhoid fever patients were aged 19–24 years and 19.05% were aged 31–36 years. The sample revealed more female patients than male patients, with a percentage of 58.33%. The length of stay for typhoid fever patients was an average of 3.06 days. The most widely used widal test results were O 1/320 titer and H 1/160 titer. Tubex examination is not often done, so not enough data were obtained. Results from routine blood tests were obtained by almost all normal patients. However, quite a lot of patients had anemia, as much as 42.86%. The most common clinical symptom experienced by typhoid fever patients was fever, which was 92.86%.

REFERENCES

Apriyadi, E. & Sarwili, I. 2018. Perilaku Higiene Perseorangan dengan Kejadian Demam Tifoid, *Jurnal Ilmiah Ilmu Keperawatan Indonesia* 8(1): 355–362.
Awa, M. E. D., Supriyadi, S., & Ka'arayeno, A. J. 2019. Hubungan Kebiasaan Mencuci Tangan Menggunakan Air Bersih Dan Sabun Dengan Kejadian Demam Typhoid Pada Orang Dewasa Di Wilayah Kerja Puskesmas Dinoyo. *Nursing news: Jurnal Ilmiah Keperawatan* 4(1): 269–278.

Brooks, G. F., Carroll, K. C., Butel, J. S., Morse, S. A., & Mietzner, T. A. 2010. *Medical microbiology*. New York: Jawetz, Melnick and Adelbergs.

Daramuna. 2018. Overview of clinical laboratory symptoms and provision of antibiotics in adult patients with typhoid fever in inpatient [treatment] at Jakarta Cempaka Putih Islamic Hospital in the January–June period 2018 [thesis]. Jakarta: Faculty of Medicine and Health, University of Muhammadiyah Jakarta.

David, O. P. 2018. Bappenas: Quality of 13 rivers in Jakarta is poor, see use of black water drones. [Online] [October 29, 2018]. Available from: https://megapolitan.kompas.com/read/2018/02/01/ 20171221/bappenas-kualitas-13-sungai-di-jakarta-buruk-lihat-pakai-drone- the water

Depkes, R. I. 2009. Undang-undang Republik Indonesia nomor 36 tahun 2009 tentang Kesehatan. *Lembaran Negara Republik Indonesia Nomor* 144.

Fithria, R. F., Damayanti, K., & Fauziah, R. P. 2017. Perbedaan Efektivitas Antibiotik pada Terapi Demam Tifoid di Puskesmas Bancak Kabupaten Semarang tahun 2014. *JIFFK: Jurnal Ilmu Farmasi dan Farmasi Klinik*, 1–6.

Garna, H. (ed.). 2012. *Textbook for the Division of Infections and Chronic Diseases*. Jakarta: Sagung Seto, 729–751.

Habte, L., Tadesse, E., Ferede, G., & Amsalu, A. 2018. Typhoid fever: Clinical presentation and associated factors in febrile patients visiting Shashemene referral hospital, southern Ethiopia. *BMC research notes* 11(605): 1–6.

Handayani, N. P. D. P. & Mutiarasari, D. 2017. Karakteristik Usia, Jenis Kelamin, Tingkat Demam, Kadar Hemoglobin, Leukosit dan Trombosit Penderita Demam Tifoid pada Pasien Anak di Rsu Anutapura Tahun 2013. *Medika Tadulako: Jurnal Ilmiah Kedokteran Fakultas Kedokteran dan Ilmu Kesehatan* 4(2): 30–40.

Indang, N., Guli, M. M., & Alwi, M. 2013. Uji resistensi dan sensitivitas bakteri Salmonella thypi pada orang yang sudah pernah menderita demam tifoid terhadap antibiotik. *Biocelebes* 7(1) 27–34.

Melarosa, P. R., Ernawati, D. K., & Mahendra, A. N. 2019. Pola Penggunaan Antibiotika Pada Pasien Dewasa Dengan Demam Tifoid Di Rsup Sanglah Denpasar Tahun 2016–2017. *E-Jurnal Medika Udayana* 8(1): 12–16.

Purnama, B. E. 2016. Poor sanitation, basic issues of DKI Jakarta. [Online] [cited October 29, 2018]. Available from: http://mediaindonesia.com/read/detail/34535-sanitasi-buruk-pers issue-dasar-dki-jakarta.

RI Minister of Health. 2006. *Guidelines for typhoid fever control*. Jakarta: RI Ministry of Health, 6–8.

Seran, E. R., Palandeng, H., & Kallo, V. 2015. Hubungan Personal Hygiene dengan Kejadian Demam Tifoid di Wilayah Kerja Puskesmas Tumaratas. *Jurnal Keperawatan* 3(2):1–8.

Setiati, S., Alwi, I., Sudoyo, A. W., Stiyohadi, B., & Syam, A. F. 2014. *Internal medicine textbook*. Vol. 1, 6th edition. Jakarta: InternaPublishing, 549–552.

Todar, K. 2008. *Todar's online textbook of bacteriology*: Salmonella and salmonellosis.

Velina, V. R., Hanif, A. M., & Efrida, E. 2016. Gambaran Hasil Uji Widal Berdasarkan Lama Demam pada Pasien Suspek Demam Tifoid. *Jurnal Kesehatan Andalas* 5(3): 687–691.

Widoyono, M. P. H. 2008. *Penyakit Tropis Epidemiologi. Penularan, Pencegahan, dan Pemberantasannya*. Jakarta: Erlangga,.

World Health Organization. 2018. *World health statistics 2018: Monitoring health for the SDGs*. Geneva: World Health Organization.

171

Medical Technology and Environmental Health – Abdullah, Widiaty & Abdullah (eds)
© 2020 Taylor & Francis Group, London, ISBN 978-0-367-86053-0

The relationship between intensity of visits of children under 5 years to the integrated health post (POSYANDU) and toddler nutritional status in Padasuka Bandung Regency

D.S. Maulidiyyah & F.A. Yulianto
Islamic University of Bandung, Bandung, Indonesia

K. Dwiastuti & N.E. Thamrin
Dinas Kesehatan Kabupaten Bandung, Bandung, Indonesia

ABSTRACT: According to the United Nations' Sustainable Development Goals (SDGs) (2018), 51 million children under 5 years of age suffer from starvation and 38 million are overweight. The number of malnourished children in Indonesia has risen 0.1% compared to the previous year. Determinants of children's health status can be seen in nutritional status and monitored regularly at the Integrated Health Post, the *Posyandu*. The weight gain of toddlers in *Posyandu* at Padasuka was only 60.85% in 2018. The aim of this study was to determine the relationship between visit intensity to *Posyandu* and toddler nutritional status. The study was conducted in Padasuka Bandung during the period January 14–15, 2019. This study was an analytical study with a cross-sectional research design. The study was conducted on 210 subjects with sample selection by cluster sampling. The visit intensity of children under 5 years of age to *Posyandu* was assessed using questionnaires and the weight was measured and plotted in a weight chart according to age. A statistical analysis test was carried out using the Fisher's exact and Spearman analysis methods. The result of the analysis showed a relationship between visit intensity to *Posyandu* and toddlers' nutritional status (p = 0.004). Furthermore, there was a correlation between visit intensity to *Posyandu* and nutritional status (p = 0.013) of 17.06%. *Posyandu* as community-based support also play an important role in increasing mothers' awareness of the nutritional status of their toddlers.

1 INTRODUCTION

Human resources are the asset of a nation. The quality of human resources is determined by the health of the next generation. One of the determinants of a child's health status is nutrition. When the nutritional status of a child is good, it is expected that he or she will grow and develop optimally (Departemen Kesehatan Republik Indonesia 2012). Conversely, if the child's nutritional status is poor, it will also affect growth and development (UNICEF 2012).

Development, growth, and nutritional status can be monitored in the Integrated Health Post (*Posyandu*). *Posyandu* is community-based health program where the community receives basic health services. *Posyandu* is managed by the community for the community (Kemenkes Republik Indonesia 2018). This community-based health service plays an important role in Indonesian healthcare. Some surveys, health promotions, and other health services are often offered in *Posyandu*.

According to the United Nations' Sustainable Development Goals (SDGs) (2018), nutrition is one of the most important concerns worldwide. All countries agree on the need to end hunger. Fifty-one million children under 5 years of age suffer from starvation and 38 million are overweight (United Nations 2018).

Basic health research data (Riskesdas) in 2018 showed the number of malnourished children under 5 was 3.9% and the number who were underweight was 13.8%. It actually escalated by

0.1% compared to the previous year. West Java itself had a proportion of malnourished children as high as 13% in 2018 (Kementrian Kesehatan Republik Indonesia 2018). One of the community health centers (*Puskesmas*) in Bandung Regency saw 16 (0.3%) children who were underweight in 2018. Nutritional program data from one of the *Puskesmas* of Bandung Regency showed that toddlers' weight gain in 2018 was only 60.85%, which meant 39.15% of them did not gain weight in some periods (UPT Cimenyan 2017).

Many factors can influence nutritional status both directly and indirectly. Direct causes include food intake and chronic infection, while indirect causes include economic status, lack of monitoring children to measure their weight periodically, and feeding them inappropriate food (Merryana & Kes 2016).

Sugiyarti and colleagues (2014) found a relationship between *Posyandu* visits and children's nutritional status (Sugiyarti et al. 2014). Children's nutritional status can be monitored periodically by visiting *Posyandu* monthly so that children with malnutrition can be treated as soon as possible.

This study aimed to establish the relationship between visit intensity to *Posyandu* of children under 5 years of age and the nutritional status of toddlers in Padasuka, Bandung Regency.

2 METHODS

The subjects of this study consisted of mothers of toddlers aged 0–59 months in Padasuka. The exclusion criteria for this study included not living in Padasuka and declining to participate in this study.

The design of this study was cross-sectional and included 210 children under 5 years of age who lived in Padasuka. The subjects in this study were determined by cluster sampling in each region of Padasuka. This study used a questionnaire that contained questions about the intensity of the mothers' visits to *Posyandu* and then the toddlers' weights were plotted into the World Health Organization's (WHO) weight and growth chart by age.

This study was held in January 14–15, 2019, in each region. The subjects received an explanation of how to fill out the questionnaires and signed informed consents. Data analysis included univariate and inferential statistics stated in the form of frequencies and percentage of each variable. The results of this study were processed by Fisher's exact and Spearman's analytical correlation study by computer program to determine the relationship between visit intensity to *Posyandu* of children under 5 and toddlers' nutritional status in Padasuka.

3 RESULTS

This study revealed that the majority of mothers do not go to *Posyandu* routinely: 20% always go, 20.96% often go, the majority sometimes go, and 3.3% never go (Table 1).

Table 2 shows that the majority of the children's nutritional status was normal (73.34%), but the number of underweight toddlers was also high (21.90%). Fortunately, the percentage of overweight toddlers was low (4.75%).

Table 1. Distribution of respondents based on the intensity of mothers' visits to *Posyandu*.

Intensity of Visits	Number	Percentage (%)
Never	7	3.33
Sometimes	105	50.0
Often	54	25.71
Always	44	20.96
Total	210	100

Table 2. Nutritional status of children under 5 years of age in *Posyandu*.

Status	Number	Percentage (%)
Malnourished < −3 SD	0	0
Underweight −3 to < 2 SD	46	21.90
Normal −2 to + 2 SD	154	73.34
Overweight > +2 SD	10	4.76
Total	210	100.00

Table 3. Spearman's correlation.

Nutritional Status	P	Rho
Intensity of Visits	0.013	0.17

Table 4. Relationship between the intensity of mothers' visits to *Posyandu* and toddlers' nutritional status.

		Nutritional Status						
		Underweight		Normal		Overweight		
		n	%	n	%	n	%	P-value
Intensity of Visits	Never	1	14.29	4	57.14	2	28.57	0.004
	Sometimes	31	29.52	72	68.57	2	1.90	
	Often	11	20.37	40	74.07	3	5.56	
	Always	3	6.82	38	86.36	3	6.82	
	Total	46	21.90	154	73.33	10	4.76	

Based on Table 4, mothers who seldom visit *Posyandu* had a higher number of underweight children (29.52%), in contrast with mothers who always visit *Posyandu*, who had a higher number of toddlers categorized as of normal weight (86.36%). This means that the more often mothers visit *Posyandu*, the lower the trend to have an undernourished toddler. Fisher's exact test in this study was $p < 0.004$, indicating a relationship between visit intensity and nutritional status.

Table 3 shows Spearman's correlation analysis result, which was $p = 0.013$. This suggests a relationship between visit intensity and nutritional status. *Posyandu* visits have a correlation of 17.06% with nutritional status.

4 DISCUSSION

This study shows that the intensity of visits to *Posyandu* of children under 5 years of age was still low, as proven by the 50% of mothers who only sometimes visit *Posyandu*. This was caused by various factors such as the lack of mothers' enthusiasm and information about the importance of *Posyandu* visits and other factors not included in this study. Low education, mothers' attitude, time limit, distance to *Posyandu*, moving house, health services, and vaccine composition influence the visit of a mother to *Posyandu* (Sihotang & Rahma 2017). Those factors can influence the supervision of toddlers' weight gain, which is recorded on their *Kartu Menuju Sehat* (KMS) card each month.

This study shows that the number of underweight toddlers in Padasuka is still high (21.90%). This could present a risk for developmental or cognitive disorders, and underweight children are more likely to be sick in the future (Katona & Katona-Apte 2008).

This study shows a relationship between the intensity of visits to *Posyandu* and toddler nutritional status in Padasuka. The higher the visit intensity, the lower the risk of malnutrition. In contrast, the mothers who has lower intensity to visit *Posyandu* saw a higher risk of malnutrition in their toddlers (p = 0.004). Another study has shown that community-based health services such as feeding centers also can increase weight gain in underweight children (Paramashanti & Sulistyawati 2019). This study showed that mothers' visits to integrated health care units can increase their children's nutritional status through health promotion and regular evaluation of their children. Health promotion can be effectively done in *Posyandu* (Susanto et al. 2019).

Other factors can influence the nutritional status of toddlers directly, so the significance of the relationship between visit intensity and nutritional status is low (17.06%, p = 0.013). Some studies show an association between maternal feeding behavior with children's nutritional status (Spill et al. 2019).

As a community-based health service, *Posyandu* control toddlers' development and growth so they will not be delayed, and offer the advantage of early diagnosis of malnutrition so that prompt treatment can be provided and complications can be avoided.

The community-based health service plays an important role in toddlers' nutritional status as it also plays an important role in mothers' increasing awareness of their children's development and growth, because they know their children's progress each month and whether their development is normal. It is also a place where they can meet healthcare providers and may have a consultation with them when there is a problem.

A study by Costa and colleagues showed that mothers' education influences eating behavior and that breastfeeding duration influences the child's weight gain (Comba et al. 2019; da Costa et al. 2019). Weight gain is directly influenced by food intake; a study by Magriplis and colleagues proved that other factors influence meal patterns such as sleeping patterns, family meals, and study hours (Magriplis et al. 2019).

5 CONCLUSION

There was a correlation between visit intensity and toddlers' nutritional status in Padasuka. *Posyandu* as community-based medicine increase the awareness of mothers about their toddlers' development and growth. Other determinants directly influence toddlers' nutritional status. To increase the number of visits to *Posyandu*, an evaluation of human resources, information provided, and goals set at *Posyandu* is needed.

REFERENCES

Comba, A., Demir, E., & Eren, N. B. 2019. Nutritional status and related factors of schoolchildren in Çorum, Turkey. *Public health nutrition* 22(1): 122–131.

da Costa, M. P., Durão, C., Lopes, C., & Vilela, S. 2019. Adherence to a healthy eating index from pre-school to school age and its associations with sociodemographic and early life factors. *British journal of nutrition* 122(2): 220–230.

Departemen Kesehatan Republik Indonesia. 2012. Pedoman Geakan Masyarakat Sadar Gizi. Direktorat Gizi Masyarakat. Jakarta: Departemen Kesehatan Republik Indonesia.

Katona, P. & Katona-Apte, J. 2008. The interaction between nutrition and infection. *Clinical infectious diseases* 46(10): 1582–1588.

Kemenkes Republik Indonesia. 2018. *Ayo ke Posyandu Setiap Bulan*. Jakarta: Pusat Promosi Kesehatan Kemenkes RI.

Kementrian Kesehatan Republik Indonesia. Riset Kesehatan Dasar 2018. *Badan Penelitian dan Pengembangan Kesehatan*. Jakarta: Kementrian Kesehatan Republik Indonesia.

Magriplis, E., Farajian, P., Panagiotakos, D. B., Risvas, G., & Zampelas, A. 2019. The relationship between behavioral factors, weight status and a dietary pattern in primary school aged children: The GRECO study. *Clinical nutrition* 38(1): 310–316.

Merryana Adriani, S. K. M. & Kes, M. 2016. *Peranan Gizi Dalam Siklus Kehidupan*. Jakarta: Prenada Media.

Paramashanti, B. A. & Sulistyawati, S. 2019. Pengaruh integrasi intervensi gizi dan stimulasi tumbuh kembang terhadap peningkatan berat badan dan perkembangan balita kurus. *Jurnal Gizi Klinik Indonesia* 15(1): 16–21.

Sihotang, H. M. & Rahma, N. 2017. Faktor Penyebab Penurunan Kunjungan Bayi Di Posyandu Puskesmas Langsat Pekanbaru Tahun 2016. *Jurnal Endurance* 2(2): 168–177.

Spill, M. K., Callahan, E. H., Shapiro, M. J., Spahn, J. M., Wong, Y. P., Benjamin-Neelon, S. E., ... & Mennella, J. A. 2019. Caregiver feeding practices and child weight outcomes: A systematic review. *American journal of clinical nutrition* 109(Supplement_1): 990S–1002S.

Sugiyarti, R., Aprilia, V., & Hati, F. S. 2014. Kepatuhan Kunjungan Posyandu dan Status Gizi Balita di Posyandu Karangbendo Banguntapan, Bantul, Yogyakarta. *Jurnal Ners dan Kebidanan Indonesia* 2(3): 141–146.

Susanto, T., Yunanto, R. A., Rasny, H., Susumaningrum, L. A., & Nur, K. R. M. 2019. Promoting children's growth and development: A community-based cluster randomized controlled trial in rural areas of Indonesia. *Public health nursing* 36(4): 514–524.

UNICEF. 2012. *Ringkasan kajian kesehatan ibu dan anak*. Jakarta: UNICEF Indonesia, dilihat 12.

United Nations (UN). *Sustainable Development Goals Report 2018*. New York: United Nations.

UPT Cimenyan. 2017. *Buku Laporan Tahunan 2018*. Bandung: UPT Cimenyan.

Tutorials or *Santri* health pocketbooks: Which one is more effective?

W.F. Sanad, N.L. Rohmatika, D. Oktaviani, Y. Triyani & T. Respati
Universitas Islam Bandung, Bandung, Indonesia

ABSTRACT: The level of cleanliness, difficulty of accessing clean water, and density of dwellings are factors that drive the high incidence of scabies in developing countries. One factor contributing to the incidence of scabies, especially in Islamic boarding schools, is the lack of clean and healthy living behavior education programs that are most likely to help students avoid scabies. This study aimed to compare the effectiveness of clean and healthy living behavior tutorials with that of a "*santri* health pocketbook." This study was a quasi-experiment with pretest and posttest approaches. The results showed that the score of tutorials are the following: the pretest: 89.24, posttest: 89.26, while those of a *santri* health pocketbook are the following: pretest: 76.89 posttest: 89.63. These results indicate that the better method for increasing students' knowledge is the *santri* health pocketbook.

1 INTRODUCTION

The groups most vulnerable to scabies are children and adults at low economic levels and in densely populated areas. Scabies has a prevalence rate of 5–10% in children across the world (World Health Organization 2015). In Indonesia, scabies ranks third out of the 12 most common skin diseases (Nuraini & Wijayanti 2016). Scabies is usually transmitted from person to person through close skin contact with infected individuals. The level of cleanliness, difficulty accessing clean water, and density of dwellings are factors that drive the high numbers of cases of scabies in developing countries. Islamic boarding schools or *pesantren* are known for the problem of scabies, especially in Indonesia (Tresnasari et al. 2019; Triyani et al. 2019).

Factors contributing to the prevention of scabies include clean and healthy living behavior (CHLB) programs. The government set a target of 80% of households and communities in Indonesia practicing CHLB by 2019 (Kementrian Kesehatan Republik Indonesia 2019). The programs represent all personal and family health efforts and play an active role in community health. In line with one of the priorities of the Ministry of Health programs in 2019, healthy Islamic boarding schools are encouraged. Since the boarding school is a place where students spend most of their time, it is imperative that students have understanding in keeping themselves healthy and their living environment clean.

Health officials can use multiple approaches to convey their messages to students (Fikriah et al. 2019; Mayzufli et al. 2013). A study on CHLB programs compared two methods – *kasugi* card games and lectures. *Kasugi* card games proved to be the more effective method (Sutriyanto et al. 2016). However, *kasugi* card games require time and skills that not all schools have, especially Islamic boarding schools.

This study aimed to determine which of two methods of teaching CHLB – a "*santri* health pocketbook" and tutorials – is the most effective.

2 METHOD

This study was a quasi-experiment study using pretest and posttest approaches. Respondents in this study were all students who lived in *pesantren* from May to September 2019, so we used

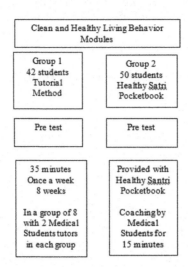

Figure 1. Chart of the processes.

total sampling. The students were divided into two groups with a first group of 42 students receiving tutorials and a second group of 50 students receiving a *santri* health pocketbook. Figure 1 is the chart of the processes.

The data presented are the mean difference in value from each answer. Statistical analyses were performed using SPSS version 19.

3 RESULTS AND DISCUSSION

In both groups, the data were not normally distributed so the Wilcoxon test was used.

In Tables 1 and 2, the results of the study found the pretest result of group A (who received the tutorials) was 89.24. There was an increase on the posttest to 89.26. Based on the results of statistical analysis, it can be seen that $p > 0.05$ ($p = 0.952$), which means that the tutorials had no significant effect on the level of knowledge. On the other hand, the pretest results of group B (who received the *santri* health pocketbooks) was 76.89. In the posttest, there was an improvement to 89.63. Based on the results of statistical analysis, it was found that $p < 0.05$ ($P = 0.012$), which means that the pocketbooks had a significant influence on the level of knowledge.

The results of this study showed an influence on the giving of CHLB tutorials and the giving of *santri* health pocketbooks, with an increase in the value of students measured

Table 1. Pretest and posttest results for the tutorial group.

Variable	Mean	N	Mean difference
Pretest	89.24	43	0.02
Posttest	89.26	43	

Table 2. Pretest and posttest results for the pocketbook group.

Variable	Mean	N	Mean difference
Pretest	76.89	24	12.74
Posttest	89.63	24	

through pretests and posttests. However, statistical analysis showed that the data were not normally distributed, because each method took weeks or months to complete and many students dropped out. This research was conducted just before Lebaran and there was a pause during Lebaran, so students who had left the *pesantren* could not be included because their data were incomplete. This issue was very influential on the tutorial method, which took 3 months, causing the Wilcoxon test results stating that the tutorials had no effect on increasing CHLB knowledge.

In this tutorial method, 12 groups were formed. Each group consisted of eight students with two tutors who were medical students of the Faculty of Medicine, Islamic University of Bandung. Each tutor was responsible for providing material and also sharing CHLB programs related to scabies. The material consisted of personal hygiene from head to toe, scabies screening, bedroom environment cleanliness, public toilet cleanliness, and cleanliness of the *pesantren* yard and waterways. The sessions were conducted in four meetings. At the first meeting, a pretest was conducted and at the last meeting, a posttest was conducted.

When the researchers gave the students the *santri* health pocketbooks, the data were also not normally distributed, but the Wilcoxon test results stated that the pocketbooks resulted in increased CHLB knowledge. The pocketbooks featured information about scabies including signs of scabies, ways of transmitting scabies, an explanation of scabies, and ways of preventing scabies. The research using this method was carried out for 1 month.

In 2018, scabies screening was conducted, and 88% of the students tested positive for scabies in the Manarul Huda *pesantren*. Various methods were used to reduce the incidence of scabies via the CHLB programs approach, including the tutorials and health pocketbooks. The results obtained in 2019 showed the incidence of scabies had decreased to 45%.

4 CONCLUSION

It can be concluded that providing *santri* health pocketbooks is the better method in comparison with the tutorials.

ACKNOWLEDGMENTS

This work was supported by a grant from Kemenristekdikti.

REFERENCES

Fikriah, N. L. N., Yulianto, F. A., Yusroh, Y., Irasanti, S. N., & Rosady, D. S. 2019. Perilaku Hidup Bersih dan Sehat dan Diare Akut di SMP Plus Pesantren Baiturrahman Bandung. *Jurnal Integrasi Kesehatan & Sains* 1(2): 170–173.

Kementrian Kesehatan Republik Indonesia. 2019. *Tingkatkan kesehatan santri, kemenkes bina pesantren sehat.*

Mayzufli, A., Respati, T., & Budiman, B. 2013. Pengetahuan, Sikap, dan Perilaku Mengenai Kesehatan Reproduksi Siswa SMA Swasta dan Madrasyah Alliyah. *Global medical & health communication* 1(2): 46–51.

Nuraini, N. & Wijayanti, R. A. (2016). Faktor Risiko Kejadian Scabies di Pondok Pesantren Nurul Islam Jember (Scabies risk factors in Pondok Pesantren Nurul Islam Jember). *Jurnal Ilmiah Inovasi* 16(2).

Sutriyanto, K., Raksanagara, A. S., & Wijaya, M. 2016. Pengaruh Permainan Kartu Kasugi terhadap Peningkatan Pengetahuan Perilaku Hidup Bersih dan Sehat pada Siswa. *Jurnal Sistem Kesehatan* 1(4).

Tresnasari, C., Respati, T., Maulida, M., Triyani, Y., Tejasari, M., Kharisma, Y., & Ismawati, I. 2019. Understanding scabies in religious boarding school (pesantren). In *Social and Humaniora Research Symposium (SoRes 2018)* 307: 520–522.

Triyani, Y., Yuniarti, L., Tejasari, M., Purbaningsih, W., Ismawati, I., & Respati, T. 2019. A journey to a better community service in religious boarding school pesantren. In *Social and Humaniora Research Symposium (SoRes 2018)* 307: 497–499.

World Health Organization. 2015. *Neglected tropical diseases: Scabies* 4–6.

Medical Technology and Environmental Health – Abdullah, Widiaty & Abdullah (eds)
© 2020 Taylor & Francis Group, London, ISBN 978-0-367-86053-0

Risk of musculoskeletal injury in the back area on small industrial workers

Y. Feriandi, B. Budiman, T. Respati & N. Romadhona
Universitas Islam Bandung, Bandung, Indonesia

ABSTRACT: The main production problem of the small industries is Manual Material Handling (MMH) that will raise risks of musculoskeletal injuries. This research was conducted to assess the risk of musculoskeletal injury among Small scale industry using the Ovako Working Posture Assessment System (OWAS) approach as a method of identifying the potential of musculoskeletal injuries. This is a case study of twelve workers from nine small industries in Bandung Indonesia. Ergofellow 3.0 software was used to measure OWAS scores. The results of the study showed that six of the workers were in category two, and the rest were in category three. Data showed that eight from twelve workers have score four, which is high risk on their back posture. These findings confirmed the high risk of musculoskeletal injury, especially diseases involving the back region. Stakeholder of the occupational health needs to take corrective action to minimize the risk of musculoskeletal injury.

1 INTRODUCTION

Ergonomics comes from the Greek word meaning "laws of work.". It is also described as 'fitting the task to man,' which means to fit the human abilities and limitations to the work activities, tools, equipment, and work environment (Jagadish et al. 2018). Manual material handling, especially lifting, is one of the significant health and safety hazards in industry (Dormohammadi et al. 2012). Musculoskeletal diseases are a spectrum of health problems related to work activities involving muscles, nerves, tendons, bones, joints, and cartilage (Ramadhani et al. 2018). One major cause of musculoskeletal diseases (MSDs) is unergonomic working postures while doing working activities. Sain & Mena (2016) concluded that musculoskeletal disorders are the most common injuries related to poor ergonomics.

The World Health Organization describes MSDs as "health problems of the locomotor apparatus, involving muscles, tendons, bone, cartilage, ligaments, and nerves. The symptoms of MSDs include a wide range of complaints, from slight transitory discomforts to irreversible and incapacitating injuries. (Gómez-galán et al. 2017) Work-related musculoskeletal disorders (WMSDs) are a major concern in the industry, including the micro, small, and medium enterprises (MSMEs). The OECD estimates that the proportion of small and medium enterprises account for 90% of firms and employ almost 63% of the workforce in the world (Pula & Berisha 2015). WMSDs has a high impact on workers' health, productivity, and economic outcome (Pheasant et al. 2019). Because of the growing number of repetitive motion injuries, as well as other disorders that are related to poor ergonomic design, employers must keep abreast of any problems in their companies that may continue to cause such injuries (Chad 2001).

The Ovako Working Posture Assessment System (OWAS) approach is a method of analyzing work posture and fast to identify the potential risk of musculoskeletal injuries. It identifies risky work postures that cause musculoskeletal diseases or injuries, such as low back pain, through analyzing and evaluate work postures used. O. Karhu from Finland founded this method in 1981. It has been used and developed for more than three decades to analyze postural stress in manual work conditions. This method identifies which body postures might be responsible for musculoskeletal problems. Its purpose is to assess the poor ergonomic postures of workers during work to improve working performance (Ramadhani et al. 2018).

The OWAS method is one of the best tools to correct poor ergonomic posture by fixing the back position of workers. The results identify the most common work postures for the back (four postures), arms (three postures), and legs (seven postures), and the weight of the load handled (three categories) and indicates whether postures are ideal or if there is a need for corrective measures (Gómez-galán et al. 2017, Sarkar et al. 2016).

2 METHOD

This research is a cross-sectional descriptive case study of twelve workers from different work processes in nine different small-scale industries in Bandung, West Java Indonesia. The observed workers have been working for more than five years with manual material handling working conditions. The researcher performed an interview and recorded the work activities through direct observation and photographs documentation of the work process.

We used the Ergofellow® software version 3.0 to calculate the OWAS score. Final OWAS results were grouped into four categories: Category 1: no corrective measures required; Category 2: corrective measures required in the near future; Category 3: corrective measures required as soon as possible; and Category 4: corrective (Sarkar et al. 2016). The working attitudes were coded on the back, hands, feet, and weight of the load based on the history and photos of work attitude. Data were exported into Microsoft excel to be analyzed descriptively.

3 RESULTS AND DISCUSSION

The worker conducts their working process in a poor ergonomic circumstance. During working, the worker tends to choose the most personally-suitable position to perform their tasks without complying to the ergonomic principles as described by the Health and Safety Authority Headquarters, Dublin (2018) such as neutral position of the joints, keep work close to the body, avoid bending forward, avoid twisted posture of the back, avoid static posture for long period, weight limit, and avoid carrying heavy loads with one hands.

Table 1 shows the possible risk of injury to each part of the body caused by work position and does not pay attention to ergonomic devices or tools. The results of the study showed that six of the workers were in category 2, and the rest were in category 3. Table 1 showed that eight from twelve workers have score four, which is categorized as high risk on their back posture. These findings confirmed the high risk of musculoskeletal injury in small scale industry, especially diseases involving the back region. Other

Table 1. OWAS score and category.

No	Back	Arms	Legs	Load	Result	Category
1	4	1	1	1	Slightly harmful	2
2	2	1	2	1	Slightly harmful	2
3	4	2	2	2	Distinctly harmful	3
4	2	2	2	1	Slightly harmful	2
5	2	2	3	1	Slightly harmful	2
6	4	1	3	2	Slightly harmful	2
7	2	1	7	2	Distinctly harmful	3
8	4	1	3	3	Distinctly harmful	3
9	4	2	2	2	Distinctly harmful	3
10	4	2	2	1	Slightly harmful	2
11	4	2	7	2	Distinctly harmful	3
12	4	2	6	1	Distinctly harmful	3

poor ergonomic postures of the back studied from this case study are bending forward of the back at neck and lower part of the back, over-rotation, and repetitive movement of the waist. These postures contribute to the high OWAS score and become further main risks for back pain.

The most frequent injuries related to poor ergonomics are known as MSDs (Musculo-skeletal Disorders), also called RMIs (Repetitive Motion Injuries) or CTDs (Cumulative Trauma Disorders). Musculoskeletal disorders (MSDs) are impairments of body structures such as muscles, joints, tendons, ligaments, nerves, bones, and the localized blood circulation system. These include tendonitis, carpal tunnel syndrome, lower back pain, and similar afflictions. These injuries result from repeated motions and exertions over a period of time and can affect many parts of the body in many different ways (Chad 2001, Wami et al. 2019).

Moreover, back pain is defined as chronic or acute pain, aches, or trouble in the lumbar or buttock area, sometimes called lumbago, or in the upper leg region, which is a major work-related disorder in almost all physically demanding jobs. Non-specific low back pain is the highest morbidity among manual handling workers.

Low Back Pain (LBP) is a common problem among workers. Many preventable risk factors have been identified, such as back trauma, Obesity, and lack of physical exercise. Research in static sitting posture performed by Bontrup et al. (2019) show a possible trend towards more static sitting behavior among workers with chronic LBP pain and pain-related disability (Alnaami et al. 2019, Bontrup et al. 2019). Sitting in a static position during work time could be one of the main focus for the workers in the small industry. The industry should have encouraged and reminds the workers to perform a 'break' relaxation regularly to prevent static posture.

To prevent back diseases related to poor ergonomic manual material handling work processes, an attempt to control these diseases should focus on assessing and reengineering the tasks of manual material handling and tools/machines used. In a developing country like Indonesia, the problem of work-related musculoskeletal disorders are extremely serious, and the working environment and behavior of the workers are not as good as and have many differences with developed countries (Dormohammadi et al. 2012).

The employer should perform policy on regular medical check-up for their workers to identify the most severe back diseases or, as we know, as red flags in back pain. Kuiper et al., through the Joint Programme for Working Life Research in Europe (SALTSA) from the Coronel Institute of Occupational Health the Netherlands, have formulated follow up tools to assess the relatedness of non-specific back pain among workers. The relatedness tools can be combined with Ergonomic measurement methods such as OWAS to enhance the input for more appropriate case management (Kuiper et al. 2004).

4 CONCLUSION

This finding confirmed the high risk of musculoskeletal injury in the small-scale industry. There is a balanced proportion between the slightly harmful and distinctly harmful groups. The back is the most impoverished ergonomic affected body region. Half of the cases have a high score on the back posture. There was variation in the Leg Score, which is influenced by the type and mobility of the tasks. Two other parameters (arms and load) were quite similar among all cases. This finding shows that the future risk of back pain disease is a serious risk of small-scale industry workers. Stakeholders of the occupational safety and health need to take corrective action to minimize the risk of musculoskeletal injuries such as generating safety and health policies, training, and development in ergonomic and manual material handling for the employer of SSIs especially modify the risk for the back working posture and hierarchy control of the manual lifting of heavy loads. Ergonomic intervention in SSIs reduces MSDs among workers. The well-being of workers would increase productivity, revenue and reduce rejection cost, which would greatly help the economy of a country.

ACKNOWLEDGEMENT

We acknowledge the head of Bandung Health District Office for the permission of this research. This study was sponsored by The Research Unit of Medical Faculty Universitas Islam Bandung.

REFERENCES

Alnaami, I., N.J. Awadalla, M. Alkhairy, S. Alburidy, A. Alqarni, A. Algarni, R. Alshehri, B. Amrah, M. Alasmari, and A.A. Mahfouz. 2019. Prevalence and Factors Associated with Low Back Pain among Health Care Workers in Southwestern Saudi Arabia. *BMC Musculoskeletal Disorders* 5: 1–7.

Bontrup, C., W.R. Taylor, M. Fliesser, R. Visscher, T. Green, P. Wippert, and R. Zemp. 2019. Low Back Pain and Its Relationship with Sitting Behaviour among Sedentary Office Workers. *Applied Ergonomics* 81: 102894.

Chad, C. 2001. New Developments in Ergonomics. Ed. Kleiner Brian H. *Management Research News* 24 (3/4): 114–117.

Dormohammadi, A., H. Amjad-Sardrudi, M. Motamedzade, R. Dormohammadi, and S. Musavi. 2012. Ergonomics Intervention in a Tile Industry: A Case of Manual Material Handling. *Journal of Research in Health Sciences* 12(2): 109–113.

Gómez-galán, M., J. Pérez-alonso, A. Callejon-Ferre, and J. Lopez-Martinez. 2017. Musculoskeletal Disorders: OWAS Review. *Industiral Health* 55: 314–337.

Health and Safety Authority Headquarters. 2018. *Ergonomics in the Workplace.* Dublin: Headquarters, Head and Safety Authority.

Jagadish, R., A. Ansari, S. Quraishi, A. Sultana, and S.M. Qutubuddin. 2018. Ergonomic Risk Assessment of Working Postures in Small Scale Industries. *Grenze International Journal of Engineering and Technologyy*no.

Kuiper, J.I., A. Burdof, M.H.W. Frings-Dresen, P.P.F.M. Kuijer, F. Lotters, D. Spreeuwers, and H. S. Miedema. 2004. *Criteria for Determining the Work-Relatedness of Nonspesific Low-Back Pain.* Ed. Monique H.W. Frings-Dresen. Coronel Institute of Occupational Health.

Pheasant, S., C.M. Haslegrave, S. Pheasant, and C.M. Haslegrave. 2019. Introduction to Ergonomic Design. In *Bodyspace*, 3–15.

Pula, J.S., and G. Berisha. 2015. Defining Small and Medium Enterprises: A Critical Review. *Academic Journal of Business, Administration, Law and Social Science* 1(1): 16–28.

Ramadhani, M., Rukman, D. Prayogo, and D.A. D.P. 2018. Assessment Analysis of Ergonomics Work Posture on Wheel Installation With Ovako Work Posture Analysis System (OWAS) Method and Rapid Entire Body Assesment (REBA) Method Preventing Musculoskeleal Disorders at Perum PPD Jakarta. *IOSR Journal Of Humanities And Social Science (IOSR-JHSS)* 23(10): 1–11.

Sain, M.K., and M.L. Meena. 2016. Occupational Health and Ergonomic Intervention in Indian Small Scale Industries: A Review. *International Journal of Recent Advances in Mechanical Engineering (IJMECH)* 5(1): 13–24.

Sarkar, K., S. Dev, T. Das, S. Chakrabarty, and S. Gangopadhyay. 2016. Examination of Postures and Frequency of Musculoskeletal Disorders among Manual Workers in Calcutta, India. *International Journal of Occupational and Environmental Health* 22(2): 151–158.

Wami, S.D., G. Abere, A. Dessie, and D. Getachew. 2019. Work-Related Risk Factors and the Prevalence of Low Back Pain among Low Wage Workers: Results from a Cross-Sectional Study. *BMC Public Health* 1–9.

Medical Technology and Environmental Health – Abdullah, Widiaty & Abdullah (eds)
© *2020 Taylor & Francis Group, London, ISBN 978-0-367-86053-0*

Initial gamification project to increase mental health awareness for Indonesian youth

E. Nurhayati, T. Respati, F.A. Yulianto, B. Budiman & Y. Feriandi
Universitas Islam Bandung, Bandung, Indonesia

E. Nugroho & A. Shandriasti
Kummara, Bandung, Indonesia

ABSTRACT: This paper presents a research design and the preliminary results of our initiatives to explore a gamification approach to increasing mental health awareness among Indonesian youth. Considering the current knowledge about mental health among youth in Indonesia, this study proposed a mixed method of research using both qualitative and quantitative approaches. This approach is believed to best explain the stigma o people with mental health disorders among youth.

1 INTRODUCTION

The World Health Organization defines mental health as "a state of wellbeing in which every individual realizes his or her own potential, can cope with the normal stresses of life, can work productively and fruitfully, and is able to make a contribution to her or his community" (World Health Organization 2011). Mental health disorder refers to a broad range of disorders that affect a person's personality, thought processes, or social interactions. Unlike physical illnesses, they can be difficult to clearly diagnose.

According to the Basic Health Research Report of the Ministry of Health of the Republic of Indonesia (Kementrian Kesehatan Republik Indonesia 2013), in 2013, the prevalence of emotional mental illness in Indonesia was indicated by depression and anxiety symptoms in 14 million people (6%) of the 15+ age group. Meanwhile, the prevalence of severe mental illness, such as schizophrenia, was 1.7 per 1,000 population, or about 400,000 people. Out of these numbers, 57,000 people (1.43%) have been shackled. The proportion of shackling in rural areas is 18.2% while in urban areas it is only 10.7%. Unfortunately, the extent to which Indonesian society was prejudicial against people with mental disorders varied, ranging from positive to negative attitudes. Societal stigma, fear, and ignorance about mental disorders form barriers between those who live with such disorders and the rest of society. Evidence is growing that public stigma against people with mental health disorders (PWMD) is one of the most significant barriers to their well-being.

The most harmful effects of the stigma that may occur are (1) barriers to community engagement and social relationships such as rejection and mistreatment by friends and family; (2) avoidance of the diagnosis and corresponding treatment; (3) work, school, or housing discrimination that causes poor outcomes; (4) internalization of prejudicial attitudes leading to self-stigma, shame, and losing self-confidence; and (5) decrease in self-esteem thereby undermining well-being and meaning in life; (6) increase possibility to get involved in substance abuse and criminal activity (Corrigan & Rao 2012; Gruttadaro & Crudo 2012; Kakuma et al. 2010; Livingston & Boyd 2010; Salzer 2012).

In order to address this issue, the government of Indonesia has implemented a number of initiatives such as (1) implementing a comprehensive, integrated, and sustained mental health-care services for society; (2) providing facilities, infrastructure, and resources for mental healthcare services across Indonesia, including medicine, medical devices, and trained health and non-health workers; (3) encouraging society to perform preventive and promotional efforts and early detection of mental illness and to undertake rehabilitation efforts that include the reintegration of PWMD in society; and (4) passing Law No. 18/2014 on mental health, approved on August 8, 2014. These initiatives aim to ensure each person can achieve good quality of life, as well as providing integrated, comprehensive, and sustained healthcare services through promotion, prevention, cures, and rehabilitation.

More recently, a global shift has occurred on promoting awareness of mental health stigmas to youth for several reasons: (1) the youth population has steadily increased; (2) youth of today are generally healthier, better educated, and more urbanized, enjoy greater access to knowledge, and are more connected with the rest of the world than the preceding generations; (3) youth has an enormous influence in the future; and (4) youth is a vulnerable group that needs to be educate about mental health. Although no universal standard exists for defining the concept of youth or the age range to which youth pertains, Indonesian Law No. 40/2009 on Youth defines "youth" as persons in the 16–30-year age range.

Consequently, information technology has become important for the delivery of awareness messages to youth. Fortunately, evidence suggests that the widespread deployment of technology in healthcare services provides a substantial opportunity for health promotion among youth or teenagers.

Although research into stigma and discrimination has resulted in a plethora of publications, most of those studies involve societies from high-income countries and, to the best of our knowledge, scholars have said little about addressing this problem in Indonesia. Following a method proposed by Brown and colleagues (2016), this research aimed as a preliminary study to increase mental health awareness among youth in Indonesia using gamification as the main media.

The guiding question of this research was whether a gamification or creative educational game implemented in a classroom setting will change the stigmas of youth toward people with mental illness.

2 METHODS

The research process began with quantitative methods by gathering pretest and posttest data to measure stigma and discrimination against PWMD among youth using a questionnaire. This first step also involved a policy and literature review. All the data gathered in the first step provided initial information and measurements of stigma and discrimination against PWMD.

The next steps involved qualitative methods using focus group discussion (FGD) to obtain in-depth and wider information about background, underlying issues, and determinants of stigma and discrimination against PWMD. The results were combined with a policy and literature review. This step produced the final measurement of stigma and discrimination against PWMD.

The process then included developing tools to measure the knowledge, attitude, and behavior of stigma and discrimination against PWMD. This process also involved a policy and literature review to provide the final product, which was educational tools such as gamification that increase mental health awareness.

2.1 Respondents

The respondents in this research comprised senior high school students in three different public schools.

2.2 Sampling technique

Frame samples were taken from each high school, consisting of students from the first to the third grades. Stratified random sampling was applied to the subjects due to the sheer number

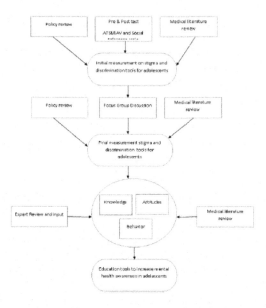

Figure 1. Research framework.

of students at each school. Chosen participants from each school were treated as a control or a treatment based on a randomly allocated cluster definition. The research framework can be seen in Figure 1.

2.3 Data-collecting instrument

The tools used in this research included pretest and posttest questionnaires developed by Indonesian mental health experts. The questionnaire was tested for its validity and reliability in a preliminary study in a group with similar characteristic (different high school). Content and construct validity were created in cooperation between researchers and the Indonesia National Health Institute, continued by testing the instruments, and analyzed by Pearson's correlation score. Reliability was assessed after the second pooling, with Kappa's coefficient agreement as the analyzing tool. Questions with poor agreement were edited and tested again.

2.4 Data processing and analysis

All the data collected were analyzed to compare the stigma and discrimination against PWMD before and after the interventions in the second and third groups. Then this comparison was compared again with the control group, which had no intervention at all. Questionnaire scores were tested for distribution by Shapiro-Wilk's test on each group of intervention and on each pretest or posttest. If normal distribution assumptions were met, repeated ANOVA were applied to see the difference within and between group treatments.

3 PRELIMINARY RESEARCH RESULT AND DISCUSSION

Gamification intervention has been considered to have a positive impact on health behavior. It may increase interest and improve the experience of the user (Brown et al. 2016). Brown and colleagues (2016) studied the role of gamification features on adherence to web-based mental health interventions. The research was done by conducting peer-reviewed randomized controlled trials (RCTs) designed to manage mental disorder.

The research reviewed 8 of 10 popular gamification features designed to manage depressive disorder using cognitive behavioral therapy. It showed that users commonly used one feature with a maximum of three features. The research also revealed that the higher the duration of intervention, the greater the adherence.

Another study, conducted by Pinfold and colleagues (2003), showed that educational workshops on mental health had a small yet positive impact on students' point of view on PWMD. The study was done by giving health awareness workshop to 472 secondary school students.

All the respondents were asked to complete pre and post questionnaires to measure mental health knowledge, attitudes, and behaviors. The workshop was done twice. The first one was delivered by mental health professionals and supported by leaflets, while the second workshop was done by persons who had experiences of mental health disorders.

4 CONCLUSION

This research chose gamification as a mode of intervention to increase mental health awareness in Indonesian youth. This method is considered effective in reducing stigma and discrimination against PWMD among youth due to enjoyment and interest. However, researchers should compare gamification and education to measure the most effective mode of interventions.

ACKNOWLEDGMENT

We acknowledge the Research Unit of Faculty Medicine, the Islamic University of Bandung (UNISBA), Indonesia, and the Institute of Mental Health Singapore for supporting this research.

REFERENCES

Brown, M., O'Neill, N., van Woerden, H., Eslambolchilar, P., Jones, M., & John, A. 2016. Gamification and adherence to web-based mental health interventions: A systematic review. *JMIR mental health* 3(3): 1–15.

Corrigan, P. W. & Rao, D. 2012. On the self-stigma of mental illness: Stages, disclosure, and strategies for change. *Canadian Journal of Psychiatry* 7(8): 464–469.

Gruttadaro, D. & Crudo, D. 2012. *College students speak: A survey report on mental health.* Virginia: National Alliance on Mental Illness, 1–24.

Kakuma, R., Kleintjes, S., Lund, C., Drew, N., Green, A., & Flisher, A. J. 2010. Mental health stigma: What is being done to raise awareness and reduce stigma in South Africa? *African Journal of Psychiatry* 13(2): 116–24.

Kementerian Kesehatan Republik Indonesia. 2013. *Riset Kesehatan Dasar.* Jakarta: Badan Penelitian dan Pengembangan Kesehatan Departemen Kesehatan Republik Indonesia.

Livingston, J. D. & Boyd, J. E. 2010. Correlates and consequences of internalized stigma for people living with mental illness: A systematic review and meta-analysis. *Social science & medicine* 71(12): 2150–2161.

Pinfold, V., Toulmin, H., Thornicroft, G., Huxley, P., Farmer, P., & Graham, T. 2003. Reducing psychiatric stigma and discrimination: Evaluation of educational interventions in UK secondary schools. *British Journal of Psychiatry* 182(4): 342–346.

Salzer, M. S. 2012. A comparative study of campus experiences of college students with mental illnesses versus a general college sample. *Journal of American College Health* 60(1): 1–7.

World Health Organization. *Mental health atlas 2011.* Geneva: WHO Press. Retrieved from http://whqlib doc.who.int/publications/2011/9799241564359eng.pdf?ua=1.

Curriculum development and evaluation

Medical Technology and Environmental Health – Abdullah, Widiaty & Abdullah (eds)
© 2020 Taylor & Francis Group, London, ISBN 978-0-367-86053-0

Intrinsic factors in learning success and passing the Computer-Based Test (CBT) of competency of graduate medical program (UKMPPD)

B. Budiman, M. Kusmiati, C. Tresnasari, C.C. Supriadi, L.D. Mulyani & R.A. Nurmaini
Universitas Islam Bandung, Bandung, Indonesia

ABSTRACT: The Uji Kompetensi Mahasiswa Program Profesi Dokter (UKMPPD) or the Test of Competency Graduate Medical Program is a national board that determines the recognition of a medical student as a doctor. Theoretically, passing a test is determined by intrinsic and extrinsic factors. Among them are the ability to think metacognitively, self-reflection, and learning approaches. This study aimed to determine the association between metacognitive thinking ability, self-reflection, and learning approaches and passing the computer-based test (CBT) of the UKMPPD. Respondents in this study comprised 107 students of the Medical Education Study Program (PSPD) at the professional stage in one of the medical faculties in West Java who underwent the UKMPPD in May 2019. Data collection was carried out via a questionnaire containing questions about the ability to think metacognitively, self-reflection, and learning approaches administered through Google Forms. The UKMPPD CBT results were collected in the May 2019 period. The design used in this study was a cohort study. Data analysis was performed by a chi-squared test. The results showed no significant association between metacognitive thinking ability, self-reflection, and learning approaches and passing the UKMPPD CBT. The ability to think metacognitively, self-reflection, and learning approaches still need to be improved to increase students' learning abilities.

1 INTRODUCTION

The Competency Standards of Indonesian Doctors (SKDI) are prepared by the government to ensure the competence of all medical graduates to meet the expectations of the community. After medical students successfully pass all examinations in the academic and clinical phases, they must take a national examination (Holmboe et al. 2010). The purpose of the government in carrying out this national exam is to ensure that graduates of all medical faculties in Indonesia are qualified for medical practice in accordance with the SKDI. The national medical exam conducted by the government is called the Test of Competency Graduate Medical Program (UKMPPD), which serves as an exit exam. The UKMPPD is carried out through computer-based tests (CBT) and objective structural clinical examinations (OSCE).

Whether students pass the UKMPPD is certainly influenced by their success in the learning process. Learning success is influenced by various internal and external factors (Chisholm et al. 1995). Among the various internal factors are the ability to think metacognitively (Schraw & Dennison 1994), self-reflection, and learning approaches (Cox et al. 2006). Faculty can take into consideration knowledge about the magnitude of the influence of various factors affecting passing the UKMPPD when determining the priority of planned intervention actions (Duncan & McKeachie 2005). The hope is to improve the optimum conditions for every student that can support the success of learning, which ultimately also supports success in passing the UKMPPD.

Based on this background, the formulation of the problem in this study was whether there is an association between metacognitive thinking, student reflection, and learning approaches and passing the UKMPPD.

2 METHODS

Respondents in this study included 107 students of the Medical Education Study Program (PSPD) at the professional stage in one of the medical faculties in West Java who underwent the UKMPPD in May 2019. The students were informed about the objective of the study, and verbal consent was obtained from them. Data collection was carried out via a questionnaire containing questions about the ability to think metacognitively, self-reflection, and learning approaches administered through Google Forms. The questionnaire was adapted from the Metacognitive Awareness Inventory, a modification of the Motivated Strategies for Learning Questionnaire (Soemantri et al. 2018), and the Revised Two Factor Study Process Questionnaire (R-SPQ-2F). The questionnaire included 52 questions on metacognitive thinking (Schraw & Dennison 1994), 23 questions on students' reflection (Duncan & McKeachie 2005), and 10 questions on learning approaches (Struyven et al. 2005). The UKMPPD CBT results were collected in the May 2019 period. The design used in this study was a cohort study. Data were analyzed with a chi-squared test.

3 RESULTS AND DISCUSSION

3.1 *Association between metacognitive ability and passing the UKMPPD CBT exam*

Chi-squared test results on the association of metacognitive abilities with the UKMPPD CBT exam can be seen in Table 1.

The effect of metacognitive abilities on passing the UKMPPD CBT in the low, medium, and high categories was not very different. The analysis showed no significant association between metacognitive abilities and success on the UKMPPD CBT exam, with a value of $p = 0.731$ ($p > 0.05$). This is likely due to other factors that influence passing the final exam, besides that metacognitive abilities are more reflective of self-regulation and management of ways of thinking (Lai 2011).

The results of this study are in line with those in the research conducted by Alaka Das in 2017 at Pub Camrup College, India, which indicated that metacognitive abilities are not related to academic achievement, in this case passing the final exam (Das 2017). Metacognitive abilities better reflect cognitive processes in managing thought processes, so they are not directly related to academic performance or success in passing the final exam (Soemantri et al. 2018).

3.2 *Association between student reflection capability and passing the UKMPPD exam*

Chi-squared test results on the association between student reflection and passing the UKMPPD CBT exam can be seen in Table 2.

In the group of students with high self-reflection ability, 10% had passed the UKMPPD CBT. This figure is higher when compared to groups of students with low and moderate self-reflection abilities. However, this percentage difference is not statistically significant because

Table 1. Association between metacognitive ability and passing the UKMPPD CBT exam.

	UKMPPD CBT			
	Fail n (%)	Pass n (%)	Total n (%)	P-value
Metacognitive				
Low	3 (9.4)	29 (90.6)	32 (100)	0.731
Medium	7 (12.5)	49 (87.5)	56 (100)	
High	3 (15.8)	16 (84.2)	19 (100)	

Table 2. Association between student reflection ability and passing the UKMPPD CBT exam.

	UKMPPD CBT			
	Fail n (%)	Pass n (%)	Total n (%)	P-value
Student reflection				
Low	6 (9.8)	55 (90.2)	61 (100)	0.427
Medium	7 (17.9)	32 (82.1)	39 (100)	
High	0 (0.0)	7 (100.0)	7 (100)	

the percentage difference is not too far away. Based on the results of the analysis, it was concluded that there was no significant association between the ability of student reflection with the results of the UKMPPD CBT exam, with a p-value = 0.427 (p > 0.05). This is likely due to student reflection better reflecting the intellectual and affective activities of individual learners to engage in exploring their experiences to lead to new understanding and appreciation (Struyven et al. 2005). Therefore, it puts more focus on new understanding and appreciation and is not directly related to achievement or success on the final exam.

The results of this study are not in line with the results of research from Devi and colleagues (2012) at Manipal University in India, which states that the lower the student reflection ability, the higher the failure in the final block test (Devi et al., 2012). There is a significant difference in the ability of reflection in students who fail the exam with first-grade students with an average score above 65%. In this research, the candidates for doctors who graduated in several levels of achievement were not categorized, only divided into groups of graduated and not graduated. The categorization of prospective doctors who passed the UKMPPD CBT exam therefore needs to be elaborated.

In a study conducted at Abulyatama University in 2016, UKMPPD graduation was influenced by many factors, including the characteristics of students as adult learners, learning material during educational programs, teaching processes, resources, and evaluation of learning outcomes (Andriaty et al. 2016). Nonacademic factors that affect UKMPPD graduation included power of concentration, examination methods, and learning environment (Franklin & Melville 2015). This study did not consider the characteristics of learners or external factors, so that the ability of student reflection is already good, but at the time of the exam other factors can affect students, such as lack of concentration, illness, and the presence of anxiety.

3.3 Association of learning approaches with passing the UKMPPD CBT exam

Chi-squared test results on the association of learning approaches with passing the UKMPPD CBT exam can be seen in Table 3.

Table 3. Association between learning approaches and passing the UKMPPD CBT exam.

	UKMPPD CBT			
	Fail n (%)	Pass n (%)	Total n (%)	P value
Learning Approach				
Low	1 (33.3)	2 (66.7)	3 (100)	0.251
Medium	10 (13.5)	64 (86.5)	74 (100)	
High	2 (6.7)	28 (93.3)	30 (100)	

Student graduation with a high learning approach is 93.3% higher than that of students with a low or moderate learning approach. However, no significant association was found between learning approach and success on the UKMPPD CBT exam, with a p-value = 0.251 (p > 0.05). This is likely due to various factors that influence success on the UKMPPD CBT exam, not just the learning approach. In addition, the learning approach better reflects the ways individuals engage in a learning activity and is not directly related to performance on a final exam (Struyven et al. 2005).

This is in line with research conducted by Lisiswati and colleagues (2015) at the University of Lampung in 2015, which concluded that the deep learning approach is not related to a passing score on the "learning skills and professionalism" test blocks, but the graduation rates for students who used a deep approach was higher than for those who used a superficial approach (Lisiswati et al. 2015).

4 CONCLUSION

Based on the results of this study, the researcher concludes that there is no significant association between metacognitive skills, self-reflection, and learning approaches and passing the UKMPPD CBT.

Further analysis is needed by categorizing the results of the UKMPPD CBT into several levels of achievement to see more closely the association between metacognitive abilities, self-reflection, and learning approaches (McCune et al. 2004). Qualitative data collection is needed to explore more deeply the factors that most contribute to UKMPPD graduation. Mixed-method quantitative and qualitative data analysis will enhance subsequent research results.

ACKNOWLEDGMENTS

We thank the research and public services organizations of the Islamic University of Bandung (Unisba) for a research grant. We would also like to thank the students who participated in this research.

REFERENCES

Andriaty, S. N., Findyartini, A., & Werdhani, R. A. 2016. Studi Eksplorasi Kemungkinan Penyebab Kegagalan Mahasiswa dalam Uji Kompetensi Dokter Indonesia, Studi Kasus di Fakultas Kedokteran Universitas Abulyatama. *Serambi PTK* 3(2): 1–12.
Chisholm, M. A., Cobb, H. H., & Kotzan, J. A. 1995. Significant factors for predicting academic success of first-year pharmacy students. *American journal of pharmaceutical education* 59(4): 364–370.
Cox, W. C., Smith, H., & Blalock, S. J. 2012. Predictors of academic difficulty in a doctor of pharmacy program. *International journal of pharmacy education & practice* 9(1):1–7.
Das, A. 2015. Relationship between metacognitive ability and academic achievement of B. Ed. students: A study. *International journal of science and research* 6(5): 1639–1642.
Devi, V., Mandal, T., Kodidela, S., & Pallath, V. 2012. Integrating students' reflection-in-learning and examination performance as a method for providing educational feedback. Journal of postgraduate medicine, 58(4),270–274.
Duncan, T. G. & McKeachie, W. J. 2005. The making of the Motivated Strategies for Learning Questionnaire. *Educational psychologist* 40(2): 117–128.
Franklin, N. & Melville, P. 2015. Competency assessment tools: An exploration of the pedagogical issues facing competency assessment for nurses in the clinical environment. *Collegian* 22(1): 25–31.
Holmboe, E. S., Sherbino, J., Long, D. M., Swing, S. R., Frank, J. R., & International CBME Collaborators. 2010. The role of assessment in competency-based medical education. *Medical teacher* 32(8): 676–682.
Lai, E. R. 2011. Metacognition: A literature review. *Always learning: Pearson research report* 24: 1–40.

Lisiswanti, R., Saputra, O., Carolia, N., & Malik, M. M. 2015. Hubungan pendekatan belajar dan hasil belajar mahasiswa Fakultas Kedokteran Universitas Lampung. *Jurnal Kedokteran dan Kesehatan: Publikasi Ilmiah Fakultas Kedokteran Universitas Sriwijaya* 2(1): 79–84.

McCune, V., & Entwistle, N. 2000. The deep approach to learning: Analytic abstraction and idiosyncratic development. In *Innovations in Higher Education Conference*. Helsinki: University of Helsinki, 30: 1–17.

Schraw, G., & Dennison, R. S. 1994. Assessing metacognitive awareness. *Contemporary educational psychology* 19(4): 460–475.

Soemantri, D., Mccoll, G., & Dodds, A. 2018. Measuring medical students' reflection on their learning: Modification and validation of the Motivated Strategies for Learning Questionnaire (MSLQ). *BMC medical education* 18(274): 1–10.

Struyven, K., Dochy, F., & Janssens, S. 2005. Students' perceptions about evaluation and assessment in higher education: A review. *Assessment & evaluation in higher education* 30(4): 325–341.

Drug discoveries and development

Medical Technology and Environmental Health – Abdullah, Widiaty & Abdullah (eds)
© 2020 Taylor & Francis Group, London, ISBN 978-0-367-86053-0

A biomechanical study using porcine knees for posterior root medial meniscus repair using arthroscopic direct meniscal extrusion reduction surgery

R. Prasetia, R. Priscilla, G.A. Utoyo & H.N. Rasyid
Hasan Sadikin General Hospital, Bandung, Indonesia
Universitas Padjadjaran, Bandung, Indonesia

R. Aditya
Universitas Padjadjaran, Bandung, Indonesia

ABSTRACT: In meniscal extrusion (ME), the meniscus cannot play its biomechanical role in load distribution and force absorption by dispersing tibiofemoral axial load. Significant ME will increase contact forces that contribute to the advancement of knee osteoarthritis. Arthroscopic direct meniscal extrusion reduction (ADMER) to reduce persistent ME in posterior root medial meniscus (PRMM) repair has been recommended and showed a promising outcome. The aim of this study was to compare the biomechanical properties of conventional PRMM repair with PRMM repair + ADMER technique. A total of eighteen (18) porcine tibia with attached intact medial meniscus were used and the specimens were randomly assigned to 1 of 3 groups (n = 6 each). The groups included: Control group (C) with healthy medial meniscus, conventional PRMM repair (P) group, and PRMM repair with ADMER technique (A) group. All specimens underwent testing for static pull-out strength with a servo-hydraulic material testing machine. Means and standard deviations of pullout strength were compared. Mean pull-out strength was 1047.19 N (±64.4) for the control group, 256.41 N (±54.26) for C group and 367.48 N (±40.5) for ADMER group. Maximum pull-out strength for ADMER was significantly higher compared to C group (p-value 0.006). PRMM Repair combined with ADMER technique had superior pullout strength compared to conventional PRMM repair. However, both techniques did not reach the strength of the native meniscus.

1 INTRODUCTION

The human meniscus is two semi-circular structures act as load sharers and shock absorbers in the knee joint and responsible for reducing axial load to the knee cartilage. Both lateral and medial meniscus play an important role to carry approximately 40%-60% of body weight. (Li) In meniscal extrusion (ME), the meniscus cannot play its biomechanical role in load distribution and force absorption by dispersing the tibiofemoral axial load. Significant ME will increase contact forces and decrease contact surface areas that contribute to the advancement of knee osteoarthritis (McDermott 2011). A systematic review showed only 56% of reduced meniscal extrusion were obtained after meniscal root repair. On the second look, the healing status was completed in only 62% of cases, 34% was healed partially, and 3% showed failed healing (Li et al. 2019).

Arthroscopic direct meniscal extrusion reduction (ADMER) to reduce persistent ME in posterior root medial meniscus (PRMM) repair has been recommended and showed a promising outcome (Moatshe et al. 2016). However, according to our knowledge, we have not found any study about the biomechanical properties of both conventional repair and ADMER technique. The aim of this study was to compare the biomechanical properties of conventional PRMM repair with PRMM repair + ADMER technique.

2 MATERIALS AND METHODS

2.1 *Specimen preparation*

A total of eighteen (18) porcine tibia with attached intact medial meniscus were used and the specimens were randomly assigned to 1 of 3 groups (n = 6 each). The groups included: Control group (C) with healthy medial meniscus, conventional PRMM repair (P) group, and PRMM repair with ADMER technique (AD) group. The knees were harvested and stored at -20° and thawed for 12h at room temperature before testing. The tibia was disconnected from the femur continued with the removal of the soft tissues, only the meniscus was left attached (Figure 1).

2.2 *Surgical technique*

2.2.1 *Conventional PRMM repair*
A PRMM tear model was created by cutting the peripheral margin of the medial meniscus root area with approximately 5mm gap left near the insertion using a no. 21 blade scalpel. The anatomical PRMM attachment was identified and a transtibial tunnel was drilled using a 2.5 mm diameter guide pin from the mid-anterior aspect of the proximal tibia through the meniscal root anatomical attachment, followed by reaming of the tibial tunnel with an Endo-button 4.5 mm drill. Sutures in the root of the medial meniscus was done using no. 0 Ethibond and pulled out to the transtibial tunnel (Figure 2).

2.2.2 *PRMM repair with ADMER technique*
A PRMM tear model was created by cutting the peripheral margin of the medial meniscus root area with approximately 5mm gap left near the insertion using a no. 21 blade scalpel. A 2.5 mm diameter guide pin was drilled from the anteromedial aspect of the proximal tibia through a 2 mm medial to the medial margin of tibial plateau at the estimated meniscus extrusion site, followed by reaming of the tibial tunnel with an Endobutton 4.5 mm drill. Subsequently, the no.2 Ethibond suture was pulled out to the transtibial tunnel (Figure 3).

The anatomical PRMM attachment was identified and a transtibial tunnel was drilled using a 2.5 mm diameter guide pin from the mid-anterior aspect of the proximal tibia through the

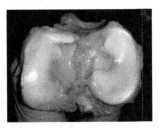

Figure 1. Porcine meniscus specimen after removal of all muscles and soft tissue.

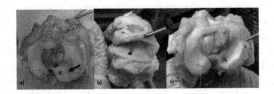

Figure 2. Surgical techniques on porcine model. a) Conventional PRMM repair by direct suture at injured site. b) & c) Axial and sagittal view of PRMM repair combined with ADMER technique.

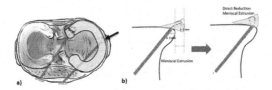

Figure 3. A) Direct reduction meniscal extrusion with transtibial pullout extrusion reduction stitch (black arrow) and free tension PRMM tear repair. b) Coronal view of direct meniscal extrusion reduction.

meniscal root anatomical attachment, followed by reaming of the tibial tunnel with an Endo-button 4.5 mm drill. Sutures in the root of the medial meniscus was done using no. 0 Ethibond and pulled out to the transtibial tunnel.

2.3 Biomechanical testing

All specimens underwent testing for static pull-out strength with a servo-hydraulic material testing machine (A&D RTF-2430 Tensilon® Universal material testing) with a metal wire hooked at the PMMR in a horizontal position (Figure 4). A maximum pullout-load was applied to the PRMM until failure. Means and standard deviations of pullout strength were compared.

A Mann-Whitney U-test, ANOVA and post-hoc test was performed to evaluate group differences in mean maximum load-to-failure strength, yield point, elastic modulus, and the ultimate strength. The significance level was set to $p < 0.05$. For all statistical analyses, SPSS 24.0 (IBM-SPSS, Armonk, NY, USA) was used.

3 RESULTS

The native/healthy medial meniscus achieved a significantly higher pull-out strength ($P = 0.002$) compared with both repair techniques. Mean pull-out strength was 1047.19 N (±64.4) for control group, 256.41 N (±54.26) for P group and 367.48 N (±40.5) for AD group (Table 1). Maximum pull-out strength for PRMM + ADMER was significantly higher compared to P group (p-value 0.006). Summary of the biomechanical properties for the two techniques after static loading tests is presented in Table 2. Compared to P group, the AD technique had significantly lower elastic modulus ($p = 0.047$) and higher ultimate strength ($p = 0.009$) (Figure 5).

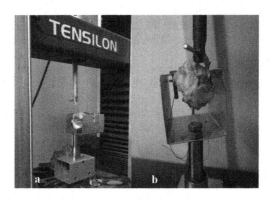

Figure 4. Biomechanical testing set-up. a) The specimen was placed horizontally in a servo-hydraulic material testing machine (A&D RTF-2430 Tensilon® Universal material testing). b) a metal wire was hooked at the insertion of the PRMM and the tibia was potted in a meta cylinder.

Table 1. Descriptive analysis of the maximum pullout strength between study groups.

	Control	Conventional PRMM	PRMM + ADMER
Mean max. Load (N)	1047.19	256.41	367.48
Minimum Load	814	152.09	302.35
Maximum Load	1341.5	309.65	418.62
SD	250.06	54.26	40.56

Table 2. Comparison of the biomechanical properties between study groups.

	Control	Conventional PRMM repair	PRMM + ADMER
Maximum Point Load (Pa)	1047.19	256.418	369.828
Yield Point (Pa)	425.935	284.158	350.774
Elastic Modulus (Pa)	79.098	107.836	86.332
P-value	<0.05		

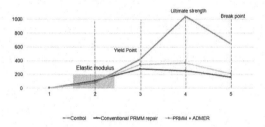

Figure 5. Graphic lines above showed PRMM + ADMER technique have better biomechanical properties than conventional PRMM repair, however both technique had lower ultimate strength compared to control group.

4 DISCUSSION

The purpose of this study was to compare the biomechanical properties of conventional PRMM repair and PRMM with ADMER technique for PRMM tears. The results confirm our hypothesis that the ADMER technique provides superior biomechanical properties under static loading testing compared with the conventional technique. However, both techniques are still inferior to the native healthy PRMM. Similar biomechanical study by Okimura et al. (2019) found that there is no significant difference strength between three meniscus root repair techniques.

Failure to reduce ME in PRMM tears could lead to several serious consequences including failed healing, osteonecrosis and rapid progression of osteoarthritis. Arthroscopic surgical repair is currently seen as the treatment of choice and showed promising clinical results (Kim et al. 2011, Pache et al. 2018). However, a study by Feucht et al. (2015) reported that only 56% of patients had successful reduction of the meniscal extrusion (Feucht et al. 2015). Another study by Moon et al. found that only in 26% patients reduced meniscal extrusion was found. Surprisingly, even with the good clinical outcome, on the second look arthroscopic surgery, the healing status was completed in only 62% of cases, 34% was healed partially, and 3% showed failed healing (Li et al. 2019).

The biomechanical properties of the repair technique may play an important role. In the conventional PRMM repair, tying of the sutures far distal at the anteromedial or anterolateral tibial cortex creates a long meniscus-suture construct. This might result in micromotion of the meniscus root fixation during the early post-operative period because of decreased stiffness and increase in elongation of the meniscus-suture complex and possibily compromising the meniscal healing (Feucht et al. 2015). Similar biomechanical study by Okimura et al. found that there is no significant difference in the ultimate load between three meniscus root repair techniques. Our results thus showed a higher pull-out strength compared to the repair techniques mentioned by Okimura et al. (2019). Therefore, another factors must be put in consideration such as the size of the specimen, types of repair technique and the materials.

There are several causes of failure to reduce ME in conventional PRMM repair mentioned by Chernchujit & Prasetia (2017): First, PRMM repair only facilitates indirect ME reduction and fixation so as to create the ineffective distance and direction force to reduce ME. Secondly, the rigid anchorage of medial capsule and meniscotibial ligament maintain extruded medial meniscus. Another reason conventional PRMM repair had failed was that meniscus has tensile capacity to elongate so that the indirect force of PRMM repair only creates elongation of the meniscus rather than reducing the ME.

A biomechanical study by Cerminara et al. (2014) showed the transtibial pull-out repair technique had substantial displacement compared to button-bone interface after a cyclic loading test (Cerminara et al. 2014). PRMM repair with ADMER technique has some advantages over conventional PRMM repair in reducing ME. This technique facilitates direct ME reduction by creating effective distance and directional force to reduce ME. The tension of the posterior medial meniscus during root repair was minimize because of the reduction of the meniscus (Chernchujit & Prasetia 2017).

In this present study, we did not analyze the biomechanical properties of all specimens under cyclic-loading, thus making this the study limitation. Because it is difficult to reproduce this study in human knees, we decide to use porcine knee models as its size, shape, soft tissue quality and bone quality are similar to humans (Rosslenbroich 2013). Although the study results of the AD group were promising, this might not enough to reflect the forces subjected during real kinematics. It is necessary to obtain more biomechanical and biological properties of the specimens in the future.

5 CONCLUSION

PRMM Repair combined with ADMER technique had biomechanical properties of higher superior pullout strength compared to conventional PRMM repair and lower elastic modulus. However, we did not test its biomechanical properties under cyclic loading and both techniques did not reach the strength of the native healthy medial meniscus.

REFERENCES

Cerminara, A. J., LaPrade, C. M., Smith, S. D., Ellman, M. B., Wijdicks, C. A. & LaPrade, R. F. 2014. Biomechanical evaluation of a transtibial pull-out meniscal root repair: challenging the bungee effect. *The American journal of sports medicine* 42(12): 2988–2995.

Chernchujit, B. & Prasetia, R. 2017. Both Posterior Root Lateral-Medial Meniscus Tears with Anterior Cruciate Ligament Rupture: The Step-by-Step Systematic Arthroscopic Repair Technique. *Arthroscopy techniques* 6(5): e1937–e1943.

Feucht, M. J., Kühle, J., Bode, G., Mehl, J., Schmal, H., Südkamp, N. P. & Niemeyer, P. 2015. Arthroscopic transtibial pullout repair for posterior medial meniscus root tears: a systematic review of clinical, radiographic, and second-look arthroscopic results. *Arthroscopy: The Journal of Arthroscopic & Related Surgery* 31(9): 1808–1816.

Kim, S. B., Ha, J. K., Lee, S. W., Kim, D. W., Shim, J. C., Kim, J. G. & Lee, M. Y. 2011. Medial meniscus root tear refixation: comparison of clinical, radiologic, and arthroscopic findings with medial meniscectomy. *Arthroscopy: The Journal of Arthroscopic & Related Surgery* 27(3): 346–354.

Li, L., Yang, X., Yang, L., Zhang, K., Shi, J., Zhu, L., ... & Jiang, Q. 2019. Biomechanical analysis of the effect of medial meniscus degenerative and traumatic lesions on the knee joint. *American journal of translational research* 11(2): 542.

McDermott, I. 2011. Meniscal tears, repairs and replacement: their relevance to osteoarthritis of the knee. *British journal of sports medicine* 45(4): 292–297.

Moatshe, G., Chahla, J., Slette, E., Engebretsen, L. & Laprade, R. F. (2016). Posterior meniscal root injuries: A comprehensive review from anatomy to surgical treatment. *Acta orthopaedica* 87(5): 452–458.

Okimura, S., Mae, T., Tachibana, Y., Iuchi, R., Nakata, K., Yamashita, T. & Shino, K. 2019. Biomechanical comparison of meniscus-suture constructs for pullout repair of medial meniscus posterior root tears. *Journal of experimental orthopaedics* 6(1): 1–6.

Pache, S., Aman, Z. S., Kennedy, M., Nakama, G. Y., Moatshe, G., Ziegler, C. & LaPrade, R. F. (2018). Meniscal root tears: current concepts review. *Archives of Bone and Joint Surgery* 6(4): 250.

Rosslenbroich, S. B., Borgmann, J., Herbort, M., Raschke, M. J., Petersen, W. & Zantop, T. 2013. Root tear of the meniscus: biomechanical evaluation of an arthroscopic refixation technique. *Archives of orthopaedic and trauma surgery* 133(1): 111–115.

Medical Technology and Environmental Health – Abdullah, Widiaty & Abdullah (eds)
© 2020 Taylor & Francis Group, London, ISBN 978-0-367-86053-0

The use of ethanolic extract of cogongrass roots to reduce triglyceride absorption in male mice

M.R.A.A. Syamsunarno, G.R. Mukarromah, A. Achadiyani & D.D. Djunaedi
Universitas Padjadjaran, Bandung, Indonesia

M. Putri
Universitas Islam Bandung, Bandung, Indonesia

ABSTRACT: The purpose of this study was to explore the effects of cogongrass (*Imperata cylindrica L)* roots ethanolic extract (CGRE) in blood triglyceride. Mice were given CGRE intragastrically with dose 90 and 115 mg/kg body weight (BW) for two weeks. The non-treatment group was used as a control. Triglyceride absorption assay and triglyceride serum level were measured at the end of treatment day. Triglyceride absorption were reduced 60% in mice that were pre–treatment with 90 and 115 mg/kg BW CGRE compare to control (p<0.05). As conclusion, CGRE has an effect to reduce triglycerides absorption in the small intestine.

1   INTRODUCTION

Lipids and carbohydrates are the main energy source for humans and rodents (Whitney & Rolfes 2016). The ability of a body to use lipids and carbohydrates as an energy source influences the adapting ability of a body to encounter extreme environmental conditions such as starvation and hypothermia (Putri et al. 2015). However, high consumption of lipids and carbohydrates increase the risk of metabolic and cardiovascular diseases (National Cholesterol Education Program (US) 2002, Harikumar et al. 2013, Nelson 2013). Triglyceride, the dominant fat in the diet is efficiently hydrolyzed into fatty acids and absorbed by the enterocytes (Putri et al. 2015, Iqbal & Hussain 2009). Fatty acids are processed into neutral lipids and utilized within the endoplasmic reticulum to form lipoproteins, and secreted from the basolateral side of the enterocyte into circulation (Iqbal & Hussain 2009).

Cogongrass (*Imperata cylindrica*) is commonly used as traditional medicine to cure fever, muscle ache, asphyxia, and nosebleed. It also has the ability to lower fat and glucose concentrations in blood (Roosita et al. 2008, VH et al. 2012). We previously showed that cogongrass roots ethanolic extract (CGRE) has an anti-hypertension effect in rats (Asmawi et al. 2013). CGRE contains tannin, saponin, flavonoid, alkaloid, and terpenoid (Krishnaiah et al. 2009). Some of those compounds are known to have effect on lipid and glucose metabolism (Deepika et al. 2013). Nevertheless, there is no study regarding the effect of cogongrass on blood triglyceride. The purpose of this study was to explore the effects of CGRE in blood triglycerides.

2   MATERIALS AND METHODS

2.1   *Preparation of plant extract*

Cogongrass was purchased from Solo, Central Java, Indonesia. Cogongrass roots were washed, dried and pounded to powder (Specimen no: 3561/10/11.CO2.2/PL/2016). Cogongrass roots powder were macerated by ethanol 95% for 72 hours, filtrated with a vacuum filter, and concentrated in a vacuum evaporator. The concentrated extracts were suspended with carboxymethylcellulose 0.5% and separated to concentration of 90 mg/kgBW and 115 mg/kgBW (Asmawi et al. 2013).

2.2 Mice treatment and sample collection

Health Research Ethics Committee Faculty of Medicine Universitas Padjadjaran approved all study protocols (No: 974/UN6.C1.3.2/KEPK/PN/2016). Eight to ten weeks old male mice (*Mus musculus*) were purchased from Biology Laboratory, Faculty of Mathematics and Natural Sciences, Universitas Padjadjaran. Mice were housed in a room with a condition of 12/12 h light and dark cycle and adequate air circulation and had unrestricted access to water and standard chow.

Mice were given CGRE intragastrically with dose 90 mg/kg (dose 1) and 115 mg/kgBW (dose 2). The non-treatment group was used as a control. Control and treated mice had unrestricted access to water and standard chow. Body weight was measured after overnight fasted every three days. At the end of the experiment, mice were sacrificed.

2.3 Determination of lipid absorption

Control and treated mice were given an olive oil (10 µl/g) intragastrically after overnight fasting. Blood was collected from the retro-orbital plexus at indicated time point.

2.4 Measurement of blood parameters

Blood samples were centrifuged at $1,200 \times g$ for 15 minutes to separate the serum. Triglyceride (Triglyceride 5 Reagent Kit, PT. Akurat Intan Madya, Jakarta), were measured according to the manufacturer's protocols.

2.5 Statistical analysis

Statistical analysis was performed using one-way ANOVA at the same indicating time point. Bonferroni's post hoc multiple comparison tests were performed to evaluate differences between the control and experimental groups. A p-value of 0.05 was considered statistically significant.

3 RESULTS AND DISCUSSION

Body weight of control and treated mice were measured every three days for two weeks. Body weight was not different among groups (Figure 1).

After two weeks, blood from retro-orbital was collected to measure triglyceride in serum. Triglyceride in serum was not different among groups (Figure 2a) ($p<0.05$). We further analyzed lipid absorption capacity in the small intestine by giving olive oil acutely. As shown in

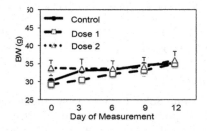

Figure 1. Body weight of control and treated mice during two weeks experiment.

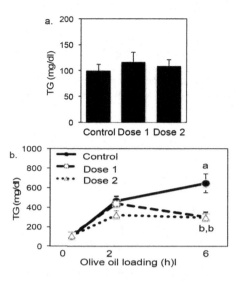

Figure 2. Triglyceride and glucose in serum. Serum was collected from retroorbital blood after overnight fasting, and (a) triglyceride was measured. (b) Mice were given olive oil intragastrically and blood was collected from retroorbital before and 2 and 6 hours after olive oil gavage. Serum was collected and triglyceride was measured. TG = Triglyceride; * = p<0.05; Different letter indicate significant result (p<0.05); BW = Body weight.

Figure 2b, triglyceride in serum was increased two hours after olive oil gavage. After six hours of olive oil gavage, triglyceride serum in control mice was two times higher compared to dose 1 and dose 2 (p<0.05). Our results showed lipid metabolism was disturbed after acute olive oil gavage, partially due to attenuation of lipid absorption in small intestine.

We hypothesized CGRE might reduce blood triglyceride at least by two mechanisms. The first mechanism is by disturbing lipid digestion process in the small intestine lumen. Degradation of lipid mainly catalyzed by pancreatic lipase, an esterase that hydrolyzed dietary triacylglyceride into β-monoglyceride and long-chain saturated and polyunsaturated fatty acids. A study in China shown that quercetin, a class of flavonoid found in fruits and vegetable, expose anti-lipase activity (Zheng et al. 2010). We speculated that flavonoid in cogongrass roots might reduce triglyceride level in blood.

The second possibly mechanism is by reducing triglyceride formation in enterocyte. Lipid absorption affected by three main components; absorption of hydrolyzed lipid into the enterocyte, intracellular processing, and exocytosis into the mesenteric lymph and vein (Mansbach & Gorelick 2007). A study using 3T3-L1 cell culture showed that flavonoids inhibit the activity of glycerol-3-phosphate dehydrogenase (GPDH), a cytosolic enzyme to convert glycerol into triglyceride (Hsu & Yen 2007). Flavonoids also regulate the activity of sterol regulatory element-binding proteins SREBP-1, and SREBP-2 and other relevant enzymes in lipid metabolism (Horton et al. 2002).

4 CONCLUSION

CGRE has an effect to lowering lipid absorption in male mice.

ACKNOWLEDGEMENTS

This project was supported by Grant from Ministry of Research, Technology and Higher Education of the Republic of Indonesia.

REFERENCES

Asmawi, M. Z., Rianse, U., Sahidin, I., Dhianawaty, D., Soemardji, A. A. & Amalia, L. (2013). Anti-hypertensive activity of Alang–Alang (Imperata cylindrica (L.) Beauv. root methanolic extract on male Wistar rat. *International Journal of Research in Pharmaceutical Sciences* 4(4): 537–542.

Deepika, D., Satish, S. K. & Lalit, S. 2013. Current updates on anti-diabetic therapy. *Journal of Drug Delivery & Therapeutics* 3(6): 121–126.

Harikumar, K., Althaf, S. A., Ramunaik, M. & Suvarna, C. H. 2013. A review on hyperlipidemic. *International Journal of Novel Trends in Pharmaceutical Sciences* 3(4).

Horton, J. D., Goldstein, J. L. & Brown, M. S. 2002. SREBPs: activators of the complete program of cholesterol and fatty acid synthesis in the liver. *The Journal of clinical investigation* 109(9): 1125–1131.

Hsu, C. L. & Yen, G. C. 2007. Effects of flavonoids and phenolic acids on the inhibition of adipogenesis in 3T3-L1 adipocytes. *Journal of agricultural and food chemistry* 55(21): 8404–8410.

Iqbal, J. & Hussain, M. M. 2009. Intestinal lipid absorption. *American Journal of Physiology-Endocrinology and Metabolism* 296(6): E1183–E1194.

Krishnaiah, D., Devi, T., Bono, A. & Sarbatly, R. 2009. Studies on phytochemical constituents of six Malaysian medicinal plants. *Journal of medicinal plants research* 3(2): 67–72.

Mansbach, C. M. & Gorelick, F. 2007. Development and physiological regulation of intestinal lipid absorption. II. Dietary lipid absorption, complex lipid synthesis, and the intracellular packaging and secretion of chylomicrons. *American Journal of Physiology-Gastrointestinal and Liver Physiology* 293 (4): G645–G650.

National Cholesterol Education Program (US). Expert Panel on Detection, & Treatment of High Blood Cholesterol in Adults. (2002). *Third report of the National Cholesterol Education Program (NCEP) expert panel on detection, evaluation, and treatment of high blood cholesterol in adults (adult treatment panel III)* (No. 2). National Cholesterol Education Program, National Heart, Lung, and Blood Institute, National Institutes of Health.

Nelson, R. H. 2013. Hyperlipidemia as a risk factor for cardiovascular disease. *Primary Care: Clinics in Office Practice* 40(1): 195–211.

Putri, M., Syamsunarno, M. R. A., Iso, T., Yamaguchi, A., Hanaoka, H., Sunaga, H., ... & Tsushima, Y. 2015. CD36 is indispensable for thermogenesis under conditions of fasting and cold stress. *Biochemical and biophysical research communications* 457(4): 520–525.

Roosita, K., Kusharto, C. M., Sekiyama, M., Fachruroci, Y. & Ohtsuka, R. (2008). Medicinal plants used by the villagers of a Sundanese community in West Java, Indonesia. *Journal of ethnopharmacology* 115(1): 72–81.

VH, E. S., Utomo, S. B., Syukri, Y. & Redjeki, T. 2012. Phytochemical screening and analysis polyphenolic antioxidant activity of methanolic extract of white dragon fruit (Hylocereus undatus). *Indonesian Journal of Pharmacy* 60–64.

Whitney, E. & Rolfes, S. R. 2016 Understanding Nutrition. 14th ed. *Nutrition Reviews*. United States of America: Cengage Learning.

Zheng, C. D., Duan, Y. Q., Gao, J. M. & Ruan, Z. G. 2010. Screening for anti-lipase properties of 37 traditional Chinese medicinal herbs. *Journal of the Chinese Medical Association* 73(6): 319–324.

Medical Technology and Environmental Health – Abdullah, Widiaty & Abdullah (eds)
© 2020 Taylor & Francis Group, London, ISBN 978-0-367-86053-0

Gynura divaricata: Natural source for carcinoma mammae therapy agent development

M. Tejasari, H. Muflihah, Z. Zulmansyah & W. Purbaningsih
Universitas Islam Bandung, Bandung, Indonesia

ABSTRACT: Cytotoxic testing is needed as an initial screening to assess the effect of a natural substance inhibiting tumor cell growth in the development of new drugs for cancer therapy agents. A flavonoid derivatives extracted from gynura divaricata is one of the compounds being investigated as an anticancer. The aim of this study was to assess the cytotoxic activity of gynura divaricata leaf in the culture of carcinoma mammae cells and to compare them with standard drugs. Cytotoxic activity of gynura divaricata leaf water extract and doxorubicin was investigated using the MTT method on T47D cell line culture. The absorbance of each well was measured at a wave length 595 nm using a spectrophotometer. Using Probhit analysis, the absorbance results were calculated to create a cytotoxic curve. The result showed inhibition concentration (IC_{50}) value 102.32 µg/mL for gynura divaricata water exact and 10,30 µg/mL for doxorubicin. This result indicated that gynura divaricata leaf water extract could inhibit 50% tumor cell growth at concentration 102. 32 µg/mL. It was concluded that gynura divaricata leaf has potential anti-cancer activity ($IC_{50} < 200$ µg/mL) against carcinoma mammae cell, but not as strong as doxorubicin.

1 INTRODUCTION

The incidence of breast cancer has increased sharply throughout the world. Until now the efficacy of chemotherapy for cancer has not been optimal because it has side effects and high toxicity and resistance occurs, therefore the discovery of anticancer drugs continues to be developed to obtain more sensitive and specific therapeutic agents (Jemal et al. 2014, Solomon & Lee 2012, Kim et al. 2010, Osborne & Schiff 2011)

One effort to find new drugs is through the exploration of bioactive compounds from natural materials. Herbal medicines have been widely used by people in Indonesia, one of them is *Gynura divaricata* or Dewa leaf which has long been believed by the community to be used to treat various diseases including cancer. Several studies have revealed that the leaves of the deity contain important compounds such as flavonoids, phenolic acids, cerebrosides, polysaccharides, alkaloids, terpenoids, and sterols. Antioxidant compounds contained in the leaves of the gods namely phenolic acids and flavonoids are relatively high so that these leaves can prevent the formation of free radicals and lead to inhibition of cancer growth. Dewa leaves have been widely used by the public, but there are no standards regarding the dosage of use and there are no studies that review the side effects of their use on organ function. (Xu & Zhang 2017, Li et al. 2018, Li et al. 2009, Jiangseubchatveera et al. 2015).

Based on the above background, researchers of this study are interested in exploring the potential of god leaf preparations in cancer therapy in particular breast cancer, comparison with standard drugs and test the effect of the combination of the leaves of the god with standard drugs. General purpose of this study was to explore the potential, effectiveness, and selectivity of preparations of Dewa leaves as candidates for new drugs from natural ingredients in cancer therapy.

2 MATERIAL AND METHOD

2.1 *Subject and material*

The subject in this study was breast cancer cell culture T47D, 100μL in 16 wells with a density of 1.5 x 10^6 cells/mL. Cytotoxic activity test was conducted to determine IC_{50} using MTT Method.

Material used were RPMI Medium (Sigma Chem Co. St. Louis, USA), *serum fetal bovine* (Sigma Chem Co. St. Louis, USA), penicillin, streptomycin (Gibco), fungizon (Gipco), sodium bicarbonate (E Merck) and 0.2 ųm (Whatman) filter paper, *phosphate buffer saline*, (PBS), MTT reagents and 10% SDS stopper. Dewa leaf water extract and standard drug used doxorubicin (Sigma Chem Co. St. Louis, USA).

2.2 *Material preparation*

Dewa leaf (Gynura divaricata) was taken from Bogor experimental garden, Indonesia. Simplicia was extracted using a water solvent and repeated 3 times. The extract obtained was concentrated by a rotary vacuum. The yield of dry extract was calculated against the weight of dry simplicia. The preparations used are the dewa leaves in various concentrations. While for the comparative preparations used standard drugs for cancer therapy which are dissolved in aqua.

2.3 *Cytotoxic activity examination*

After being incubated the culture media are discarded and washed with PBS and added 100 μL of MTT solution (1 mL MTT in 10 mL, culture media) into each well and then re-incubated at 37°C 5% CO_2 for 4 hours. After 4 hours, 100 μL of added stopper was 10% SDSto 0.1 N HCl into each well (to dissolve purple formazan). Then it was locked out for 5 minutes and sealed tightly left at room temperature overnight. After overnight at room temperature, the absorption reading was done with a microplate ELISA reader at a wavelength (λ) of 595 nm. From the absorbance analysis results, cell viability can be calculated with the formula.

$$\% \text{ live (viability)} = (cb)/(ab) \times 100.$$

The letter a is the absorbance of cell control; b is the absorbance of the media control and c is the absorbance of the sample. The cytotoxic activity of Dewa leaf extract and doxorubicin is expressed by the value of Inhibition Concentration (IC_{50}) which were a concentration capable of inhibiting cell growth to 50% which is calculated using probhit analysis based on the relationship between levels of cell growth inhibitors.

3 RESULTS AND DISCUSSION

In the development of new anticancer drugs as candidates for cancer therapy agents, preclinical evaluation is one of the important things to know the potential for cytotoxic activity. This evaluation is not only used for anticancer drugs, but also other drugs, cosmetics, food additives, pesticides and others. Standardized evaluations to determine whether a material contains biologically toxic material is called a cytotoxic test (Zhou et al. 2014, Acquaviva & Lauk 2010, Rodríguez et al. 2013).

Cytotoxic tests are used as an initial screening to determine the effect of a natural substance in inhibiting tumor cell growth. A compound is considered active if it can inhibit the growth of 50% of tumor cell populations at certain concentrations. Requirements that must be met for the cytotoxicity test system are that the test system must be able to produce a reproducible dose-response curve with low variability, the response criteria must show a linear relationship with the number of cells and the information obtained from the dose-response curve must be in line with the effect appear. One method commonly used to assess this cytotoxic activity is the MTT method (Batra & Sharma 2013, Fraga & Oteiza 2011)

Based on the cytotoxic curve in T47D cell culture in this study, IC_{50} values obtained from dewa leaf water extract were 102.32 μg/mL. This showed that the dewa leaf water extract has anticancer activity against breast cancer cells. To assess its comparison strength with a standard drug doxorubicin, the cytotoxic activity test was done. Based on the cytotoxic curve in T47D cell culture, the IC_{50} doxorubicins values were 10.30μg/mL. This showed that the anticancer activity of water extracts from the dewa leaves was not as strong as doxorubicin.

An extract is said to have strong cytotoxic activity if it can inhibit the growth of 50% of the cell population at concentrations below 200 μg/mL (IC_{50}: 200 μg/mL) (Meiyanto et al. 2007).

In this study, the results of cytotoxic tests dewa leaf water extract against breast cancer *T47D cell line cells* with the MTT method produces an $ICvalue_{50}$ of 102.32 μg/mL. With this $ICvalue_{50}$ it can be concluded that the flavonoid compounds from clove leaf oil are categorized as active compounds with strong anticancer potential so that these compounds can be used in subsequent tests in preclinical evaluation to examine anticancer potentials and analyze their mechanism of action using reference values The IC50. The IC_{50} value was weaker than the IC_{50} value of the compounds isolated from green tea (57.53) as well as other studies with IC_{50} value 65.7 in Hs579T human breast cancer cells. (Elfita et al. 2009).

The results of this comparison do not affect the ability of the water extract of dewa leaves as a potential anti-cancer candidate especially in terms of selectivity because doxorubicin caused normal cell damage.

4 CONCLUSION

It was concluded that the dewa leaves (gynura divaricata) were a potential natural source to be developed as an anticancer agent for breast cancer therapy.

ACKNOWLEDGEMENTS

This research was made possible by the full support of the Institute of Research and Community Service (LPPM) Universitas Islam Bandung which supported the funding of this study (PDU grant).

REFERENCES

Acquaviva, R., & Iauk, L. 2010. Natural polyphenols as anti-inflammatory agents. *J Front Biosci*, 2: 318–31.

Batra, P., & Sharma, A. K. 2013. Anti-cancer potential of flavonoids: recent trends and future perspectives. *3 Biotech*, 3(6): 439–459.

Fraga, C. G., & Oteiza, P. I. 2011. Dietary flavonoids: role of (−)-epicatechin and related procyanidins in cell signaling. *Free Radical Biology and Medicine*, 51(4): 813–823.

Jemal, A., Bray, F., Center, M. M., Ferlay, J., Ward, E., & Forman, D. 2014. Global cancer statistics CA Cancer J Clin. 2011; 61: 69–90. *Erratum in: CA Cancer J Clin*, 61.

Jiangseubchatveera, N., Liawruangrath, B., Liawruangrath, S., Korth, J., & Pyne, S. G. 2015. The chemical constituents and biological activities of the essential oil and the extracts from leaves of Gynura divaricata (L.) DC. growing in Thailand. *Journal of Essential Oil Bearing Plants*, 18(3): 543–555.

Li, J., Feng, J., Wei, H., Liu, Q., Yang, T., Hou, S., ... & Yang, C. 2018. The aqueous extract of Gynura divaricata (L.) DC. improves glucose and lipid metabolism and ameliorates type 2 diabetes mellitus. *Evidence-Based Complementary and Alternative Medicine*, 2018.

Kim, T. H., Seo, W. D., Ryu, H. W., Seo, H. R., Jin, Y. B., Lee, M., ... & Lee, Y. S. 2010. Anti-tumor effects by a synthetic chalcone compound is mediated by c-Myc-mediated reactive oxygen species production. *Chemico-biological interactions*, 188(1), 111–118.

Li, W. L., Ren, B. R., Min-Zhuo, Hu, Y., Lu, C. G., Wu, J. L., … & Sun, S. 2009. The anti-hyperglycemic effect of plants in genus Gynura Cass. *The American journal of Chinese medicine*, 37(05): 961–966.

Meiyanto, E., Susilowati, S., Tasminatun, S., & Murwanti, R. (2007). Chemopreventive effect of ethanolic extract of Gynura procumbens (Lour), Merr on the carcinogenesis of rat breast cancer development. *Indonesian Journal of Pharmacy*, 154-161.Osborne, C. K., & Schiff, R. 2011. Mechanisms of endocrine resistance in breast cancer. *Annual review of medicine*, 62: 233–247.

Zhou, Y., Zhang, A. H., Sun, H., Yan, G. L., & Wang, X. J. 2014. Plant-derived natural products as leads to antitumor drugs. *Plant Science Today*, 1(2): 46–61.

Rodríguez, M. L., Estrela, J. M., & Ortega, Á. L. 2013. Natural polyphenols and apoptosis induction in cancer therapy. *J Carcinog Mutag S*, 6.

Solomon, V. R., & Lee, H. 2012. Anti-breast cancer activity of heteroaryl chalcone derivatives. *Biomedicine & Pharmacotherapy*, 66(3): 213–220.

Xu, B. Q., & Zhang, Y. Q. 2017. Bioactive components of gynura divaricata and its potential use in health, food and medicine: A mini-review. *African Journal of Traditional, Complementary and Alternative Medicines*, 14(3), 113–127.

Elfita, E., Muharni, M., Latief, M., Darwati, D., Widiyantoro, A., Supriyatna, S., … & Foubert, K. 2009. Antiplasmodial and other constituents from four Indonesian Garcinia spp. *Phytochemistry*, 70(7), 907–912.

Copper nanoparticles synthesis optimization using melinjo (*Gnetum gnemon* L.) leaves extract and beta cyclodextrin as a stabilizer

H.A. Wisnuwardhani, R.D. Shafira, Y. Lukmayani & A. Arumsari
Universitas Islam Bandung, Bandung, Indonesia

ABSTRACT: Copper nanoparticles are one type of metal nanoparticles that have not been widely studied. Copper nanoparticles have some uniqueness that affects their optical, catalysis and conductive properties. In addition, nanoparticles can also be applied in the field of microbiology because of their antibacterial activity. This study aims to determine the optimum conditions for the synthesis of copper nanoparticles, using melinjo leaves extract as bioreductor. This research used copper (II) sulfate as a precursor. Melinjo leaves are extracted with water at 60°C. Beta cyclodextrin was used as a stabilizer. The results showed that the optimum conditions for the synthesis of copper nanoparticles were using a ratio of extracts: $CuSO_4$ solution: beta cyclodextrin 10 mg/mL = 1: 2: 2 (v/v/v). Nanoparticles were synthesized by reflux at 90°C for 4 hours. The optimization results show that the sonication process for 4 hours after synthesis affects the stability of the formed nanoparticles. Copper nanoparticles have an average particle diameter of 682.8 nm and polydispersity index of 0.497. Copper nanoparticle synthesis conditions can be used routinely in the laboratory for further investigation as a topical antibacterial candidate.

1 INTRODUCTION

A nanoparticle is a particle that has a size of 1-100 nm. The small size of the nanoparticles will affect the properties and characteristics compared to its larger size. At present, one of the nanoparticles that has been developed is nanoparticles which are used as antimicrobials (Ayu 2015). Copper nanoparticles are a potential material, in the non-food and health fields, nanoparticles can be used as wound dressing and biocides (Borkow et.al. 2010). In the pharmaceutical field, the use of nanoparticles can be for topical ointments and creams used to prevent burns and open wound infections (Elumalai et al. 2011).

According to Dewi et al. (2012), melinjo leaves contain flavonoids which can act as bioreducers. The content of flavonoids in melinjo leaves amounted to 0.187 mg/ml. The aim of this research is to synthesize copper nanoparticles using melinjo leaves extract as a bioreductor. The benefit of this research is that it can provide information on green synthesis of copper nanoparticles. Those copper nanoparticles are expected to become alternatives antimicrobial material as an active pharmaceutical ingredient for disinfectant preparation.

The reaction between copper (II) sulphate with plant metabolite (flavonoid or polyphenolic which are act as a bioreductor) be (Kuppusamy et al. 2016):

$$CuSO_4.5H_2O + \text{Plant metabolite}$$

$$\downarrow$$

$$Cu°NPs + \text{Side Product}$$

Nanoparticles have physical properties including magnetic, electrical, optical, mechanical and thermal properties that are examined for later use in various fields such as material engineering, electronics, magnetic, and biomedical (Kuppusamy et al. 2016).

Melinjo (*Gnetum gnemon* L.) is one of the local commodities that has several benefits. Melinjo is widely cultivated in Indonesia, but the usage of this plant is limited to its usage as vegetables and as a material of *emping* (kind of melinjo chips in Indonesia). According to Pudjiatmoko (2007), the antioxidant activity of the phenolic content in melinjo is equivalent to the synthetic antioxidant Butylated Hydroxytolune (BHT). Chemical content contained in melinjo seeds and leaves are saponins, flavonoids, and tannins while melinjo skins contain saponins and flavonoids.

Natural antioxidants contained in plants are generally phenolic or polyphenolic compounds which can be in the form of flavonoids, cinnamic acid derivatives, coumarin, tocopherol, and polyfunctional acids. Flavonoids that have antioxidant activity include flavones, flavonoids, flavanones, isoflavones, catechins, and chalcon (Markham 1988).

In this research, the synthesis of copper nanoparticles is assisted by the content of secondary metabolites which have antioxidant activity. Certain types of plants contain certain chemical compounds that can act as reducing agents (Handayani 2011).

The development of the emergence of microbial resistance to a particular antimicrobial, making the discovery of a new drug (antimicrobial) that has a broad spectrum is very important. Copper nanoparticles are one of the most widely studied metal nanoparticles as a promising antimicrobial candidate besides silver nanoparticles. Copper nanoparticles themselves have been widely studied as antibacterial against bacteria *E. coli* and *S. aureus* (Prabhu et al. 2017), *E. coli* (Lee et al. 2013, Chatterjee et al. 2014). Synthesis of metal nanoparticles using plant extracts as bioreductors, most of which are still focused on the synthesis of silver nanoparticles, which are already widely studied as having antibacterial activity. One of the studies on the synthesis of copper nanoparticles using plant extracts is the synthesis of copper nanoparticles using clove flower extract (*Syzigium aromaticum*) (Rengga et al. 2017), and even then, not to test its antibacterial activity. Other researchers are mostly from abroad, including using *Nag Champa* (Kathad & Gajera, 2014), *mangos teen* leaves (Prabhu et al. 2017), tea leaves (Mandaya, et al. 2017), *Magnolia Kobus* (Lee et al. 2013) leaves, *Asparagus adscendens* leaves (Thakur et al. 2018), *Citrus medica* fruit (Shende et al. 2015). According to previous research, there are no studies that have carried out the synthesis of copper nanoparticles using extracts.

This research aims to optimize the synthesis of copper nanoparticles using melinjo leaves extract as a bioreductor. The results of this synthesis optimization are expected to be used as a reference in terms of the best copper nanoparticles synthesis conditions, so that the copper nanoparticles produced can be continued for research on its antibacterial ability.

2 MATERIAL AND METHODS

2.1 *Material*

Copper (II) sulphate, beta cyclodextrin, citric acid, aquabidest, crude drugs melinjo (*Gnetum gnemon* L.) leaves. All reagent that was used in this research was pro analytical grade.

2.2 *Methods*

Melinjo leaves extract was prepared by infusa method using water as extract ant. $CuSO_4$ solution and liquid extract from melinjo leaves are used for nanoparticles synthesis through bio reduction process. To synthesize nanoparticles, the concentration of $CuSO_4$ solution used is 1 mM. 1 mM $CuSO_4$ solution was mixed with melinjo leaves extract with various ratio then the stabilizer was added consisting of 5 mM citric acid and 10 mM betaclodextrin with various concentration. Copper nanoparticles characterization was done using UV-Visible spectrophotometer and Particle Size Analyzer. The optimization conditions performed include Table 1-4:

Table 1. Optimization of copper nanoparticles synthesis time.

No	Condition	Synthesis methods	Time	Melinjo leaves extract: $CuSO_4$ solution
1	A1	Reflux	4 h	1:4
2	B1	Reflux	1 h	1:4
3	C1	Reflux	4 h	1:4
4	D1	Magnetic stirer	24 h	1:4

Table 2. Optimization of copper nanoparticles synthesis (citric acid as a stabilizer).

No.	Condition	Synthesis methods	Time	Melinjo leaves extract: $CuSO_4$ solution: citric acid
1	A2	Reflux	4 h	1:1:1
2	B2	Reflux	4 h	1:2:2
3	C2	Reflux	4 h	1:4:4
4	D2	Reflux	4 h	2:1:1

Table 3. Optimization of copper nanoparticles synthesis (without stabilizer).

No.	Condition	Synthesis methods	Time	Melinjo leaves extract: $CuSO_4$ solution
1	A3	Reflux	4 h	1:1
2	B3	Reflux	4 h	1:2
3	C3	Reflux	4 h	1:3
4	D3	Reflux	4 h	1:4

Table 4. Optimization of copper nanoparticles synthesis (beta cyclodextrin as a stabilizer.

No.	Condition	Synthesis methods	Time	Melinjo leaves extract: $CuSO_4$ solution: beta cyclodextrin
1	A4	Reflux	4 h	1:2:2
2	B4	Magnetic stirer	5 h	1:2:2
3	C4	Reflux	4 h	1:2:2
4	D4	Reflux	4 h	1:2:2

3 RESULTS AND DISCUSSION

The Surface Plamin Resonance (SPR) of Cu nanoparticles shows peak wavelengths in the range of 400-700 nm (Putri et al. 2019). Synthesis is carried out until there is color change in nanoparticles solution. The nanoparticles color change was occurred after 4 hours of reflux. The color changes from light green to light brown. The change in color indicates that the nanoparticles have formed. The best results on optimization of synthesis conditions are the synthesis conditions with a ratio between Melinjo leaves extract: $CuSO_4$: betacyclodextrin = 1: 2: 2 (v/v/v) using reflux for 4 hours. The results show the SPR in copper nanoparticles range. The results can be seen in Figure 1.

After that the characterization is done using Particle Size Analyzer. The results showed that the size of the nanoparticles formed was 1734.4 nm with a polydispersity index of 0.479. The results can be seen in Figure 2.

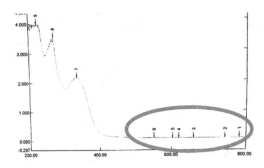

Figure 1.   Copper nanoparticles solution spectrum.

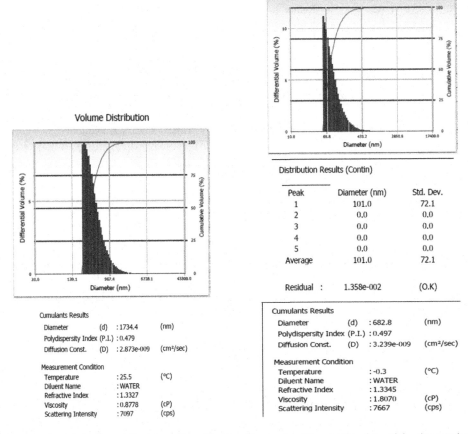

Figure 2.   The result of particle size analyzer for the
best synthesis condition.

Figure 3.   The result of particle size analyzer
after 4 hours of sonication.

In order to improve the particle size of nanoparticles, the sonication process was carried out
for 4 hours. Sonication utilizes mechanical vibrations from ultrasonic waves, ultrasonic waves
produce high energy and can make large-sized particles into nanometer-sized (Gupta & Kompella 2006). In addition, the sonication process is thought to also be able to prevent the agglom-
eration of copper nanoparticles that are formed so that its stability can be increased. The results
show that there is a decrease in particle size after sonication. The measurement results after sonication were 682.8 nm with a polydispersity index of 0.497. The results can be seen in Figure 3.

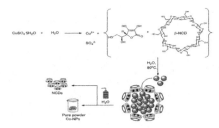

Figure 4. General reaction for the synthesis and stabilization of CuNPs with native cyclodextrin.

Based on Suarez-Cerda, et.al. (2016), general reaction for the synthesis and stabilization of CuNPs with native cyclodextrin can be seen in Figure 4.

From Figure 4 we can find out the mechanism of native cyclodextrin as a stabilizer of copper nanoparticles. Copper nanoparticles that are formed will initially be trapped inside the structure of the native cyclodextrin, then after passing through washing with water, the copper nanoparticles that are formed will separate from the structure of the cyclodextrin. Copper nanoparticles that are formed will be more stable because the stabilization by cyclodextrin will prevent the fusion of the components of nanoparticles that are formed into a non-nano form.The reaction for the synthesis and stabilization of CuNPs with beta cyclodextrin was suspected to be the same as with native cyclodextrin.

4 CONCLUSION

Based on the results of the study, it can be concluded that the synthesis of copper nanoparticles using melinjo leaves extract as bioreductor produces the most optimum conditions in the D4 treatment, which uses a ratio between Melinjo Leaf Extract: CuSO4: betaclodextrin = 1: 2: 2 (v/v/v) with 4 hours of sonication before being characterized by PSA. The result produces nanoparticles, which were suspected as copper nanoparticles, with diameter of 682.8 nm and polydispersity index of 0.497.

ACKNOWLEDGEMENT

This research was funded in part by Penelitian Dosen Muda 2018/2019 by Centre for Research and Community Service, Universitas Islam Bandung, Indonesia.

REFERENCES

Ayu, H. 2015. *Kinetika Sintesis Nanopartikel Perak Dari Larutan AgNO3 Dengan Menggunakan Ekstrak Bungkil Biji Jarak Pagar (Jatropha Curcas L.) Sebagai Reduktor* (Doctoral dissertation, Tesis).

Borkow, G., Gabbay, J., Dardik, R., Eidelman, A. I., Lavie, Y., Grunfeld, Y., ... & Marikovsky, M. 2010. Molecular mechanisms of enhanced wound healing by copper oxide-impregnated dressings. *Wound repair and regeneration*, 18(2): 266–275.

Chatterjee, A. K., Chakraborty, R., & Basu, T. 2014. Mechanism of antibacterial activity of copper nanoparticles. *Nanotechnology*, 25(13): 135101.

Dewi, C., Utami, R., & Riyadi, P. N. H. (2012). Antioxidant and Antimicrobial Activity of Melinjo Extract (Gnetum gnemon L.). *Jurnal Teknologi Hasil Pertanian*, 5(2): 74–81.

Elumalai, E. K., Prasad, T. N. V. K. V., Nagajyothi, P. C., & David, E. 2011. A bird's eye view on biogenic silver nanoparticles and their applications. *Der Chemica Sinica*, 2(2): 88–97.

Gupta, R. B., & Kompella, U. B. 2006. Nanoparticle technology for drug delivery.

Handayani, W. 2011. *Pemanfaatan Tumbuhan Tropis untuk Biosintesis Nanopartikel Perak dan Aplikasinya sebagai Indikator Kolorimetri Keberadaan Logam Berat* (Doctoral dissertation, Thesis diterbitkan).

Kathad, U., & Gajera, H. P. 2014. Synthesis of copper nanoparticles by two different methods and size comparison. *Int J Pharm Bio Sci*, 5(3): 533–540.

Kiranmai, M., Kadimcharla, K., Keesara, N. R., Fatima, S. N., Bommena, P., & Batchu, U. R. 2017. Green synthesis of stable copper nanoparticles and synergistic activity with antibiotics. *Indian Journal of Pharmaceutical Sciences*, 79(5): 695–700.

Kuppusamy, P., Yusoff, M. M., Maniam, G. P., & Govindan, N. 2016. Biosynthesis of metallic nanoparticles using plant derivatives and their new avenues in pharmacological applications–An updated report. *Saudi Pharmaceutical Journal*, 24(4): 473–484.

Lee, H. J., Song, J. Y., & Kim, B. S. 2013. Biological synthesis of copper nanoparticles using Magnolia kobus leaf extract and their antibacterial activity. *Journal of Chemical Technology & Biotechnology*, 88 (11): 1971–1977.

Markham, K. R. 1988. Cara mengidentifikasi flavonoid. *Bandung: ITB*, 1–3.

Prabhu, Y. T., Rao, K. V., Sai, V. S., & Pavani, T. 2017. A facile biosynthesis of copper nanoparticles: a micro-structural and antibacterial activity investigation. *Journal of Saudi Chemical Society*, 21(2),180–185.

Pudjiatmoko. 2007. *Potensi Melinjo di Jepang*. [Online]. Retrieved from: http://id.wikipedia.org/wiki/.

Putri, S. E., Fudhail, A., & Rauf, R. 2019, April. Pengaruh PVA terhadap kestabilan nanopartikel tembaga dari CuSO4 menggunakan bioreduktor kulit buah naga merah (Hylocereus costaricensis). In *Seminar Nasional LP2M UNM*.

Rengga, W. D. P., W. P. Hapsari & D. W. Ardianto. 2017 Synthesis of Copper Nanoparticles from $CuNO_3$ Solution by Using Clove Flower Extract (Syzygium aromaticum). *Jurnal Rekayasa Kimia dan Lingkungan*, 12(1): 15–21.

Shende, S., Ingle, A. P., Gade, A., & Rai, M. 2015. Green synthesis of copper nanoparticles by Citrus medica Linn. (Idilimbu) juice and its antimicrobial activity. *World Journal of Microbiology and Biotechnology*, 31(6): 865–873.

Suárez-Cerda, J., Espinoza-Gómez, H., Alonso-Núñez, G., Rivero, I. A., Gochi-Ponce, Y., & Flores-López, L. Z. 2017. A green synthesis of copper nanoparticles using native cyclodextrins as stabilizing agents. *Journal of Saudi Chemical Society*, 21(3): 341–348.

Thakur, S., Sharma, S., Thakur, S., & Rai, R. 2018. Green synthesis of copper nano-particles using Asparagus adscendens roxb. Root and leaf extract and their antimicrobial activities. *International Journal of Current Microbiology and Applied Sciences*, 7(4): 683–694.

Medical Technology and Environmental Health – Abdullah, Widiaty & Abdullah (eds)
© 2020 Taylor & Francis Group, London, ISBN 978-0-367-86053-0

Alternative drug combination to treat chronic myeloid leukemia resistance in developing countries

A.F. Sumantri
Al Ihsan Hospital, Bandung, Indonesia
Universitas Islam Bandung, Bandung, Indonesia

A. Oehadian & M.H. Bashari
Hasan Sadikin Hospital, Bandung, Indonesia
Padjadjaran University, Bandung, Indonesia

ABSTRACT: The use of Tyrosine Kinase Inhibitors (TKI) is still the main therapy for Chronic Myeloid Leukemia (CML) but in some cases, there is high resistance to TKI due to an autophagy mechanism in resistant stem cells. Chloroquine is a drug that has an anti-autophagy function. The objective of this study was to analyze the synergism of a combination therapy of chloroquine and nilotinib in causing cell death in nilotinib-resistant CML. The research method used in this study was an in vitro laboratory experimental test using peripheral blood mono-nuclear cells from CML patients that failed with TKI. The results indicated that the combination of chloroquine and nilotinib had synergistic results in increasing the death of CML cells that failed with TKI. This study concludes that the combination of nilotinib with chloroquine is synergistic in causing the death of nilotinib-resistant CML cells and it can increase the death of CML cells compared to nilotinib alone. However, further research needs to be done on the anticancer activity of the combination of chloroquine with nilotinib in clinical trials.

1 INTRODUCTION

Chronic Myeloid Leukemia (CML) is a myeloproliferative malignancy with an estimated incidence of 1–2 cases per 100,000 adults. On average, about 15% of new CML cases are diagnosed as adults. In 2015, 7,000 new cases of CML were diagnosed in the USA and around 1,100 cases were found dead due to CML (Jabbour & Kantarjian 2016).

CML is caused by translocation t (9; 22), also known as the Philadelphia (Ph) chromosome, which is a fusion of genes between Abelson Murine Leukemia (ABL1) on chromosome 9 and Breakpoint Cluster Region (BCR) on chromosome 22, and then produces genes that can express oncoprotein, namely BCR-ABL1, which has the ability to activate Tyrosine Kinase (TK) so that it can trigger leukemogenesis through cytokine-independent cell cycle 2 (Lussana et al. 2018, Hamad 2013). The management of CML significantly changed with the development of the Tyrosine Kinase Inhibitors (TKI) molecule which has the potential for intervention in the interaction process between BCR-ABL1 oncoprotein and Adenosine Triphosphate (ATP), which will inhibit cellular proliferation of malignancies (Hamad 2013, Reksodiputr et al. 2015).

The use of TKI in CML treatment is still the main therapy because it is considered to be highly effective. However, at present, up to one-third of new cases of CML have shown TKI resistance (Lussana et al. 2018). In a multi-center Indonesian study, there was higher TKI resistance, 47.69% compared to 24% in Europe (Jabbour 2013, Druker 2006).

TKI resistance can be interpreted as the inability of TKI drugs to reach pharmacological targets as a consequence of reduced bioavailability, in vivo inactivation, and negative interaction with other substances, and it can be evaluated through hematological, cryptogenic, and molecular responses within a certain period (Hamad 2013, Jabbour 2013).

Various molecular and genetic mechanisms in CML cases are basic for the formation of TKI resistance mechanisms: one of them is the mechanism of CML stem cells. One of the mechanisms in CML stem cells causing resistance is the autophagy mechanism. Autophagy is a process of recycling a cell under conditions of stress (hypoxia, malnutrition, drug therapy, etc.), thus causing a cell to survive in these conditions. One drug that is known to function as an anti-autophagy agent is chloroquine (Hamad 2013, Mukhopadhyay 2016).

Various studies have shown an increase of cell death in CML cells in the first-line TKI resistance, imatinib, when combined with an anti-autophagy agent, chloroquine (Mukhopadhyay 2016, Crowley 2013, Carew 2007).

In Indonesia, the only available TKIs are imatinib as the first line and if there is further resistance, only nilotinib as the second line. Besides, there are not yet available third-line TKIs that can overcome second-line TKI resistance. Under these conditions, the researchers tried to find alternative combination therapies to overcome second-line resistance. It is hoped that through this study, researchers can provide alternative combination therapies in both first-line imatinib and second-line nilotinib–resistant CML cases.

2 METHOD

This research was an in vitro laboratory experimental test wherein the subject was a mononuclear cell that isolated from peripheral blood of CML patients with TKI Nilotinib resistance who sought treatment at the Hemato-Oncology Polyclinic of Hasan Sadikin Hospital Bandung. Nilotinib resistance is determined based on treatment evaluation reviewed through the absence of a major molecular response after one year with an BCR-ABL result > 0.1%.

Assessment of patients' characteristics as cell donors includes gender, age, BMI, duration of CML, initial BCR value, duration of imatinib therapy, duration of nilotinib therapy, the latest BCR-ABL value, hematologic response of TKI therapy, and examination of palpable palpation.

Procedure for mononuclear cell isolation was carried out using Ficoll-Paqque media and then inserted into centrifugation tubes of 3 ml and carefully poured 4 ml of peripheral blood specimens to form layers of blood samples and Ficoll-Paqque media. Centrifugation was carried out at a speed of 400 g for 30–40 minutes at a temperature of 18–20°C. After that, the researchers moved the mononuclear cell layer into a sterile centrifugation tube and then calculated viability of the cells before the cells were treated, and then analyzed by researchers.

All data from this study were processed using CompuSyn® software using the Chou-Talalay method to determine test results of drug combinations that provide a theoretical basis for a combination index (CI) equation that allows the quantitative determination of drug interactions. Meanwhile, to determine the effect of nilotinib combined with chloroquine on CML death cell, a one-way ANOVA test was used, which was then continued with a post hoc test using the Tukey test if the ANOVA test results gave meaningful results.

3 RESULTS AND DISCUSSION

3.1 *Results*

The study was conducted over a period of four months, September–December 2018, and has received a letter of ethics from the Medical Research Ethics Committee of Padjajaran University, Dr. Hasan Sadikin Hospital Bandung.

From medical records for 2017–2018, 45 CML patients received nilotinib therapy at the Hemato-Oncology Polyclinic of Department of Internal Disease Faculty of Medicine Padjajaran University/Hasan Sadikin Hospital Bandung. Eleven of them showed resistance to nilotinib therapy as indicated by the results of the examination that still showed leukocytosis, thrombocytosis, and BCR-ABL detected after nilotinib treatment for 12 months. Four people were chosen randomly to participate in the study.

Based on the study subjects, all patients had previously used imatinib for three to six years. The BCR-ABL value of the study subjects were still quite high, ranging from 34% – 60.69% after receiving nilotinib second-line TKI therapy. Two of the four study subjects had received nilotinib therapy for four years, while the other two patients were one and two years. All patients still have not shown hematologic response.

Results of cell viability calculations showed the presence of contamination in the sample of thesecond patient, so the researchers conducted a retrial. However, the results of retesting the sample of second patient still found contamination in the cell, so the researchers decided to exclude the sample of the second patient from this study. Therefore, this study only used patient's samples 1, 3, and 4 (see Figure 1).

When the cell is in the log-growth phase, the cell is harvested and the cell is counted. The recommended optimal calculation of the cell is 1×10^4 cells/ml (cell density). To adjust to plate 96, the test drug is added to nilotinib or chloroquine to get the results of drug testing with the cell viability test.

Based on Figure 2, results of tests with the use of the single nilotinib drug have a minor effect on increasing cell death. The use of a nilotinib-and-chloroquine combination drug increases the rate of cell death in line with the increasing concentration of the drug nilotinib and chloroquine that is given.

Based on Figure 3, the use of a combination of nilotinib and chloroquine has synergistic results in increasing cell death (CI <1). So, in this case, the use of a combination of nilotinib with 10 mg and 25 mg of chloroquine can increase the rate of cell death as the concentration of the nilotinib drug and chloroquine that is given.

The average cell death upon administration of a single dose of nilotinib 10μ, a combination of nilotinib 10μ + chloroquine 10μ, and of nilotinib 10μ + chloroquine 25μ, were 8.08% (SD = 1.49), 19.61% (SD = 10.27) and 23.50% (SD = 1.99) respectively. Whereas, the average cell death between treatment of a single dose of nilotinib 25μ, the combination of nilotinib 25μ + chloroquine 10μ, and nilotinib 25μ + chloroquine 25μ, were 10.63% (SD = 2.06), 30.21% (SD =

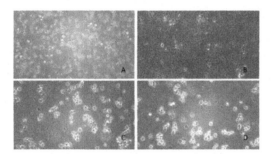

Figure 1. Result of cell viability calculation showed the presence of contamination in the sample of second patient (B).

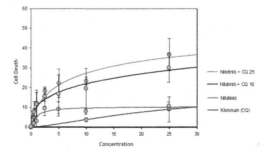

Figure 2. Average results of cell viability test with single nilotinib drug and nilotinib combination with chloroquine.

221

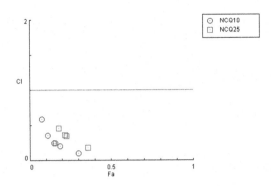

Figure 3.   Test results of synergism combination of nilotinib + chloroquine.

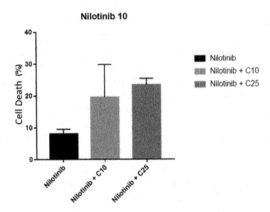

Figure 4.   Average cell mortality in treatment of single 10μ nilotinib drug combined with chloroquine 10μ and 25μ.

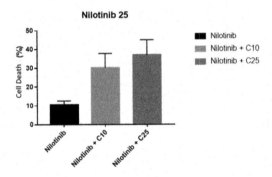

Figure 5.   Average of cell death in treatment of 25μ nilotinib combined with 10μ and 25μ chloroquine.

7, 55) and 36.84% (SD = 8.13) respectively. Analysis of these data showed significant differences (p <0.05) of cell death between single nilotinib with the combination of nilotinib + chloroquine at doses of 10μ and 25μ (see Figure 5). Because ANOVA test results showed significant difference, post-hoc test was conducted to see which groups were different (see Figure 4).

3.2   *Discussion*

This study aimed to analyze the synergism of chloroquine combination therapy with nilotinib in causing the death of nilotinib-resistant CML cells and it compares the differences in the percentage of CML cell death between single drug nilotinib and the combination of nilotinib with chloroquine.

This study used nilotinib (0.5 µg, 1 µg, 2.5 µg, 5 µg, 10 µg, and 25 µg) drug tests combined with chloroquine (10 µg and 25 µg) in vitro because they are considered potent in producing CML resistant cell death.

Nilotinib is designed as a BCR-ABL inhibitor, potentially 30 times more potent than imatinib in vitro, with specification against BCR-ABL. Nilotinib is used in CML patients with resistant or intolerant imatinib in chronic and acceleration phases (not for the blast and Ph + LLA phases) at a dose of 400 mg twice daily (O'Dwyer et al. 2009).

Chloroquine is a drug known to function as an anti-autophagy agent, with weak basic properties so that it can enter an environment with acidic conditions (such as endosomes and lysosomes) during the protonization process and works in the final stages of the autophagy process by inhibiting the degradation process of autolysosomes that are formed from merging auto phagosomes with lysosomes (Mukhopadhyay 2016, Maes 2016).

Through the process of intracellular examination, the use of TKI therapy can induce the autophagy process (drug-induced autophagy). In CML cells, the autophagy process induced by TKI has a role in the mechanism of the cell's defense, and this process can be suppressed by chloroquine (Salomoni & Calabretta 2009).

CML stem cells have characteristics similar to those of hematopoietic stem cells in normal cells, both of which can regenerate and grow populations of heterogeneous cells: these differences are found in BCR-ABL genetic markers that are specific in CML disease. With the presence of CML stem cells, future treatment is expected to be carried out. Based on this, the study subject is primary nucleated cell (peripheral blood mononuclear cell) isolated from the peripheral blood of CML patients with nilotinib TKI resistance. This cell was chosen by the researchers because it can increase physiological relevance of data obtained from in vitro tests or in vivo models.

After obtaining nucleated primary cells, a procedure of mononuclear cell isolation is carried out using Ficoll-Paqque media and followed by cleaning the isolated mononuclear cells to obtain the desired mononuclear cells and then calculating cell concentration (viable cells/ml), which is followed by cell implantation in the medium and then observed for the growth after two to three days. After that the harvest was carried out and cell counts were performed in log phase with Alamar Blue's reagent before being tested using test drugs.

Alamar Blue's reagent in cell viability assay quantitatively measures mammalian, bacterial, and fungal cell proliferation. Dyes incorporate good, oxidative-reduction (REDOX) indicators, fluorescence, and discoloration in response to chemical reduction in growth media due to cell growth (Sun et al. 2018).

Resauzin is the active ingredient of Alamar Blue's reagent, a non-toxic permeable cell compound that is blue and almost non-fluorescent. When entering cells, resazurin is reduced to resorufin, which is a red compound and highly fluorescent. Proper cells continuously convert resazurin to resorufin, increasing the overall fluorescence and cell color around the media (Rampersad 2012).

Previous research has been carried out by Mishima using the combination of imatinib drug and chloroquine to confirm that the autophagy mechanism can regulate systems cell defense in K562 cells, and imatinib is used to induce cell death in K562 cells. From his research data, Mishima concluded that autophagy is strongly associated with cell defense systems and autophagy inhibition, which accelerates cell death induced by TPA or imatinib so that the mechanism of this autophagy block can become a new strategy in CML treatment (Mishima 2008).

4 CONCLUSION

In this study, the combination of nilotinib with chloroquine is synergistic in causing the death of nilotinib-resistant CML cells and increasing cell death of CML compared to single-drug nilotinib. This study can lead to new strategies in CML's treatment with nilotinib resistance so that cases of CML patients with nilotinib resistance can continue their treatment especially in a country with limited access to third-line TKI.

ACKNOWLEDGEMENTS

We acknowledge the Hematology Medical Oncology Division, Department of Internal Medicine, Hasan Sadikin Hospital Bandung and Laboratory of Culture and Cytogenetic, Faculty of Medicine, Padjadjaran Univerity, Indonesia. We thank Mikoriza Mustofanny and Eliza Nurazizah for their help in collecting data.

REFERENCES

Carew, J. S., Nawrocki, S. T., Kahue, C. N., Zhang, H., Yang, C., Chung, L., … & Cleveland, J. L. 2007. Targeting autophagy augments the anticancer activity of the histone deacetylase inhibitor SAHA to overcome Bcr-Abl–mediated drug resistance. *Blood: The Journal of the American Society of Hematology*, 110(1): 313–322.

Crowley, L. C., O'Donovan, T. R., Nyhan, M. J., & McKENNA, S. L. 2013. Pharmacological agents with inherent anti-autophagic activity improve the cytotoxicity of imatinib. *Oncology Reports*, 29(6): 2261–2268.

Druker, B. J., Guilhot, F., O'Brien, S. G., Gathmann, I., Kantarjian, H., Gattermann, N., … & Cervantes, F. 2006. Five-year follow-up of patients receiving imatinib for chronic myeloid leukemia. *New England Journal of Medicine*, 355(23): 2408–2417.

Hamad, A., Sahli, Z., El Sabban, M., Mouteirik, M., & Nasr, R. 2013. Emerging therapeutic strategies for targeting chronic myeloid leukemia stem cells. *Stem Cells International*, 2013.

Jabbour, E. J., Cortes, J. E., & Kantarjian, H. M. 2013. Resistance to tyrosine kinase inhibition therapy for chronic myelogenous leukemia: A clinical perspective and emerging treatment options. *Clinical Lymphoma Myeloma and Leukemia*, 13(5): 515–529.

Jabbour, E., & Kantarjian, H. 2016. Chronic myeloid leukemia: 2016 update on diagnosis, therapy, and monitoring. *American Journal of Hematology*, 91(2): 252–265.

Lussana, F., Intermesoli, T., Stefanoni, P., & Rambaldi, A. 2018. Mechanisms of resistance to targeted therapies in chronic myeloid leukemia. In *Mechanisms of Drug Resistance in Cancer Therapy* (pp. 231–250). Springer, Cham.

Maes, H., Kuchnio, A., Carmeliet, P., & Agostinis, P. 2016. Chloroquine anticancer activity is mediated by autophagy-independent effects on the tumor vasculature. *Molecular & Cellular Oncology*, 3(1): e970097.

Mishima, Y., Terui, Y., Mishima, Y., Taniyama, A., Kuniyoshi, R., Takizawa, T., … & Hatake, K. 2008. Autophagy and autophagic cell death are next targets for elimination of the resistance to tyrosine kinase inhibitors. *Cancer Science*, 99(11): 2200–2208.

Mukhopadhyay, S., Sinha, N., Das, D. N., Panda, P. K., Naik, P. P., & Bhutia, S. K. 2016. Clinical relevance of autophagic therapy in cancer: Investigating the current trends, challenges, and future prospects. *Critical Reviews in Clinical Laboratory Sciences*, 53(4): 228–252.

O'Dwyer, M. C., Kent, E., Parker, M., Swords, R. T., Giles, F., Coutre, P. L., … & Egan, K. 2009. Nilotinib 300 mg twice daily is effective and well tolerated as first line treatment of Ph-positive chronic myeloid leukemia in chronic phase: Preliminary Results of the ICORG 0802 Phase 2 Study, *Blood Joournals*, 114(22): 3294 .

Rampersad, S. N. (2012). Multiple applications of Alamar Blue as an indicator of metabolic function and cellular health in cell viability bioassays. *Sensors*, 12(9): 12347–12360.

Reksodiputro, A. H., Tadjoedin, H., Supandiman, I., Acang, N., Kar, A. S., Bakta, I. M., … & Salim, N. 2015. Epidemiology study and mutation profile of patients with chronic myeloid leukemia (CML) in Indonesia. *J Blood Disord Transfus*, 6(3): 1000271.

Salomoni, P., & Calabretta, B. 2009. Targeted therapies and autophagy: New insights from chronic myeloid leukemia. *Autophagy*, 5(7): 1050–1051.

Sun, Y., Wang, P., Li, H., & Dai, J. 2018. BMAL1 and CLOCK proteins in regulating UVB-induced apoptosis and DNA damage responses in human keratinocytes. *Journal of cellular physiology*, 233(12): 9563–9574.

Medical Technology and Environmental Health – Abdullah, Widiaty & Abdullah (eds)
© *2020 Taylor & Francis Group, London, ISBN 978-0-367-86053-0*

Total polyphenol and flavonoid content comparation of Kertasari Arabica coffee (*coffea arabica* L.) leaves, pulp, and beans

E.R. Sadiyah, L. Purwanti, S. Hazar, S.O. Sasmita & A. Yuniarti
Universitas Islam Bandung, Bandung, Indonesia

ABSTRACT: Kertasari Arabica coffee (*Coffea arabica*, L.) has been known for its specific characteristics and has international sale value. Many research studies have shown that coffee has antioxidant activity, which is related to the content of polyphenols and flavonoid substances. Polyphenol substances contained in coffee plant parts are one of largest secondary metabolite groups produced by plant. Phenol and polyphenol have an important role in long life health care, reduce the risk of chronic and degenerative diseases. The roles may be related to antioxidant activity, antitumor, antiviral, and antimicrobe. Besides the beans, coffee leaves and pulp are also known to have polyphenols and flavonoids, but there is no information about which plant part of Kertasari Arabica coffee has the highest level. Kertasari Arabica coffee leaves, pulp, and beans were extracted by maceration with 96% ethanol. The next step was total polyphenols and flavonoids content measurement from each extract collected. The results showed that the level of total polyphenols content in coffee leaves, peels, and beans consecutively were 181.07; 37.05; and 112.39 mg GAE/g extract. Furthermore, based on the measurement results, the level of total flavonoids in coffee leaves, peels, and beans consecutively were 7.98; 5.43; and 6.85 mg QE/g extract. To be concluded, the leaves part had the highest level of polyphenols and flavonoids content in Kertasari Arabica coffee.

1 INTRODUCTION

Free radicals are molecules with non-paired electrons. Free-radical substances are very unstable and readily react with other substances, trying to catch their electrons to become stable (Sarma et al. 2010). Free radicals are naturally formed as by-product from metabolism. When free radicals contained in the human body reach beyond the limit, then it will become a "stress" condition. High levels of free radicals in the biological system might trigger degenerative diseases (Giusti & Jing 2007). Antioxidants may inhibit or dampen the negative oxidants or free radicals in the body (Winarsi 2007). These substances are found in many plants. A plant may have antioxidants if it contains the substances that can catch free radicals, such as phenolics, flavonoids, and carotenoids (Nurung 2016). One of the flavonoid that functions as an antioxidant is anthocyanin (Shipp & Abdel-Aal 2010, Giusti & Jing 2007).

Coffee is a drink that contains the highest level of antioxidants, followed by red wine and green tea (Carelsen et al. 2010). Coffee beans, leaves, and peels contain polyphenols and flavonoids, both of which are phenolic substances that have antioxidant potency (Apak et al. 2007).

Arabica coffee (*Coffea arabica* L.) is a kind of coffee and the first that was cultivated, especially in Indonesia (Aak 2006). West Java is the area that produces the most kinds of coffee and has been acknowledged in the world. One of the products is Arabica coffee, which has more consumers and the price is higher (Budiman 2017). One of the areas in West Java that became the Arabica coffee cultivation center is Kertasari Sub-district, Bandung District (Lestari 2017a), located on 1,700 m from sea level (Bandung District Goverment 2018). Kertasari Arabica coffee is known for its unique characteristics and has international-grade selling value (Lestari 2017b). Potentially, Kertasari Arabica coffee can be further developed, including in pharmacology and health, especially in terms of the data about polyphenolic and flavonoid content as antioxidants in the coffee leaves, pulp, and beans.

Based on the information, this research aimed to measure and compare total polyphenolic and flavonoid content of Arabica coffee (*Coffea arabica* L.) leaves, peels, and beans that collected from Kertasari Sub-district, at Bandung District.

2 METHODS

2.1 *Plant sample collection, preparation, phytochemical screening, and extraction*

This research was conducted in stages, beginning with plant sample collection and preparation, and continuing with phytochemical screening using leaves, pulp, and beans. The next step was extraction by maceration using 500 g from each sample and 96% ethanol as solvent. After having concentrated extract, the yield and the content of total polyphenols and flavonoid were measured from each sample extract.

2.2 *Total polyphenol content measurement*

The measurement of total polyphenol content was done according to Chun et al. (2003). Standard curve was made using six concentrations of gallic acid solution (5, 15, 30, 50, 70, and 100 ppm). Then 0.5 ml of each solution was mixed with 2.5 ml of Folin-Ciocalteu reagent (7.5% in distilled water) and incubated at room temperature for 8 minutes. After that, 2 ml of NaOH solution was added. The mixture was then homogenized and incubated for 1 hour at room temperature. The absorbance of each solution was then measured using a UV-visible spectrophotometer at 732 nm wavelength. A linear regression equation was derived from the standard curve produced.

For sample measurement, about 10 mg from each extract was dilluted with 10 ml ethanol (pro analytical grade) to get an initial solution (1,000 ppm). Then 10 ml of methanol (pro analytical grade) was added to 5 ml of initial solution to get a 500 ppm extract solution. The solutions were then mixed and incubated with Folin-Ciocalteu and NaOH using the same procedure as the gallic acid standard. The absorbance of each sample was then measured using a UV-visible spectrophotometer at 732 nm wavelength. Using the linear regression equation from the gallic acid standard curve, the sample absorbances were then calculated to get the level of total polyphenols from each extract.

2.3 *Total flavonoid content measurement*

In order to measure the level of total flavonoids, first the standard curve using quercetin was made, by diluting 25 mg of standard quercetin in 25 ml 96% ethanol to produce the initial solution (100 ppm). This initial solution was then diluted to make 20, 30, 40, 50, and 60 ppm quercetin solution. Then 0.5 ml of each solution was mixed with 0.1 ml AlCl$_3$ (10%), 0.1 ml Na-acetate (1 M), and 2.8 ml of distilled water. The mixture was then incubated at room temperature for 30 minutes before the absorbances were measured with a UV-visible spectrophotometer at 437.5 nm wavelength.

For the sample measurement, 20 mg from each extract was diluted in 10 ml ethanol (pro analytical grade) to get the extract solution (2,000 ppm). The next step was similar to the procedures conducted for standard quercetin preparation, including the absorbance measurement using a UV-Vis spectrophotometer. As in total polyphenol content, the level of total flavonoid was calculated using the linear regression equation that resulted from the standard quercetin calibration curve.

3 RESULTS AND DISCUSSION

3.1 *Plant sample collection, preparation, phytochemical screening, and extraction*

Fresh Kertasari Arabica coffee leaves (6.5 kg), peels (3.8 kg), and green beans (4.8 kg) were dried using an artificial dryer with lamp (36–40°C, 7–10 days). The beans were soaked in water for 18 hours to release the sticky left-over peels enzymatically by natural fermentation. After the drying process, the resultant dry leaves were 610 g, while the dry peels were 580 g, and the beans were 1.08 kg. Based on the result of extraction, the highest yield was from leaves (19.57%), followed by peels (3.55%), and beans (2.63%).

Table 1 showed the results of phytochemical screening from each crude drug and extract. From these results, the three samples showed the same content whether in the crude or the extract. The result was relevant to the results explained by Harahap (2017) for peels and beans (Gunalan et al. 2012). Only saponins were not detected in the samples. This finding was different from the result mentioned by Nayeem et al. (2011), that coffee leaves contain saponins.

3.2 *Total polyphenol content measurement*

Figure 1 showed the standard gallic acid calibration curve, with the resultant linear regression equation y = 0.0083x + 0.0309 (R^2 = 0.9912).

Folin-Ciocalteu reagent was used to measure the total polyphenol content in the samples. This method is based on hidroxy phenolic residue oxidation by the reagent in base condition that would produce a blue molybdenum-tungstate complex (Zhang & Hamauzu 2004). According to the calculation results, the level of total polyphenol content in leaves, peels, and beans of Kertasari Arabica coffee were 180.98, 37.13, and 112.31 mg GAE/g extract, respectively.

Phenolic substances hava an aromatic ring with one or more hidroxyl residues (OH-) and other residues. Most phenolic substances have more than one hidroxyl residue so they are named polyphenols. Rice-Evans et al. (1997) explained that polyphenols may act as antioxidants because the substances have the ability to reduce or have a role as hydrogen donor.

Table 1. Results of phytochemical screening.

Phytochemicals	Crude drug & Extract		
	Leaves	Peels	Beans
Flavonoids	+	+	+
Polyphenols	+	+	+
Alkaloids	+	+	+
Tanins	+	+	+
Saponins	–	–	–
Monoterpenes/Sesquiterpenes	+	+	+
Triterpenes/Steroids	+	+	+

(+) Identified, (–) Not identified

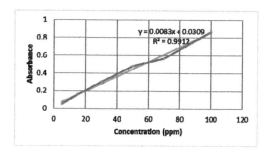

Figure 1. Linear regression calibration curve of standard gallic acid absorbance.

3.3 Total flavonoid content measurement

The results of standard quercetin (20–60 ppm) absorbance measurement showed in Table 2.

Based on the absorbances in Table 2, a linear regression curve was obtained with the equation $y = 0.0088x - 0.0235$, and $R^2 = 0.999$ (Figure 2). Table 3 showed the result of average absorbances from each sample extract.

Based on the calculation result, the highest level of total flavonoid showed in leaves extract (Table 4).

The antioxidant potency was confirmed from the result of total polyphenolic and flavonoid content. The results were in line with previous findings by Sasmita et al. (2019) using the same sample that the antioxidant activity of coffee leaves ethanol extract (IC50 35.24 ppm) was higher than coffee bean extract (IC50 57.45 ppm), and coffee pulp extract (IC50 175.46 ppm).

Table 2.   Standard quercetin absorbance.

Concentration (ppm)	Average absorbance
20	0.152 ± 0.002
30	0.240 ± 0.004
40	0.326 ± 0.013
50	0.417 ± 0.019
60	0.502 ± 0.016

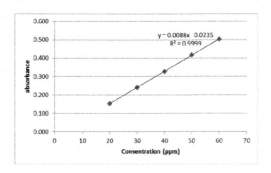

Figure 2.   Linear regression calibration curve of standard quercetin absorbance.

Table 3.   Results of average absorbance from leaves, peels, and beans extract.

Extract	Average absorbance
Leaves	0.117 ± 0.0017
Peels	0.097 ± 0.0015
Beans	0.072 ± 0.0023

Table 4.   Total flavonoid content of leaves, peels, and beans extract.

Extract	Average total flavonoid content
Leaves	15.966 ppm or 7.98 mg/g extract
Peels	10.852 ppm or 5.43 mg/g extract
Beans	13.693 ppm or 6.85 mg/g extract

4 CONCLUSIONS

Based on the results, the leaves, peels, and beans of Kertasari Arabica coffee (*Coffea arabica* L.) contained the same phytochemical substance. The leaves extract gave the highest yield (19.57%). The leaves also showed the highest polyphenol (180.98 mg GAE/g extract) and flavonoid (7.98 mg/g extract) content. The lowest polyphenol (37.13 mg GAE/g extract) and flavonoid (5.43 mg/g extract) content were found in peel extract.

ACKNOWLEDGEMENT

The authors would like to give appreciation to LPPM Universitas Islam Bandung for the research funding and supports.

REFERENCES

Aak. 2006. *Budidaya Tanaman Kopi*. Penerbit Kanisius: Yogyakarta.

Apak, R., Güçlü, K., Demirata, B., Özyürek, M., Çelik, S. E., Bektaşoğlu, B., ... & Özyurt, D. 2007. Comparative evaluation of various total antioxidant capacity assays applied to phenolic compounds with the CUPRAC assay. *Molecules*, 12(7): 1496–1547.

Bandung District Government. 2018. *Kecamatan Kertasari* [Online]. http://www.bandungkab.go.id/arsip/20180213013708-profil-kecamatan-kertasari.

Budiman, A. 2017. *Bersaing dengan Kopi Arabika, Peminat Robusta Menyusut*, [Online]. https://www.pikiranrakyat.com/ekonomi/2017/09/03/bersaing-dengan-kopi-arabika-peminat-robusta-menyusut-408618.

Carlsen, M. H., Halvorsen, B. L., Holte, K., Bøhn, S. K., Dragland, S., Sampson, L., ... & Barikmo, I. 2010. The total antioxidant content of more than 3,100 foods, beverages, spices, herbs and supplements used worldwide. *Nutrition Journal*, 9(1): 3.

Chun, O. K., Kim, D. O., & Lee, C. Y. 2003. Superoxide radical scavenging activity of the major polyphenols in fresh plums. *Journal of Agricultural and Food Chemistry*, 51(27): 8067–8072.

Giusti, M. M., & Jing, P. 2007. Natural pigments of berries: Functionality and application. *Food Science And Technology-New York-Marcel Dekker*, *168*, 105.

Gunalan, G., Myla, N., & Balabhaskar, R. 2012. In vitro antioxidant analysis of selected coffee bean varieties. *Journal of Chemical and Pharmaceutical Research*, 4(4): 2126–2132.

Harahap, M. R. 2017. Identifikasi Daging Buah Kopi Robusta (Coffea robusta) Berasal Dari Provinsi Aceh. *Elkawnie*, 3(2): 201–210.

Lestari, E. A., 2017a. *Ironis, Kopi Jawa Barat Diklaim Daerah Lain* [Online]. https://www.pikiran-rakyat.com/bandung-raya/2017/10/10/ironis-kopi-jawa-barat-diklaim-daerah-lain-411239.

Lestari, E. A., 2017b. *Kopi Kertasari Akan Disuguhkan di Sidney* [Online]. https://www.pikiran-rakyat.com/hidup-gaya/2017/09/29/kopi-kertasari-akan-disuguhkan-di-sidney-410479

Nayeem, N., Denny, G., & Mehta, S. K. 2011. Comparative phytochemical analysis, antimicrobial and antioxidant activity of the methanolic extracts of the leaves of Coffea arabica and Coffea robusta. *Der Pharmacia Lettre*, 3(1): 292–297.

Nurung, S. H. H. 2016. Penentuan Kadar Total Fenolik, Flavonoid, dan Karotenoid Ekstrak Etanol Kecambah Kacang Hijau (Vigna radiata L.) Menggunakan Spektrofotometer UV-Vis.

Rice-Evans, C., Miller, N., & Paganga, G. 1997. Antioxidant properties of phenolic compounds. *Trends in Plant Science*, 2(4): 152–159.

Sarma, A. D., Mallick, A. R., & Ghosh, A. K. 2010. Free radicals and their role in different clinical conditions: an overview. *International Journal of Pharma Sciences and Research*, 1(3): 185–192.

Sasmita, S. O., Purwanti, L., and Sadiyah, E. R. 2019. Antioxidant activities comparison between Arabica Coffee (*Coffea arabica* L.) leaves, pulp and its beans ethanol extract with DPPH-Free radical scavanging method. *Prosiding Farmasi, SPESIAL Unisba*, 5(2).

Shipp, J., & Abdel-Aal, E. S. M. 2010. Food applications and physiological effects of anthocyanins as functional food ingredients. *The Open Food Science Journal*, 4(1).

Winarsi, H. 2007. Antioksidan Alami dan Radikal Bebas Penerbit Kanisius.

Zhang, D., & Hamauzu, Y. 2004. Phenolic compounds and their antioxidant properties in different tissues of carrots (*Daucus carota* L.). *Journal of Food Agriculture and Environment*, 2, 95–100.

Phycochemical screening and standard parameter determination of *Spirulina plantesis, Chlorella vulgaris*, and *Euchema spinosum,* cultivated in Indonesia

I.T. Maulana, L. Mulqie, K.M. Yuliawati, Y. Sukarman, N.A. Suhara, N.A. Suhara &
R. Safira
Universitas Islam Bandung, Bandung, Indonesia

ABSTRACT: *Spirulina plantesis, Chlorella vulgaris*, and *Euchema spinosum* are three kinds of algae samples that have been cultivated in Indonesia. These three algae are used both as food ingredients, medicinal ingredients, and cosmetics. Until now, there has been no official release related to the applicable standard parameters of algal material. Therefore, this research study aims to analyze the content of active compounds and the standardization of algae involved in loss-on-drying value, total ash content, acid-insoluble ash content, and water-soluble and ethanol-soluble extractable matter. Phycochemical screening showed that all of the algae samples contained alkaloids, flavonoids, and steroids. Standard parameter determination showed that polar compounds contained in all algae were more dominant than semi-polar and non-polar compounds; it could be seen in the value of water-soluble extractable matter, which was greater than ethanol-soluble extractable matter. All algae were also rich in inorganic compounds, as indicated by a total ash content of more than 5%. But even so, all algae were indicated safe from dangerous heavy-metal content. This can be seen from the value of acid-insoluble ash levels lower than 1%.

1 INTRODUCTION

Algae are still not widely used as an active material of traditional medicine in Indonesia. However, some algae such as *Spirulina plantesis, Chlorella vulgaris*, and *Eucheuma spinosum* have been cultivated in Indonesia. Every natural material that would be used as medicinal active material must fulfill the specified quality requirements to ensure the safety, efficacy, and quality of the materials. An herbal material cannot be promoted as quality material if it does not meet the quality requirements of the Indonesian Herbal Pharmacopoeia (Menkes 2009).

The problem in Indonesia at this time is that there is no official standard related to algae material when used as herbal material. However, standardization has an important role in maintaining consistent quality of medicinal active materials so as to produce traditional medicines of consistent quality (Pandey et al. 2016, Torey et al. 2010). The quality of materials, especially algae, is mostly determined by the location and environment in which it grows, including altitude, temperature, and sunlight (Singh et al. 2005).

Three algae, namely *Spirulina platensis, Chlorella vulgaris*, and *Eucheuma spinosum*, are known to have many health benefits. *Spirulina plantesis* is known to be easily digested and absorbed by the human body (Choi et al. 2013), and to be rich in phycocyanin compounds, cyanocobalamin, tocopherols, carotenoids, and sulfated polysaccharide. Spirulina is also known to inhibit lipid oxidation, and have anti-aging, antitumor, antioxidants (Okechukwu et al. 2019, Choi et al. 2013), and antidiabetic (Okechukwu et al. 2019) properties. *Chlorella vulgaris* is known as a source of an organic and inorganic bioactive substance (Muszyńska et al. 2018), flavonoids (Ahmed 2016), steroids and triterpenoids, phenolic compounds (Adhoni et al. 2016), considerable chlorophyll (Chia et al. 2013, Safafar et al. 2016) and also contains Chlorellin compounds that are suspected to be antibacterial (Asadi et al. 2016, Shannon & Abu-Ghannam 2016, Ghasemi et al. 2007). However, *Chlorella vulgaris* is not a good source of minerals such as zinc, iron, magnesium (Muszyńska et al. 2018). *Chlorella vulgaris* is known to be able to inhibit the growth of *E. coli, P. aeruginosa, S. aureus*, and *S. Pyogenes* bacteria (Shannon and Abu-Ghannam 2016, Syed et al. 2015, Ghasemi et al. 2007). *Chlorella vulgaris* and *Spirulina plantesis* have been also recognized as a single-cell protein source supplement (Kose et al. 2017).

Seaweed agar of class *Rhodophyta* is known to contain more metabolite compounds than other marine agar (Balasubramaniam et al. 2016). Seaweed agar has been widely used in the health supplement, nutraceutical, and pharmaceutical industries (Cian et al. 2015, Hurtado et al. 2014). Some studies linked the consumption rates of seaweed to low cancer prevalence (Cian et al. 2015). Types of *Eucheuma* sp were known to inhibit the α-amylase enzyme, produce anti-inflammatory effects, and have high antioxidant activity (Balasubramaniam et al. 2016). Besides that, *Eucheuma spinosum* was known active to inhibit the growth of *E. Coli* and *S. Aureus* bacteria and it has also a cytotoxic activity (Sugrani et al. 2019).

The purpose of this study is to determine the parameter standard values from three types of algae cultivated in Indonesia. This type of data is very important as a form of standardization of natural products that have the potential to be used as Indonesian traditional medicine.

2 MATERIAL AND METHODS

2.1 *Material*

The material used in this research is *Spirulina plantesis* obtained from PT. Neo Alga Sukoharjo, Central Java, Indonesia; *Chlorella vulgaris* obtained from *Balai Besar Perikanan Budidaya Air Payau* (BBPBAP) Jepara, Central Java; and *Eucheuma spinosum* obtained from the Cipatujah Tasikmalaya region of West Java. Specific reagents used include $FeCl_3$ 1%, Dragendorff, Liebermann Bouchard, sitroborat and vanillin-sulfuric acid, which are made according to listings in the Indonesian Herbal Pharmacopoeia (Menkes 2009). The equipment used includes oven, furnace, and glass and ceramic equipment commonly used in laboratories.

2.2 *Methods*

This research was completely conducted from February to July 2019 in the Pharmacy Laboratory, Faculty of Mathematics and Natural Sciences, Universitas Islam Bandung (UNISBA), Indonesia. The alga was firstly determined via the secondary metabolite content based on the method conducted by (Harborne 1973, Octaviani et al. 2019, Tiwari et al. 2011) methods. The target compounds determined include alkaloids, flavonoids, terpenoids, tannins, and anthraquinones. Saponin numbers were calculated as determined by a foam index as stated in WHO guidelines (WHO 2011).

Determination of water-soluble extractable matter (WSEM) was conducted by weighing a 5-gram alga, then macerated with chloroform saturated aquadest. The mixture was stirred then allowed to stand for 24 hours. The mixture was then filtered and the filtrate obtained was evaporated at a temperature of 105°C until a dried extract was obtained. The water-soluble extract was calculated as a percentage ratio between the amount of dry extract and the material that was used as described at equation (1) (Menkes 2009, WHO 2011).

$$WSEM = \frac{Dry\ Extract}{Amount\ of\ Material} \times 100\% \qquad (1)$$

This procedure was also conducted in the determination of ethanol-soluble extractable matter (ESEM), where the solvent was replaced with ethanol and the equation of extractable matter is described at equation (2).

$$ESEM = \frac{Dry\ Extract}{Amount\ of\ Material} \times 100\% \qquad (2)$$

Total ash content (TAC) was determined by weighing 2 gr of material then ignited using a furnace at 500°C until the charcoal ran out and only ash remained. Next, ash content was calculated as a percentage ratio between the amount of ash obtained and the amount of material used as described at equation (3) (WHO 2011, Menkes 2009).

$$TAC = \frac{Ash}{Amount\ of\ Material} \times 100\% \qquad (3)$$

Ash produced from the determination of TAC, was then dissolved in 2NHCl then boiled for 5 minutes. The insoluble part was then filtered using ash-free filter paper. Insoluble ash was washed with hot water until a neutral pH was obtained. Insoluble ash and filter paper were ignited until ash was obtained. The percentage ratio between the amount of residual ash and the amount of starting material was calculated as acid-insoluble ash content (AIAC) as described at equation (4) (WHO 2011, Menkes 2009).

$$AIAC = \frac{Ash\ insoluble\ in\ acid}{Amount\ of\ Material} \times 100\% \qquad (4)$$

The water content in the material is determined using two methods, namely azeotropic distillation (AD) and loss-on-drying value (LoDV). Azeotropic distillation was conducted by mixing approximately 20 grams of material with 200 mL water-saturated toluene in a distillation flask. The mixture was then boiled until the water and toluene mixture evaporated and passed through the condenser. Distillation was stopped when the volume of water was steady or did not increase anymore. Then the AD was obtained by calculating the percentage ratio between the volume of water and the weight of the starting material as described at equation (5) (WHO 2011, Menkes 2009).

$$AD = \frac{volume\ water}{Amount\ of\ Material} \times 100\% \qquad (5)$$

Loss-on-drying value (LoDV) was determined by drying 2-gram alga using an oven at 105°C until a constant weight was obtained. Then the value of compounds that evaporate was determined by calculating the difference between the weight of starting material (SW) and the weight after heating (WaH). Then the LoDV was obtained by calculating the percentage ratio between that value and the weight of the starting material as described at below equation (6) (WHO 2011, Menkes 2009).

$$LoDV = \frac{SW - WaH}{Amount\ of\ Material} \times 100\% \qquad (6)$$

3 RESULTS AND DISCUSSION

3.1 *Screening of phycochemistry*

The phycochemical screening data of the three algal materials showed varied results that could be seen in Table 1. Alkaloids and flavonoids are secondary metabolites that exist in all three algae. This happens because those two secondary metabolites are derived from the same pathway, which is the shikimic pathway. Shikimate acid is known to be a precursor in the

Table 1. Data on phycochemical screening of algae from Indonesian aquaculture.

Metabolite Group	Chlorella vulgaris	Spirulina platensis	Eucheuma spinosum
alkaloid	+	+	+
flavonoid	+	+	+
saponin	–	+	–
tannin	+	–	–
triterpenoid	–	+	–
steroid	+	–	+
polifenolat	+	+	–
mono and sesquiterpenoid	+	+	+
antraquinon	+	+	–

+ = identified; – = unidentified

formation of amino acids that form proteins that are precursors for alkaloid formation (Dewick 2009). In addition, shikimic acid in the biosynthetic pathway also forms phenylalanine, which is a precursor to the formation of flavonoids.

Alkaloids are secondary metabolites from an amino acid precursor that have a specific structure, namely a nitrogen heterocyclic (Brindha 2016, Dewick 2009). Alkaloids are generally known to have a variety of pharmacological effects, especially influences on the nervous system. Some alkaloids have structural similarities to the neurotransmitter acetylcholine, muscarinic, and adrenergic (Dewick 2009). Several alkaloids also demonstrate as antitumor, antimicrobial, immunostimulant, and analgesic activity (Brindha 2016).

Flavonoids based on the structure of compounds that have two phenyl groups connected by three carbon atoms are often linked as an antioxidant (Brindha 2016, Dewick 2009, Harborne 1973). In fact, many flavonoids are associated with anticancer (Brindha 2016) and antibacterial properties (Linfante et al. 2018). Flavonoids are also known to be able to penetrate the film layer produced by bacterium so can be used to overcome resistance problems (Linfante et al. 2018). The interesting thing was that, although *Eucheuma spinosum* contained flavonoids, the polyphenolate test showed negative results. This is most likely due to the substitution in 5-OH or 3-OH; it could also be because the flavonoid contained is a type of isoflavone (Harborne 1973).

Tannin is a complex phenol compound that is widely found in plants. Tannins are known to have a variety of properties including astringent, anti-cancer, neuroprotector, and immunosuppressant (Shimozu et al. 2017, Brindha 2016). Table 1 shows that only *Chlorella vulgaris* algae were positive for tannins. Monoterpenoids and sesquiterpenoids are terpenoid groups as a compound of many essential oils. Some monoterpenoid and sesquiterpenoid compounds are known to have antibacterial properties.

Saponins are only known in algae *Spirulina plantesis* but it is not known with certainty what kind of saponins were contained in the *Spirulina plantesis*. Saponins are glycoside compounds that formed from secondary metabolites that are glycosylated, such as alkaloid saponins, triterpenoid saponins, steroid saponins, and so forth.

3.2 *The determination of the quality standard from algae*

The three algal materials analyzed were known to contain predominantly polar compounds compared to a sum of semi-polar and nonpolar compounds, as can be seen from the value of WSEM (Table 2) which greater than ESEM. WSEM indicates the amount of polar compound or water-soluble compounds that are extractable when the material is macerated with water solvents. ESEM indicates the number of compounds that are extractable when it is macerated with ethanol solvent. ESEM can also be used to predict yield if the material is macerated with solvent ethanol.

The comparison between WSEM and ESEM value in *Chlorella vulgaris* is almost balanced, which is 51.85:48.15. This means that most of the compounds in *Chlorella vulgaris* are polar.

Table 2. Data analysis of the quality parameters of *Chlorella vulgaris*, *Spirulina plantesis*, and *Eucheuma spinosum*.

	Chlorella vulgaris	Spirulina plantesis	Eucheuma spinosum
WSEM (b/b %)	36.44+0.89	36.60+0.57	20.78+0.13
ESEM (b/b %)	33.84+1.01	6.85+0.93	12.30+0.04
TAC (b/b %)	59.14+1.07	8.54+0.07	22.52+0.17
AIAC (b/b %)	0.22+0.06	0.28+0.01	0.81+0.34
LoDV (b/b %)	8.27+0.07	15.93+0.38	41.7+1.47
AD (v/b %)	8+0.00	7+0.71	6.75+0.87

WSEM = water-soluble extractable matter; ESEM= ethanol-soluble extractable matter; TAC = total ash content; AIAC = ash insoluble in acid content; LoDV = loss-on-drying value; AD = azeotrophic distilation

The extraction using ethanol will produce a yield that is also quite large. *Spirulina plantesis* was known to have a WSEM to ESEM ratio of 84.23:15.77, and *Eucheuma spinosum* has a ratio of 62.82:37.18. This mean that *Spirulina plantesis* will have a lower yield than the other algae when its extracted with ethanol.

Ash content indicates the number of inorganic elements contained in the material. As known, when the incandescent process at a temperature 450°C, all organic elements, such as C, H, O, N, P, K, and others, would evaporate, while the inorganic elements become residue. Inorganic elements are divided into two, namely light metals having a density of less than 4 kg/L and heavy metals having a density of more than 4 kg/L. To calculate the total content of each of these metal groups, a separation method was conducted using an acid solution, HCl. Light-metal groups such as alkali metals, alkaline earth metals, and 3A (Santana et al. 2018) react with acidic compounds to form water-soluble salts, whereas heavy metals, such as Pb, Hg, Cd and transition metals, react with acidic compounds to form complex compounds that do not dissolve. Therefore, the total ash content is related to the number of elements of inorganic compounds – both light metals and heavy metals – contained in the material, while acid-insoluble ash content indicates the number of heavy metal groups contained in the material. The smaller the content of heavy metals, the better the level of material safety.

Chlorella vugaris was known to contain the highest levels of inorganic elements, compared to *Spirulina plantesis* and *Eucheuma spinosum*. The exact cause of this was not yet known with certainty: possibly it was caused by the high content of chlorophyll in algae. Although it contained a fairly high level of metal elements, it can be ascertained that *Chlorella vulgaris* was still safe to use, due to its low heavy-metal content. That was indicated by the AIAC value that less than 1 % (Santana et al. 2018). *Spirulina plantesis* was known to contain the lowest levels of metal elements among other test materials, while *Euchema spinosum* contains the largest heavy-metal level. However, based on the value of AIAC, all materials were still safe to be used as medicinal material where the heavy-metal content levels were less than 1%.

Determining the water content in the material was done by three methods, namely Karl Fischer titration, azeotropic distillation, and gravimetry. The gravimetric method was done by measuring the amount of compound that evaporates/disappeared after the material was heated within a certain period. Water content could be determined by the gravimetric method if it was certain that there were no other metabolite compounds that could evaporate at 105°C beside water, such as essential oils and other compounds (Menkes 2009). If there were volatile compounds in the material, then the amount of water content could not be determined by the gravimetric methods but by the azeotropic distillation method.

Determination of water content was very important to ensure that the material was safe during the storage process and there was no potential for the growth of microorganisms such as bacteria, mold, and khamir (Santana et al. 2018). Determination of moisture content and loss-on-drying value provided quite interesting information. The value of LoDV using the gravimetric method was known greater than AD value using the azeotropic distillation method. This happened to all ingredients. In the *Eucheuma spinosum*, although it contained the lowest AD value compared to the two other algae, it contained the highest volatile compounds compared to both. This indicated that the *Eucheuma spinosum* contained volatile compounds such as essential oils in very large proportions of more than 40%. However, this needs to be confirmed further.

4 CONCLUSION

The three algae were known to contain alkaloids and flavonoids that were known to have a variety of pharmacological properties that are beneficial to health. *Chlorella vulgaris* was known to contain the highest ethanol-soluble compound and also contained the highest inorganic metal element among the three algae. The concentrations of heavy-metal content of the three algae were still in the low category, below 1%, so they were safe to be used as traditional medicinal ingredients. Those three ingredients are known to contain quite high levels of volatile compounds, where the *Eucheuma spinosum* was known to contain the largest volatile compounds among the three ingredients. It has not been known certainly whether these compounds were essential oils or not.

REFERENCES

Adhoni, S. A., Thimmappa, S. C., & Kaliwal, B. B. 2016. Phytochemical analysis and antimicrobial activity of Chorella vulgaris isolated from Unkal Lake. *Journal of Coastal Life Medicine*, 4(5): 368–373.

Ahmed, E. A. 2016. Antimicrobial activity of microalgal extracts isolated from Baharia Oasis, Egypt. *Glob Adv Res J Microbiol*, 5, 033–041.

Asadi, S., Doudi, M., & Darki, B. Z. 2016. An in-vitro investigation of the antibacterial effects of the methanol and aqueous extracts and the supernatant of the algae *Chlorella vulgaris* CCATM 210-1 on multiantibiotic-resistant *Staphylococcus aureus* isolates causing urinary tract infections. *International Journal of Advanced Biotechnology And Research*, 7(2): 806–814.

Balasubramaniam, V., Lee, J. C., Noh, M. F. M., Ahmad, S., Brownlee, I. A., & Ismail, A. 2016. Alpha-amylase, antioxidant, and anti-inflammatory activities of *Eucheuma denticulatum* (NL Burman) FS Collins and Hervey. *Journal of Applied Phycology*, 28(3): 1965–1974.

Brindha, P. (2016). Role of phytochemicals as immunomodulatory agents: A review. *International Journal of Green Pharmacy (IJGP)*, 10(1).

Chia, M. A., Lombardi, A. T., & Melao, M. D. G. G. 2013. Growth and biochemical composition of *Chlorella vulgaris* in different growth media. *Anais da Academia Brasileira de Ciências*, 85(4): 1427–1438.

Choi, W. Y., Kang, D. H., & Lee, H. Y. 2013. Enhancement of immune activation activities of *Spirulina maxima* grown in deep-sea water. *International Journal of Molecular Sciences*, 14(6): 12205–12221.

Cian, R. E., Drago, S. R., De Medina, F. S., & Martínez-Augustin, O. 2015. Proteins and carbohydrates from red seaweeds: Evidence for beneficial effects on gut function and microbiota. *Marine Drugs*, 13(8): 5358–5383.

Dewick, P. M. 2002. *Medicinal Natural Products: A Biosynthetic Approach*. John Wiley & Sons.

Ghasemi, Y., Moradian, A., Mohagheghzadeh, A., Shokravi, S., & Morowvat, M. H. 2007. Antifungal and antibacterial activity of the microalgae collected from paddy fields of Iran: Characterization of antimicrobial activity of *Chroococcus dispersus*. *Journal of Biological Sciences*, 7(6): 904–910.

Harborne, J. B. 1973. *Textbook of Phytochemical Methods*, 1st dn, Champraan and Hall Ltd., London, pp, 110–113.

Hurtado, A. Q., Gerung, G. S., Yasir, S., & Critchley, A. T. 2014. Cultivation of tropical red seaweeds in the BIMP-EAGA region. *Journal of Applied Phycology*, 26(2): 707–718.

Kose, A., Ozen, M. O., Elibol, M., & Oncel, S. S. 2017. Investigation of in vitro digestibility of dietary microalga *Chlorella vulgaris* and cyanobacterium *Spirulina platensi*s as a nutritional supplement. *3 Biotech*, 7(3): 170.

Linfante, A., Allawh, R. M., & Allen, H. B. 2018. The role of *Propionibacterium acnes* biofilm in acne vulgaris. *J Clin Exp Dermatol Res*, 9, 1–4.

Menkes, R. I. 2009. Farmakope Herbal Indonesia Edisi Pertama. Jakarta: Menkes.

Muszyńska, B., Krakowska, A., Lazur, J., Jękot, B., Zimmer, Ł., Szewczyk, A., ... & Opoka, W. 2018. Bioaccessibility of phenolic compounds, lutein, and bioelements of preparations containing *Chlorella vulgaris* in artificial digestive juices. *Journal of Applied Phycology*, 30(3): 1629–1640.

Octaviani, M., Fadhli, H., & Yuneistya, E. 2019. Uji Aktivitas Antimikroba Ekstrak Etanol Kulit Bawang Merah (Allium cepa L.) dengan Metode Difusi Cakram. *Pharmaceutical Sciences & Research*, 6(1): 8.

Okechukwu, P. N., Ekeuku, S. O., Sharma, M., Nee, C. P., Chan, H. K., Mohamed, N., & Froemming, G. R. A. 2019. In vivo and in vitro antidiabetic and antioxidant activity of spirulina. *Pharmacognosy Magazine*, 15(62): 17.

Pandey, R., Tiwari, R. K., & Shukla, S. S. 2016. Omics: A newer technique in herbal drug standardization and quantification. *Journal of Young Pharmacists*, 8(2).

Safafar, H., Uldall Nørregaard, P., Ljubic, A., Møller, P., Løvstad Holdt, S., & Jacobsen, C. 2016. Enhancement of protein and pigment content in two *Chlorella* species cultivated on industrial process water. *Journal of Marine Science and Engineering*, 4(4): 84.

Santana, P. M., Quijano-Avilés, M., Chóez-Guaranda, I., Barragán Lucas, A., Viteri Espinoza, R., Martínez, D., ... & Miranda Martínez, M. 2018. Effect of drying methods on physical and chemical properties of *Ilex guayusa* leaves. *Revista Facultad Nacional de Agronomía Medellín*, 71(3): 8617–8622.

Shannon, E., & Abu-Ghannam, N. 2016. Antibacterial derivatives of marine algae: An overview of pharmacological mechanisms and applications. *Marine Drugs*, 14(4): 81.

Shimozu, Yuuki, et al. 2017. Ellagitannins of *Davidia involucrata*. I. Structure of davicratinic acid A and effects of davidia tannins on drug-resistant bacteria and human oral squamous cell carcinomas. *Molecules* 22(470): 1–10.

Singh, S., Kate, B. N., & Banerjee, U. C. 2005. Bioactive compounds from cyanobacteria and microalgae: An overview. *Critical Reviews in Biotechnology*, 25(3): 73–95.

Sugrani, Andis, Hasnah Natsir, M Natsir Djide, and Ahyar Ahmad. 2019. Biofunctional protein fraction from red algae (*Rhodophyta*) eucheuma spinosum as an antibacterial and anticancer drug agent. *Int. Res. J. Pharm* 10(3): 64–69.

Syed, S., Arasu, A., & Ponnuswamy, I. 2015. The uses of *Chlorella vulgaris* as antimicrobial agent and as a diet: The presence of bio-active compounds which caters the vitamins, minerals in general. *International Journal of Bio-Science and Bio-Technology*, 7(1): 185–190.

Tiwari, P., Bimlesh, K., Mandeep, K., Gurpreet, K., & Harleen, K. 2011. Phytochemical screening and extraction: A review. *Internationale Pharmaceutica Sciencia*, 1(1): 98–106.

Torey, A., Sasidharan, S., Yeng, C., & Latha, L. Y. (2010). Standardization of *Cassia spectabilis* with respect to authenticity, assay and chemical constituent analysis. *Molecules*, 15(5), 3411–3420.

World Health Organization. (2011). *Quality Control Methods for Herbal Materials*. World Health Organization..

In Silico approach of soursop leaf for prediction of anticancer molecular target therapy

M.K. Dewi, Y. Kharisma & L. Yuniarti
Universitas Islam Bandung, Bandung, Indonesia

ABSTRACT: The success rate of standard therapies for breast cancer currently is not optimal, the side effects of breast cancer treatment have also not been overcome, and resistance frequently occurs. The aim of this study is to analyze the potential and protein target of soursop leaf as a breast cancer chemotherapy. This research is a bioinformatics study using an *in silico* method with the pathway analysis method employing PubChem software, SwissTarget-Prediction, STRING, and cystoscopy. The results show many phytochemicals present in soursop leaf and several phytochemical compounds are predicted to behave apoptoticly, inhibiting proliferations, metastasis, and angiogenesis. In conclusion, based on *in silico* analysis, soursop leaf components have anticancer properties in breast cancer.

1 INTRODUCTION

Breast cancer is the most common cancer in women and the incidence has risen sharply worldwide in the last five years (Bianco & Gévry 2012). Chemotherapy is an important therapeutic modality, despite high incidence of side effects and treatment resistance. Therefore, it is necessary to develop more sensitive and specific anticancer drugs using target therapy (Osborne & Schiff 2011).

Indonesia is rich in natural ingredients that could prevent and treat cancer, one of which is soursop (*Annona muricata* L.), a member of the Annonaceae family. Soursop contains acetogenins, tannins, and flavonoids. Acetogenin has a selective cytotoxic effect on cancer cells and multi-drug resistant cancer cells with minimal toxicity to normal cells. Flavonoids have a metastatic inhibiting effect on the culture of breast, liver, colon, lung, and ovarian cancer cells. Tannins can inhibit the growth and angiogenesis of Caco-2 colon cancer cells. Tannin derivatives have a selective cytotoxic effect on cancer cells by inducing apoptosis 3 (Rajesh & Kala 2015, Zhou et al. 2014).

Cancer treatment strategies with targeted therapy are considered more promising and can provide greater efficacy and better survival, but malignant cells have alternative survival strategies. Changes in conditions in microenvironments, systemic signals, immune cells, or stress related to cancer therapy cause changes in proteomic/genomic profiles in metastatic cells that lead to the mechanism of suppression of cancer cells or survival. Active substances contained in soursop leaves are proven to inhibit carcinogenesis by in vitro (Zhou et al. 2014, Batra & Sharma 2013, Rodríguez et al. 2013).

Annona muricata Linn (Annonaceae), commonly known as soursop or graviola, is in the Annonacea family. Some phytochemicals that are reported to have been isolated and characterized from various parts of the soursop plant are annonaceous acetogenins, lactones, isoquinoline alkaloids, tannis, coumarins, procyanidins, flavonoids, pentacyclic terpenoids saponins, p-coumaric acid, myristic acid, stepharine, reticulags, reticulins, ellicags, reticulins, and iced phytosterol (Chen et al. 2012, Liu et al. 2012). The first generation of annonaceous acetogenin mimetic (1, AA005) not only shows antitumor activity in some human cancer cells in vitro but also has high selectivity between normal cells and cancer cells (Qayed et al. 2015, Liang et al. 2009). Flavonoids have cytotoxic effects and selective apoptotic induction activity in vitro on squamous cell carcinoma HSC-2, submandibular gland carcinoma HSG. Tannins have a selective in vitro cytotoxic effect on human T cell lines, human oral squamous cell carcinomas, and salivary gland tumor cell lines rather than normal human gingival fibroblasts with activity inducing apoptosis (Watson & Preedy 2010, Chin 2009).

Computation-based methodology becomes an important component in drug discovery, ranging from identifying targets, optimization to optimization with a ligand structure approach based on virtual screening techniques. The key to this method is docking a small molecule on a protein-binding site. (Chin et al. 2009).

Based on the above background, it is necessary to investigate the anticancer and co-chemotherapy activities of soursop leaf preparation in breast cancer cells *in silico*.

2 RESEARCH METHODS

The research processes and interprets data obtained from the database. Initial step was a literature search of previous research for the soursop leaf based on active ingredient. Next step was a search for the chemical structure of active compounds with pathway analysis using PubChem Software. The compound-structure searches were performed through the PubChem Database, then by selecting the 4D structure of the compound and the SMILES canonical data, used in the next step.

The third step was tracking the target protein of aloe vera-based active compounds on the SMILES canonical structure using SwissTargetPrediction, (http://www.SwissTargetPredic tion.ch/).

SMILES is able to predict the target protein of interaction of a compound based on structure-based similarity between the structure we want to predict (query structure) with the structure of FDA-approved drugs. It is also able to predict the interaction on non-drug compounds that have been analyzed in vitro and in vivo (Gfeller et al. 2014, Dunken et al. 2008). Based on the findings, the target proteins selected were the ones above 70%. The next step is the pathway analysis of these proteins on protein targets that play a role in the melanogenesis process is done using a STRING and cystoscope.

STRING is a database that can be used to predict the interaction of a protein, its expression in a network, and trace its relationship to other proteins in the cell mechanism.

3 RESULTS AND DISCUSSION

The results of the literature search indicate soursop leaves contain several active compounds (Moghadamtousi et al. 2015, Sawant & Dongre 2014), as shown in Table 1. The table also shows the structure of canonical SMILES for the related active compound and its target protein.

Results for the protein pathway analysis on protein targets' role in apoptosis are shown in Figure 1.

Results for the protein pathway analysis on protein targets' role in proliferation are shown in Figure 2.

Search results for the pathway analysis of these proteins on protein targets that play a role in tumor suppressor gene genes are shown in Figure 3.

Search results for the pathway analysis of these proteins on protein targets that play a role invasive, metastasis, and amgiogenesis are shown in Figure 4.

Soursop leaves are known to contain a lot of compounds such as alkaloids, annonaceus acetogenin, megastigmane, flavonol triglycoside, phenolic, and cyclopeptide (Moghadamtousi et al. 2015). Soursop leaf water extracts contain alkaloids, falvanoids, quinons, saponins, tannins, and steroids, while ethanol extracts do not contain quinones, the ethyl acetate fraction does not contain saponins, and the n hexane fraction does not contain quinons and saponins (Sad & Tejasari 2016). Soursop leaves are useful to kill cancer cells and inhibit the growth of cancer cells, have antidiarrheal, analgesic, antidysentery, anti-asthmatic, antihelmitic, and blood vessel dilation

Table 1. Active compounds, canonical SMILES, and Target Protein of soursop leaf.

No	Active Compound	Canonical SMILES	Target Protein
1	Anonaine	C1CNC2CC3=CC=CC=C3C4=C2C1=CC5=C4OC05	SLC6A3 DRD2 HTR2A SLC6A4
2	Atherosperminine	CN(C)CCC1=CC(=C(C2=C1C=CC3=CC=CC=C32)OC)OC	*DRD2*
3	Coreximin	COC1=C(C=C2C3CC4=CC(=C(C=C4CN3CCC2=C1)OC)O)O	DRD2 DRD1 f4 DRD3 SIMAR1 DRD4 DRD5
4	Gentisic acid	CC1=CC2=C(C(=C1)O)C(=O)C3=C(C2=O)C=C(C=C3O)O	CA2 CA1 CA12 CA9
5	β-caryophyllene	CC1=CCCC(=C)C2CC(C)(C)C2CC1	PPARA CNR2
6	α-muurolene	CC1=CC2C(CC1)C(=CCC2C(C)C)C	PPARA CNR2
7	Isolaureline	CN1CCC2=CC3=C(C4=C2C1CC5=C4C=CC(=C5)OC)OCO3	SLC6A3 DRD2 HTR2B
8	Xylopine	COC1=CC2=C(C=C1)C3=C4C(C2)NCCC4=CC5=C3OCO5	DRD2 SLC6A3
9	Quercetin sophoroside	CC1C(C(C(C(O1)OC2C(OC(C(C2O)O)OCC3C(C(C(O3)OC4=C(OC5=CC(=CC(=C5C4=O)O)O)C6=CC(=C(C=C6)O)O)O)O)O)O)O)O	NMUR2 ADRA2A ADRA2C ACHE
10	Gallic acid	C1=C(C=C(C(=C1O)O)O)C(=O)O	CA2 CA7 CA1 CA3 CA6 CA12 CA14 CA9 FUT7 CA4 CA5B CA5A CA13
11	Chlorogenic acid	C1C(C(C(CC1(C(=O)O)O)OC(=O)C=CC2=CC(=C(C=C2)O)O)O)O	AKR1B1 AKR1B10
12	Kaempferol	C1=CC(=CC=C1C2=C(C(=O)C3=C(C=C(C=C3O2)O)O)O)O	NOX4 AKR1B1 XDH TYR FLT3 CA2 ALOX5

(Continued)

Table 1. *(Continued)*

No	Active Compound	Canonical SMILES	Target Protein
			CA7
			HSD17B2
			ABCC1
			HSD17B1
			AHR
			CA12
			ESRRA
			ABCB1
13	RETICULINE	CN1CCC2=CC(=C(C=C2C1CC3=CC(=C(C=C3)OC)O)O)OC	DRD2
			DRD1
14	COCLAURINE	COC1=C(C=C2C(NCCC2=C1)CC3=CC=C(C=C3)O)O	DRD2
15	COREXIMINE	COC1=C(C=C2C3CC4=CC(=C(C=C4CN3CCC2=C1)OC)O)O	DRD2
			DRD1
			F3
			SIGMAR1
			DRD3
			DRD4
			DRD5
16	ATHEROSPERMININE	CN(C)CCC1=CC(=C(C2=C1C=CC3=CC=CC=C32)OC)OC	DRD2

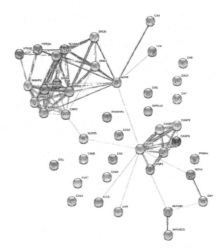

Figure 1.   Pathway analysis of apoptotic target protein from soursop leaf active compounds.

properities, stimulate digestion, and reduce depression. Previous research shows that soursop leaves have anticancer, antibacterial, anti-inflammatory, anti-asthmatic, antihelmitic, anti-obesity, and antioxidant blood vessel dilatation properties, stimulate digestion, and reduce depression (Yuniarti et al. 2016, Moghadamtousi et al. 2015, Yuniarti et al. 2014, Vijayameena et al. 2013).

Apoptosis is one of the targets of cancer treatment. The success of cancer therapy depends mainly on the ability of therapy to induce apoptosis through the mechanism of increasing expression of genes that induce apoptosis, as well as inhibiting genes that inhibit apoptosis (Hassan et al. 2014).

Antiapoptotic proteins include as Bcl-2, Bcl-xL, Bcl-W, Mcl-1, and Bfl-1/A1. If the signal originating from the activated receptor is not strong enough, help is needed via mitochondria-

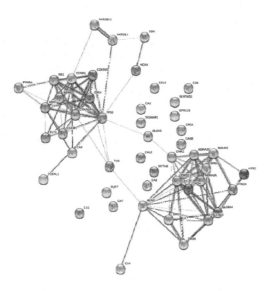

Figure 2.   Analysis of target protein pathway regulation of proliferation of soursop leaf active compounds.

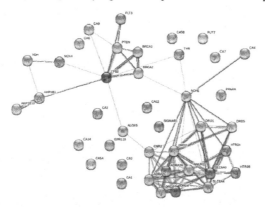

Figure 3.   Pathway analysis of protein tumors to suppressor genes from active compounds of soursop leaf.

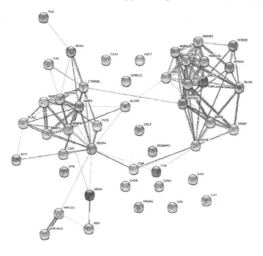

Figure 4.   Pathway analysis of target proteins for migration, metastasis and angigenesis of the active compounds of soursop leaf.

dependent apoptotic pathways (Bcl-2 family member Bid). The bid will be broken down into truncated form (tBid) and then act with the Bcl-2 family member proapoptosis Bax and Bak in the mitochondria to induce cytochrome c into the cytosol. In the cytochrome c cytosol it binds to apoptotic protease activating factor 1 (Apaf-1) to form apoptosome, which is a wheel-like complex that activates procapase 9; Caspase 9 initiates activation of the exclusionary caspase (Green, 2006).

Intrinsic pathway (mitochondria) occurs due to mitochondrial permeability that releases proapoptotic molecules into the cytosol. Growth factors and other signals can stimulate the formation of Bcl2 and Bcl-XL antiapoptotic proteins, which are present in the mitochondrial membrane and cytoplasm to prevent apoptosis. When cells are stressed, antiapoptosis is replaced by Bak and Bax proapoptosis which forms the channel through which cytochrome c and other mitochondrial proteins exit into the cytosol. Cytochrome c binds to the Apaf-1 protein and activates the caspase 9. Caspase 9 activates the exclusionary caspases 3, 6, and 7. Caspase, Bax, and Bid are proapoptotic proteins (Green 2006).

4 CONCLUSION

This research shows that the active compounds present in soursop leaves can bind to proteins that modulate proapotosis proteins, such as caspase 2, caspase 3, caspase6, caspase 8, and caspase 9.

ACKNOWLEDGEMENTS

This work was funded by the Institute for Research and Community Service, Universitas Islam Bandung, through Hibah Dosen Utama.

REFERENCES

Batra, P., & Sharma, A. K. 2013. Anti-cancer potential of flavonoids: recent trends and future perspectives. *3 Biotech*, 3(6): 439–459.

Bianco, S., & Gévry, N. 2012. Endocrine resistance in breast cancer: From cellular signaling pathways to epigenetic mechanisms. *Transcription*, 3(4): 165–170.

Chen, Y., Xu, S. S., Chen, J. W., Wang, Y., Xu, H. Q., Fan, N. B., & Li, X. 2012. Anti-tumor activity of Annona squamosa seeds extract containing annonaceous acetogenin compounds. *Journal of ethnopharmacology*, 142(2): 462–466.

Chin, Y. W., Yoon, K. D., & Kim, J. 2009. Cytotoxic anticancer candidates from terrestrial plants. *Anti-Cancer Agents in Medicinal Chemistry (Formerly Current Medicinal Chemistry-Anti-Cancer Agents)*, 9(8): 913–942.

Chin, Y. W., Yoon, K. D., & Kim, J. 2009. Cytotoxic anticancer candidates from terrestrial plants. *Anti-Cancer Agents in Medicinal Chemistry (Formerly Current Medicinal Chemistry-Anti-Cancer Agents)*, 9(8): 913–942.

Dunkel, M., Günther, S., Ahmed, J., Wittig, B., & Preissner, R. 2008. SuperPred: drug classification and target prediction. *Nucleic acids research*, 36(suppl_2): W55–W59.

Gfeller, D., Grosdidier, A., Wirth, M., Daina, A., Michielin, O., & Zoete, V. 2014. SwissTargetPrediction: A web server for target prediction of bioactive small molecules. *Nucleic acids research*, 42(W1): W32–W38.

Green, D. R. 2006. At the gates of death. *Cancer cell*, 9(5): 328–330.

Hassan, M., Watari, H., AbuAlmaaty, A., Ohba, Y., & Sakuragi, N. 2014. Apoptosis and molecular targeting therapy in cancer. *BioMed research international*, 2014.

Liang, Y. J., Zhang, X., Dai, C. L., Zhang, J. Y., Yan, Y. Y., Zeng, M. S., … & Fu, L. W. 2009. Bullatacin triggered ABCB1-overexpressing cell apoptosis via the mitochondrial-dependent pathway. *BioMed Research International*, 2009.

Liu, Y. Q., Cheng, X., Guo, L. X., Mao, C., Chen, Y. J., Liu, H. X., … & Zhou, G. B. 2012. Identification of an annonaceous acetogenin mimetic, AA005, as an AMPK activator and autophagy inducer in colon cancer cells. *PloS one*, 7(10).

Moghadamtousi, S. Z., Fadaeinasab, M., Nikzad, S., Mohan, G., Ali, H. M., & Kadir, H. A. 2015. Annona muricata (Annonaceae): a review of its traditional uses, isolated acetogenins and biological activities. *International journal of molecular sciences*, 16(7): 15625–15658.

Osborne, C. K., & Schiff, R. 2011. Mechanisms of endocrine resistance in breast cancer. *Annual review of medicine*, 62, 233–247.

Qayed, W. S., Aboraia, A. S., Abdel-Rahman, H. M., & Youssef, A. F. 2015. Annonaceous acetogenins as a new anticancer agent. *Der Pharma Chemica*, 7(6): 24–35.

Rajesh, V., & Kala, M. B. 2015. Antiproliferative and chemopreventive effect of Annona muricata Linn. on Ehrlich ascites carcinoma and Benzo [a] pyrene induced lung carcinoma. *Oriental Pharmacy and Experimental Medicine*, 15(4): 239–256.

Rodríguez, M. L., Estrela, J. M., & Ortega, Á. L. 2013. Natural polyphenols and apoptosis induction in cancer therapy. *J Carcinog Mutag S*, 6.

Sad, T., & Tejasari, M. 2016. Gardnerella vaginalis ATCC 14018 resistant to metronidazol and soursop leaves (Annona Muricata Linn) preparation.

Sawant, T. P., & Dongre, R. S. 2014. Bio-chemical compositional analysis of Annona muricata: a miracle fruit's review. *International Journal of Universal Pharmacy and Bio Sciences*, 3(2): 82–104.

Vijayameena, C., Subhashini, G., Loganayagi, M., & Ramesh, B. 2013. Original Research Article Phytochemical screening and assessment of antibacterial activity for the bioactive compounds in Annona muricata. *Int. J. Curr. Microbiol. Appl. Sci*, 2, 1–8.

Watson, R. R., & Preedy, V. R. 2010. *Bioactive foods in promoting health: Probiotics and prebiotics.* Academic Press.

Yuniarti, L., Dewi, M. K., Lantika, U. A., & Bhatara, T. 2016. Potensi Ekstrak Air Daun Sirsak sebagai Penurun Kolesterol dan Pengendali Bobot Badan. *Acta VETERINARIA Indonesiana*, 4(2): 82–87.

Yuniarti, L., Sastramihardja, H. S., Purbaningsih, W., Tejasari, M., Respati, T., Hestu, E., & Adithya, A. 2014. Soursop effect in cervical cancer apoptosys mechanism. *Global Medical & Health Communication*, 2(1): 6–14.

Zhou, Y., Zhang, A. H., Sun, H., Yan, G. L., & Wang, X. J. 2014. Plant-derived natural products as leads to antitumor drugs. *Plant Science Today*, 1(2): 46–61.

Features of lymphocyte infiltration in lungs of rats given ethanolic extract of white oyster mushrooms and exposure to cigarette smoke

S.B. Rahimah, S. Fitriyana, I.B. Akbar & N. Soetadipura
Universitas Islam Bandung, Bandung, Indonesia

ABSTRACT: Cigarette smoke is a huge source of free radicals and high levels of chemicals. Exposure to cigarette smoke can cause inflammation and lung destruction. This study aims to look at the effect of ethanolic extracts of white oyster mushrooms in preventing the development of lymphocyte infiltration in rats exposed to cigarette smoke. This experimental study used 32 rats divided into 4 groups. Group I (positive control, only given drinking water and standard food), group II (negative control) were given drinking water, standard food, and exposure to cigarette smoke 1 hour/day for 6 weeks. Group III (treatment group) were given cigarette smoke exposure 1 hour/day and ethanol extract of white oyster mushroom at a dose of 250 mg/KgBB of mice/day. Group IV (comparison group) were given cigarette smoke exposure 1 hour/day/group and were given N-acetyl cysteine at a dose of 600 mg/day. Lymphocyte infiltration can be seen from the histopathological picture of the rat lungs-stained by haematoxylin eosin staining. The results showed much of lymphocyte infiltration in rats exposed to cigarette smoke, but there was no significant difference in the proportion of lymphocyte infiltration between all groups (P= 0,14), which was assessed using fisher's exact test. It can be concluded that ethanolic extracts of white oyster mushrooms can prevent development of lymphocyte infiltration in rats exposed to cigarette smoke.

1 INTRODUCTION

Cigarettes contain huge free radicals, as many as 4,700 chemicals in each suction, and reactive oxygen species (ROS). These substances can trigger inflammation in susceptible individuals. ROS production will increase in conditions of respiratory bursts associated with the migration of large macrophage and neutrophil cells to the lungs so that the inflammatory process is increasing. ROS also influences the inflammatory response in the lungs through the activity of nuclear factor kappaB transcription factors (NF-κB), mitogen-activated protein kinases (MAPK) signal transduction pathways, chromatin remodelling, and transcription of inflammatory pro-gene mediators (Rahman 2012, Valavanidis 2009, van der Vaart 2004, Manzel 2011).

Provision of antioxidants is thought to be very beneficial in inhibiting the negative effects of cigarette smoke. Commonly used antioxidants such as thiols, glutathione, N-acetyl cysteine and carbocysteine are reported to be able to modulate various cellular and biochemical aspects of the inflammatory condition of the lungs. Antioxidants can capture and detoxify various free radicals and antioxidants, regulate the biosynthesis of glutathione, inhibit the activation of Nf-kappaB and inhibit the expression of inflammatory genes. Endogenous antioxidants are often not enough, so exogenous antioxidants are needed and one of them can come from herbs such as white oyster mushrooms (Manzel 2011, Rosenbaum 2010, El-Missiry 2012, Noori 2012).

White oyster mushroom (Pleurotus ostreatus) is a food fungus that is popular with the community, easily cultivated and affordable. This mushroom contains a variety of active substances that have antioxidant properties, including vitamin C, beta-carotene, selenium, and phenolic components. The phenolic component is the main component that affects its antioxidant activity (Jayakumar et al. 2009, Jayakumar et al. 2006, Jayakumar et al. 2008, Elmastas 2007). Phenolic antioxidant activity is mainly due to its ability as a reducing agent for hydrogen donors and oxygen quencher singlets, besides this component also has potential metal chelation effects. Phenolic antioxidants also stimulate the synthesis of. Previous research has proven that white oyster mushrooms have strong antioxidant capabilities both in vivo and in vitro (Jayakumar et al. 2009, Jayakumar et al. 2006, Jayakumar et al. 2008, Elmastas 2007, Boonsong et al. 2016, Dogan 2011)

This study's objective was to look at the effect of ethanolic extracts of white oyster mushrooms in preventing the development of lymphocyte infiltration in rats exposed to cigarette smoke.

2 METHODS

The rats were adapted to a laboratory atmosphere for 7 days. During the adaptation period, they were well cared for, fed and drunk sufficiently, and at the end of the adaptation period were weighed again and observed. Furthermore, rats were randomly divided into 4 groups each consisting of 6 rats, and given different treatments as follows: Group I: (normal group) 1. Group II (negative controle): Rats were exposed to cigarette smoke 60 minutes/day/group for 42 days. Group III (treatment group): Rats were exposed to 60 minutes of cigarette smoke/day/group and ethanol extract of white oyster mushroom dose 250 mg/kg BW orally for 42 days. Group IV (comparative group): Rats were exposed to cigarette smoke 60 minutes/day/group and N-acetyl cysteine at a dose of 600 mg/day. On day 43, rats were taken for the histopathological examination.

Lung tissue is taking by thoracotomy and fixed in 10% formalin. The tissue is then made into paraffin blocks and then sliced for each 4-mm thick sample. The preparations that have been stained with hematoxylin and eosin are then examined using a microscope by a pathobiologist. Lymphocyte infiltration uses a score of 0-4, with details 0 - no infiltration around the airway, 1- minimum random infiltration, 2 (mild) for aggregate formation, 3 (moderate) there are at least two aggregations per small visual field and 4 (severe) for more than two aggregates (Dogan 2011).

3 RESULTS

There are 4 groups of experimental animals involved in this study. Each group consisted of 6 rats. Table 1 shows the results of the lymphocyte infiltration assessment in rat lungs.

The Results of each group were compared and statistically tested using fisher's exact test. Table 2 shows the difference in the proportion from each group.

Table 2 shows that group 2 had more lymphocyte infiltrates than the other groups, but there was no statistically significant difference in the proportion of lymphocyte infiltration between all groups (P 0.14). Lymphocyte infiltration scores mostly range from score 1 and 2 for the entire treatment group.

4 DISCUSSION

In this study, it appears that administration of cigarettes to experimental animals within six weeks has been proven to cause an increase in lymphocyte infiltration in the lung tissue to form an aggregate. Previous studies, exposure to massive cigarettes or smoke for six months significantly caused massive lymphocyte infiltration compared to other groups. This can be caused by differences in the duration of exposure to cigarette smoke (Dogan 2011, Naota 2013).

Ethanol extract of white oyster mushrooms on the results of histopathological research showed an excellent antioxidant effect. There was a decrease in the number of lymphocyte

Table 1. Lymphocyte infiltration in lung rats exposed to cigarette smoke.

No	Group	Animal Testing	Score lymphocytes Infiltratration (counting the aggregate nodular)				
			0	1	2	3	4
1.	Normale group	1	-	-	√	-	-
		2	-	√	-	-	-
		3	-	-	√	-	-
		4	-	-	√	-	-
		5	√	-	-	-	-
		6	-	-	√	-	-
2.	Negative controle	1	-	-	√	-	-
		2	-	-	√	-	-
		3	-	-	√	-	-
		4	-	-	-	-	√
		5	-	-	-	√	-
		6	-	-	-	√	-
3.	Treatment by EEWOM	1	-	-	√	-	-
		2	-	-	√	-	-
		3	-	√	-	-	-
		4	-	√	-	-	-
		5	-	√	-	-	-
		6	-	√	-	-	-
4.	Competitive	1	-	√	-	-	-
		2	-	√	-	-	-
		3	-	-	√	-	-
		4	-	-	√	-	-
		5	-	√	-	-	-
		6	-	-	√	-	-

EEJTP: Ethanolic extract of white oyster mushroom
0: There is no peribronchial lymphocyte infiltration
1: There is random peribronchial infiltration
2: There is at least one nodular aggregation
3: There are two nodular aggregation
4: There are more than 2 nodular aggreagation

Table 2. Differences in the proportion of lymphocyte infiltration between groups.

Groups	0		1		2		3		4		Fisher's Exact
	n	%	n	%	n	%	n	%	n	%	
1	1	16.67	1	16.67	4	66.67	0	0	0	0	
2	0	0	0	0	3	50	2	33.33	1	16.67	
3	0	0	4	57.14	3	42.86	0	0	0	0	0.14
4	0	0	3	50	3	50	0	0	0	0	
total	1	4	8	32	13	52	2	8	1	4	

infiltration in the group treated with white oyster mushroom ethanol extract. White oyster mushrooms contain various secondary metabolites that are very much like ergotien, phenol compounds, alkaloids, flavonoids, glucan, and saponins. These compounds have a very high antioxidant effect and also have good anti-inflammatory properties (Jayakumar et al. 2006, Jayakumar et al. 2008, Elmastas 2007, Elmastas 2007).

Giving cigarettes to experimental animals will cause an increase in free radicals and high ROS in these animals, and ethanol extracts of white oyster mushrooms can prevent the development of leukocyte infiltration that occurs by administering at a dose of 250 mg/kg BW for six weeks. This effect is comparable to the antioxidant effect of GSH precursors caused by N-Acetylcysteine.

5 CONCLUSION

Having analyzed the data, this study concluded that ethanolic extracts of white oyster mushrooms can prevent development of lymphocyte infiltration in rats exposed to cigarette smoke. Ethanolic extract of white oyster mushroom have good antioxidant potential to inhibit the negative effect of cigarette smoke.

ACKNOWLEDGEMENT

We acknowledge Research Unit of Faculty of Medicine, Universitas Islam Bandung for sponsoring this research.

REFERENCES

Boonsong, S., Klaypradit, W., & Wilaipun, P. 2016. Antioxidant activities of extracts from five edible mushrooms using different extractants. *Agriculture and Natural Resources*, 50(2): 89–97.

Dogan, O. T., Elagoz, S., Ozsahin, S. L., Epozturk, K., Tuncer, E., & Akkurt, I. 2011. Pulmonary toxicity of chronic exposure to tobacco and biomass smoke in rats. *Clinics*, 66(6): 1081–1087.

Elmastas, M., Isildak, O., Turkekul, I., & Temur, N. 2007. Determination of antioxidant activity and antioxidant compounds in wild edible mushrooms. *Journal of Food Composition and Analysis*, 20(3-4): 337–345.

Elmastas, M., Isildak, O., Turkekul, I., & Temur, N. 2007. Determination of antioxidant activity and antioxidant compounds in wild edible mushrooms. *Journal of Food Composition and Analysis*, 20(3-4): 337–345.

El-Missiry, M. A. (Ed.). 2012. *Antioxidant enzyme*. BoD–Books on Demand.

Jayakumar, T., Ramesh, E., & Geraldine, P. 2006. Antioxidant activity of the oyster mushroom, Pleurotus ostreatus, on CCl4-induced liver injury in rats. *Food and Chemical Toxicology*, 44(12): 1989–1996.

Jayakumar, T., Sakthivel, M., Thomas, P. A., & Geraldine, P. (2008). Pleurotus ostreatus, an oyster mushroom, decreases the oxidative stress induced by carbon tetrachloride in rat kidneys, heart and brain. *Chemico-Biological Interactions*, 176(2-3): 108–120.

Jayakumar, T., Thomas, P. A., & Geraldine, P. 2009. In-vitro antioxidant activities of an ethanolic extract of the oyster mushroom, Pleurotus ostreatus. *Innovative Food Science & Emerging Technologies*, 10(2): 228–234.

Manzel, L. J., Shi, L., O'Shaughnessy, P. T., Thorne, P. S., & Look, D. C. 2011. Inhibition by cigarette smoke of nuclear factor-κB–dependent response to bacteria in the airway. *American journal of respiratory cell and molecular biology*, 44(2): 155–165.

Naota, M., Shiotsu, S., Shimada, A., Kohara, Y., Morita, T., Inoue, K., & Takano, H. 2013. Pathological study of chronic pulmonary toxicity induced by intratracheally instilled Asian sand dust (kosa). *Toxicologic pathology*, 41(1): 48–62.

Noori, S. 2012. An overview of oxidative stress and antioxidant defensive system. *Open access scientific reports*, 1(8): 1–9.

Rahman, I. 2012. Pharmacological antioxidant strategies as therapeutic interventions for COPD. *Biochimica et Biophysica Acta (BBA)-Molecular Basis of Disease*, 1822(5): 714–728.

Rosenbaum, C. C., O'Mathána, D. P., Chavez, M., & Shields, K. 2010. Antioxidants and antiinflammatory dietary supplements for osteoarthritis and rheumatoid arthritis. *Alternative Therapies in Health & Medicine*, 16(2).

Valavanidis, Athanasios; Vlachogianni, Thomais; Fiotakis, Konstantinos. Tobacco smoke: involvement of reactive oxygen species and stable free radicals in mechanisms of oxidative damage, carcinogenesis and synergistic effects with other respirable particles. *International journal of environmental research and public health*, 2009, 6.2: 445–462.

van der Vaart, H., Postma, D. S., Timens, W., & Ten Hacken, N. H. 2004. Acute effects of cigarette smoke on inflammation and oxidative stress: a review. *Thorax*, 59(8): 713–721.

Medical Technology and Environmental Health – Abdullah, Widiaty & Abdullah (eds)
© *2020 Taylor & Francis Group, London, ISBN 978-0-367-86053-0*

MicroRNA-16 in novel liver cancer targeted therapy by clove leaf oil

M. Tejasari, T. Respati, S.A.D. Trusda, E. Hendryanny & L. Yuniarti
Universitas Islam Bandung, Bandung, Indonesia

ABSTRACT: Over-expression of miRNA-16 inhibits proliferation, invasion, and metastasis of hepatocellular carcinoma (HCC). Detection of miRNA levels has potency for the initial diagnosis of malignancy, prognosis prediction and can monitor the therapeutic response because miRNA is very stable in the blood. It was hypothesized that flavonoid from clove leaf oil will be able to interfere MiRNA-16 expression in liver cancer cells. The aim of the study is to analyse miRNA-16 expression on liver cancer cell culture after administration of pure compound from clove leaf oil. This study used the HepG2 cell line culture and treatment groups were given pure compound isolated from clove leaf oil. Measurement of gene expression was done with real-time PCR. In this study, the measurement of miRNA-16 expression in liver cancer cell culture after administration of flavonoid pure compound from clove leaf oil showed an increase in expression more than five times the level of expression of the control group. This means that flavonoid compound from clove leaf oil can induce miRNA-16 expression on liver cancer cells culture. This result provides a preclinical proof-of-concept for a potential strategy for HCC treatment based on the regulation of miRNA-16 expression with clove leaf as novel therapy agent.

1 INTRODUCTION

MicroRNAs (miRNAs) had been involved in most cancers progression, and the focus on miRNAs with the aid of natural agents has opened opportunities for cancer treatment and drug development. miR-16 functions as a tumor suppressor and is regularly deleted or down-regulated in quite a number of human cancers, such as hepatocellular carcinoma (HCC) (Zhang et al. 2019, Chen et al. 2011). In the present study, miR-16 responsive luciferase is reported to display screen candidate com-pounds that modulate miR-16 expression from an herbal product library. These studies offer a good preclinical proof of concept for a potential HCC therapeutic plat-form based on restoring the role of miR-16 as tumor suppressor. (Zhang et al. 2019, Jelic et al, 3010).

The pathogenesis of liver cancer is related to the level of expression of several microRNAs counting miRNA-16. MiRNA is an endogenous RNA that does not encode proteins, with a generally little estimate of around 19-24 nucleotides (Calin & Croce 2006). Disturbance of miRNA expression will eventually influence the miRNA target quality. Dysregulated miRNA actuates an instrument that has suggestions for the pathogenesis of a few infections counting danger. The sum of miRNAs encoded by the genomes of distinctive life forms shifts significantly, and more than 2000 miRNAs have been recognized in humans. A few of these miRNAs have pulled in extraordinary consideration due to their inclusion in initiation, progression, and metastasis of cancer in human. One well-studied case is mir-16, one of the primary miRNAs to be related with cell malignancy. Evidence proves that miRNA16 can modify the cell cycle, decreased proliferation, stimulated cell apoptosis and inhibit tumorigenicity (Liu et al. 2008, Yan et al. 2013).

Previous studies indicate that deregulated miRNAs ought to be applied as workable bio-markers for cancer diagnosis and therapeutics. Natural compounds are a viable source of medicinal materials. More than 50% of the drugs availability are inert ingredients and their derivatives. Numerous natural remedies use different mechanisms to exert a wide variety of cytotoxic effects on cancer cells (Nobili et al. 2009).

Nowadays, HCC has not found any single or integrated chemotherapy regimen to be effective. Traditional medicine plays a vital role in the management of various liver disorders, particularly herbal medicine. With their successful therapeutic action and lack of side effects, demand for medicinal drugs from natural products is growing day by day in this era of science and technology. (Jain et al. 2016, Hassan et al. 2016). Screening program for the identification of potential antitumor agents from traditional herbs found that the clove extract showed powerful cytotoxic activity against several lines of human cancer cells. Previous experiments identified the bioactive compounds and tested the clove extract in vivo antitumor activity to explore possible biological activities (Liu et al. 2014).

2 METHODS

2.1 Subject and material

This study used HepG2 cell line from the Parasitology Laboratory, Faculty of Medicine, University of Gadjah Mada, Yogyakarta. A compound isolated from clove leaf oil used 1,2-epoxy-3 (3-(3,4-dimethoxyphenyl) -4H-1-benzopyran-4on) propane (EPI) synthetized at the MIPA Faculty of Organic Chemistry Laboratory at Gajah Mada University, Yogyakarta.

The material used were DMEM (Sigma Chem Co. St. Louis, USA), fetal bovine serum (Sigma Chem Co. St. Louis, USA), penicillin, streptomycin (Gibco), fungizone (Gipco), sodium bicarbonate (E Merck) and filter paper 0; 2 mm (Whatman), phosphate buffer saline (PBS), MTT reagent and 10 percent stopper SDS. The mini kit, miScript II RT package, SYBR green PCR kit, MiR-196b secondary MiScript and miR-16 primary myScript (Qiagen) are the materials used to test gene expression.

2.2 HepG2 cell culture

HepG2 cells embedded in media culture comprising 1.5×10^4 cells/100μL, the suspension was inserted into each 96 microwave plate and incubated for 24 hours in an incubator at 37CC and 5% CO2. The growth medium used is DMEM, a medium commonly used for the growth of mammalian cells, with the addition of 10% FBS (fetal bovine serum), 0.5% fungal, and 1% penicillin-streptomycin.

2.3 Gene expression detection using real-time PCR

This research using miR-Neasy Mini Kit (cat #217004) the maximum RNA was isolated. CDNA made with QiagenmiScript II RT Kit (cat number 218161). Qiagen SYBR Green PCR Kit paint #218073), the primary collection of miRNA, cDNA that was previously made, is the material for real-time qPCR. Work procedures are performed according to each kit's protocol. The primer used was Ctrl_miRTC_1 Primary MyScript Assay (MS00000001) as the internal control and U6 as the primer of host keeping gene. The primers used in this study was Hs_miR-16_1 Primary MiScript Assay (MS00003318): target hsa-miR-16-5p.

3 RESULTS AND DISCUSSION

The measurement of miRNA-16 expression on liver cancer cells culture following administration of clove leaf oil EPI compound showed an increase in expression more than five times compared to the control group expression rate (Figure 1).

The result showed that EPI compound can enhance the apoptosis cycle and can restore the chemistry of liver cancer cells to normal level (Rebucci et al. 2015). The results of

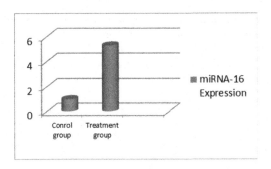

Figure. 1. MiRNA-16 expression in group treatment, five times higher than the control group, after administration of EPI compounds from clove leaf oil in liver cancer cells culture.

this study were consistent with those of Wu et al. (2015) report, which concluded that miRNA-16 overexpression prevented proliferation, invasion, and metastasis of HepG2 cells. MiRNA16 is found to be abnormally expressed in hepatocellular carcinoma (HCC) and miRNA-16 overexpression prevents multiple cancer cell proliferation, invasion, and metastasis. Excess miRNA-16 overexpression also inhibits phosphorylation of PI3 K and Akt (Wu et al 2015).

An emerging field of human cancer is the use of miRNAs for cancer therapy. In recent years, scientists have tried to change the expression of miRNAs in order to achieve the inhibition of cancer growth and the increase in resistance to chemotherapeutic drugs. MiRNAs role in cancer management strategies may include inactivation of oncogenic miRNAs, activation of miRNAs tumor suppressor, and identifying potential miRNAs to restore drug sensitivity. MiRNAs role in cancer management strategies may include inactivation of oncogenic miRNAs, activation of miRNAs tumor suppressor, and identifying potential miRNAs to restore drug sensitivity. Furthermore, in recent years, more attention has been paid to natural agents and their chemical analogs in the field of cancer research (Li et al. 2010).

Flavonoid derivative compound known as eugenol (4-allyl-2-methoxyphenol; EUG) is a bioactive phenolic component of Syzigium aromaticum (the clove). It has commonly been used as a traditional medicine in Asian countries, mainly as a disinfectant, painkiller and antimicrobial agent (Carrasco et al. 2009). EUG's cancer activity against different cancer cell lines has just been demonstrated. EUG's antiproliferative role has been attributed to its ability to generate apoptosis in cancer cells by inhibiting the Bcl-2 family proteins (Manikandan et al. 2011, Jaganathan & Supriyanto 2012, Wiirzler et al. 2016). However, EUG has different cytotoxicity in opposition to special human most cancers cell lines induced apoptosis which plays the most vital role in the chemo preventive motion of EUG on human cancer cells. Researchers cautioned that the EUG induced apoptosis impact on most cancers cell lines is related to the modulation of the Bcl-2 family proteins (Manikandan et al. 2011) and tiers of p53 (Jaganathan & Supriyanto 2012). Therefore, we hypothesized that clove leaf oil compound could induce apoptosis of liver cancer cell leading to tumor growth inhibition. In cancer, 65 which is also characterized by increased cell proliferation has been studied the anti-proliferative and molecular mechanism of eugenol-induced apoptosis (Ali et al. 2014). Natural products that induce apoptosis and has antiproliferative activity in cancer cells are valuable resources in the suppression of cancer. This make natural compound an ideal anticancer agent that do not show large obvious effects on normal cells (Bayala et al. 2014, Gautam et al. 2014).

In this study the administration of EPI compounds isolated and modified from clove leaf oil was able to stimulate increased expression of miRNA-16 in HepG2 cell culture. miRNA-16 overexpression could induce apoptosis leading to inhibit tumor progression. Therefore, this EPI compound can be developed as a therapeutic agent candidate for liver cancer with miRNA-16 as a molecular target of therapy.

4 CONCLUSIONS

The study concluded that clove leaf oil flavonoid compound stimulated miRNA-16 over-expression in liver cancer. This finding provides a scientific basis mechanism based on miRNA-16 role control for a future liver cancer treatment strategy.

ACKNOWLEDGEMENT

This research was made possible by the full support of the Republic Indonesia's Ministry of Technology Research and Higher Education (Ristekdikti), which supported the funding of this study (PDUPT grant contract number: 156/B.04/Rek/III/2018), Institute of Research and Community Service (LPPM) Universitas Islam Bandung and Research Unit at Faculty of Medicine Universitas Islam Bandung for the support of this research.

REFERENCES

Ali, S., Prasad, R., Mahmood, A., Routray, I., Shinkafi, T. S., Sahin, K., & Kucuk, O. 2014. Eugenol-rich fraction of Syzygium aromaticum (clove) reverses biochemical and histopathological changes in liver cirrhosis and inhibits hepatic cell proliferation. *Journal of cancer prevention*, 19(4): 288.

Bayala, B., Bassole, I.H.N., Scifo, R., Gnoul, C., Morel, Lobaccaro, L.J.M., & Simpore, J., 2014. Review Article Anticancer activity of essential oils and their chemical components - a review *Am J Cancer Res* 4 (6):591–607 www.ajcr.us/ISSN:2156-6976/ajcr0001130

Calin, G. A., & Croce, C. M. 2006. MicroRNA signatures in human cancers. *Nature reviews cancer*, 6(11): 857–866.

Carrasco, F. R., Schmidt, G., Romero, A. L., Sartoretto, J. L., Caparroz-Assef, S. M., Bersani-Amado, C. A., & Cuman, R. K. N. 2009. Immunomodulatory activity of Zingiber officinale Roscoe, Salvia officinalis L. and Syzygium aromaticum L. essential oils: evidence for humor-and cell-mediated responses. *Journal of Pharmacy and Pharmacology*, 61(7): 961–967.

Chen, C., Zhang, Y., Zhang, L., Weakley, S. M., & Yao, Q. 2011. MicroRNA-196: critical roles and clinical applications in development and cancer. *Journal of cellular and molecular medicine*, 15(1): 14–23.

Gautam, N., Mantha, A. K., & Mittal, S. 2014. Essential oils and their constituents as anticancer agents: a mechanistic view. *BioMed research international*, 2014.

Hassan, H. A., El-Gharib, N. E., & Azhari, A. F. 2016. Role of natural antioxidants in the therapeutic management of hepatocellular carcinoma. *Hepatoma Res*, 2, 216–23.

Jaganathan, S. K., & Supriyanto, E. 2012. Antiproliferative and molecular mechanism of eugenol-induced apoptosis in cancer cells. *Molecules*, 17(6): 6290–6304.

Jain, S., Dwivedi, J., Jain, P. K., Satpathy, S., & Patra, A. 2016. Medicinal plants for treatment of cancer: A brief review. *Pharmacognosy Journal*, 8(2).

Jelic, S., Sotiropoulos, G. C., & ESMO Guidelines Working Group. 2010. Hepatocellular carcinoma: ESMO Clinical Practice Guidelines for diagnosis, treatment and follow-up. *Annals of oncology*, 21(suppl_5): v59–v64.

Li, Y., Kong, D., Wang, Z., & Sarkar, F. H. 2010. Regulation of microRNAs by natural agents: an emerging field in chemoprevention and chemotherapy research. *Pharmaceutical research*, 27(6): 1027–1041.

Liu, H., Schmitz, J. C., Wei, J., Cao, S., Beumer, J. H., Strychor, S., ... & Zhao, X. 2014. Clove extract inhibits tumor growth and promotes cell cycle arrest and apoptosis. *Oncology Research Featuring Preclinical and Clinical Cancer Therapeutics*, 21(5): 247–259.

Liu, Q., Fu, H., Sun, F., Zhang, H., Tie, Y., Zhu, J., ... & Zheng, X. 2008. miR-16 family induces cell cycle arrest by regulating multiple cell cycle genes. *Nucleic acids research*, 36(16): 5391.

Manikandan, P., Vinothini, G., Priyadarsini, R. V., Prathiba, D., & Nagini, S. 2011. Eugenol inhibits cell proliferation via NF-κB suppression in a rat model of gastric carcinogenesis induced by MNNG. *Investigational new drugs*, 29(1): 110–117.

Nobili, S., Lippi, D., Witort, E., Donnini, M., Bausi, L., Mini, E., & Capaccioli, S. 2009. Natural compounds for cancer treatment and prevention. *Pharmacological research*, 59(6): 365–378.

Rebucci, M., Sermeus, A., Leonard, E., Delaive, E., Dieu, M., Fransolet, M., ... & Michiels, C. 2015. miRNA-196b inhibits cell proliferation and induces apoptosis in HepG2 cells by targeting IGF2BP1. *Molecular cancer*, 14(1): 79.

Wiirzler, L. A. M., Aguiar, R. P., BersaniAmado, C. A., Velázquez-Martínez, C. A., & Cuman, R. K. N. 2016. Anticancer activity of eugenol is not related to regulation of the oncogenic transcription factor Forkhead Box M1. *Acta Scientiarum. Health Sciences*, 38(2): 159–163.

Wu, W. L., Wang, W. Y., Yao, W. Q., & Li, G. D. 2015. Suppressive effects of microRNA-16 on the proliferation, invasion and metastasis of hepatocellular carcinoma cells. *International journal of molecular medicine*, 36(6),1713–1719.

Yan, X., Liang, H., Deng, T., Zhu, K., Zhang, S., Wang, N., . . . & Zhang, C. Y. 2013. The identification of novel targets of miR-16 and characterization of their biological functions in cancer cells. *Molecular cancer*, 12(1): 92.

Zhang, B., Wang, X., Deng, J., Zheng, H., Liu, W., Chen, S., . . . & Wang, F. 2019. p53-dependent upregulation of miR-16-2 by sanguinarine induces cell cycle arrest and apoptosis in hepatocellular carcinoma. *Cancer letters*, 459, 50–58.

Medical Technology and Environmental Health – Abdullah, Widiaty & Abdullah (eds)
© 2020 Taylor & Francis Group, London, ISBN 978-0-367-86053-0

In-vitro investigation of cytotoxicity of West Java *Curcuma longa* and its potential therapeutic use against breast cancer

A. Fatimah, A. Anggraeni, T. Firda & Y. Lelly
Universitas Islam Bandung, Bandung, Indonesia

ABSTRACT: Breast cancer is the second leading cause of death for women and, in Indonesia, the incidence keeps increasing each year. Previous studies have found the use of *Curcuma longa* as antioxidant, antidiabetic, antiviral, even antiangiogenic. The role as antioxidant and antiangiogenic points to *Curcuma longa* as one of the potential plants for therapy of cancer. This study aims to know potential therapeutic use of this West Java plant for breast cancer and to know the cytotoxic range. *Curcuma longa* was prepared as an ethanol extraction using reflux methods. In-vitro investigation used MCF7 cell line. We conducted the cytotoxic examination using MTT assay of the extract to the cell line, and another examination for secondary metabolites. The screening results were positive for flavonoid and tannin as secondary metabolites and the IC50 from MTT assay of MCF7 is 0.256µg/ml (95% CI 0–3.04µmol). The ethanolic turmeric rhizome extract had high anticancer activity with IC50 values of 0.256µg/ml, revealing that *Curcuma longa* inhibits MCF7 proliferation. This result indicates the potential of *Curcuma longa* for further research as a therapeutic agent for breast cancer.

1 INTRODUCTION

1.1 *Curcuma longa as therapeutic plant*

Curcuma longa (turmeric) is a rhizomatous herbaceous plant in the ginger family (*Zingiberaceae*). It originated from South Asia and was used as spice and for religious reasons almost 4,000 years ago in India (Prasad & Aggarwal 2011). Turmeric was distributed to tropical and subtropical countries of the world, including South East Asian countries. A study of *Curcuma* genus in Indonesia shows that there are eight species used as medicinal plant. One of the species is *Curcuma longa,* which is frequently mentioned by ethnic groups who use turmeric for several diseases/symptoms such as gastritis, cough, and diarrhea. The part mainly used is the rhizome of the plant. The rhizome contains several secondary metabolites that have potential for medicine (Subositi & Wahyono 2019).

1.2 *Breast cancer treatment by natural remedies*

Breast cancer has become leading cause of death in women. In Indonesia, this is another cause of death by cancer for women besides cervical cancer (International Agency for Research on Cancer 2018). Management for breast cancer mainly uses chemotherapy, radiotherapy, hormonal therapy, immunotherapy, and surgery (Ramli 2015). However, researchers today try to do studies about natural resources either from plant or other natural sources to make a co-chemotherapy or an additional treatment (Cheng et al. 2018, Zhang et al. 2018, Shareef 2016, Zheng et al. 2016). Curcumin, the active compound of curcuma, shows anticancer properties with several mechanisms for lung, nasopharyngeal, hepatobiliary, colorectal, and breast cancer (Zheng et al. 2016).

This study aims to check the possibility of *Curcuma longa* originating in Indonesia for potential treatment for breast cancer by checking for possible cytotoxicity toward the breast cancer cell line.

2 METHOD

2.1 *Preparing MCF7 cell line*

The preparation of MCF-7 breast cancer cell line was done in Parasitology Laboratorium, Faculty of Medicine, Universitas Gadjah Mada, Indonesia. The cell originated from Erasmus Medical Center (Netherlands) with growth media using DMEM (sigma). Then, it was added by fetal bovine serum (sigma) 10% (v/v) (Gibco), fungison 0.5% (v/v) (Gibco), and 1% penicillin-streptomycin (v/v) (Gibco). The concentration of the cell is 1x103 cells/mL in microplate with 12 wells, each well receiving 1 ml of cell suspension. It needed 72 hours to incubate the microplate in 5% CO_2 environment to make 80% of the cell subconfluent.

2.2 *Preparing turmeric extract*

Preparation of turmeric extract starts with collection of the wet *Curcuma longa*, cultivated from Manoko Experimental Garden, West Java. The rhizome was then dried at room temperature with direct sunlight to get 3 kg of dried rhizome (simplicia). The simplicia was brought to the pharmacy laboratory, Faculty of Mathematics and Natural Sciences, Universitas Islam Bandung (UNISBA), Indonesia, for an extraction. This research used ethanol 95% as solvent and reflux method based on previous study (Bagchi 2012).

2.3 *Plant determination and phytochemical screening of secondary metabolites*

Wet *Curcuma longa* plant collected from Manoko Experimental Garden, West Java, was identified macroscopically at School of Life Sciences and Technology. The dried simplisia and also the ethanolic extract of *Curcuma longa* underwent phytochemical screening for alkaloids, flavonoids, tannins, and saponnins using each respective reagent and methods (Tiwari et al. 2011).

2.4 *Cytotoxicity assay*

After the cell line 80% confluent by preparation of cell line, a series of extract concentrations was added to the wells. The series of concentration were divided into seven different concentrations (1000μg/ml; 500μg/ml; 250 μg/ml; 125μg/ml; 62.5μg/ml; 31.25μg/ml; 15.625 μg/ml). Each of the concentrations was added to 12 wells so there were 96 well-plates in total. The plate was incubated for another 24 hours until cytotoxic effect could be seen. Cytotoxicity assay was done to know the IC50 of the ethanolic turmeric extract by using the MTT assay. Breast cancer ($1x10^3$) cells were seeded in DMEM medium culture, then MTT reagent (50 mg) mixed with PBS (10 ml) added by 100μl to each 96 well-plate (12 wells for each extract concentration). The cells cultured 2–4 hours in a CO_2 incubator until formazan formation. After that, stopper was added to each well: 100μL SDS 10% in 0.01 NHCl (Sápi & Kovács, 2015). The cytotoxicity absorbant in each well was read using ELISA reader before calculating IC50 using probit in SPSS version 24.

3 RESULTS AND DISCUSSION

Determination of the wet *Curcuma longa* plant was done and the result was documented in determination letter 2383/11.CO2.2/PI/2019. The plant was identified as *Curcuma longa* L or *Curcuma domestic Valeton* macroscopically. Extraction processes used reflux and got a liquid extract, which was thickened by using a rotary evaporator and hot plant at 60°C. The *rendement* was 26.67% ethanolic extract. Phytochemical screening results were summarized in Table 1. The screening shows different secondary metabolites identified from the simplisia and the ethanol extract. The finding is coherent with previous studies stating that turmeric extract has phytochemical compounds such as

Table 1. Phytochemical screening secondary metabolites of *Curcuma longa*.

Secondary Metabolites	Simplisia	Extract
Alkaloid		
- Meyer	+	–
- Dragendorf	–	–
Flavonoid	–	+
Tannin	–	+
Saponnin	–	–

* + positive for secondary metabolites, – negative for secondary metabolites.

flavonoids and tannins (Oghenejobo 2017, Rajesh et al. 2013). The differences between this study and previous one is that there are also alkaloid findings but here alkaloid can be found when the rhizome still in the simplisia state. Cytotoxic assay results calculated IC50 from the extract 0.25µg/ml (95% CI 0–3.04µg).

Secondary metabolites are important and have a key role in the plant to give rise as a therapeutic agent. Both tannin and saponin are anticancer agents (Yildirim & Kutlu 2015). A study of several plants that have tannin in them shows that these phenols act as an antioxidant to cancer cells in in-vitro conditions. Tannin has a high molecular weight, found in in the root, stem, and outer layers of plants (Widsten, Cruz, & Fletcher, 2014). The mechanism of anticancer activity was found by increasing apoptotic activity to breast and prostate cancer cell line (Yildirim & Kutlu 2015). Tannins also have a subgroup, gallotanin, with structure related to Maplexin-1 that can give anti-proliverative activity to breast and colon cancer cell line (González-Sarría et al. 2012). Tannic acids originated from tannin have a role in decreased overexpression of fatty acid synthase (FAS) by inhibiting cellular activity of this synthesis that is mainly overexpressed in breast cancer (Nie et al. 2016).

Flavonoids are another phenolic compound mainly studied for anticancer potential, with characteristics of C6-C3-C6 structure. Groups of flavonoids can be found from dietary sources (Abdotaleb et al. 2019). Flavonoids in the rhizome of *Curcuma longa* have been identified from different varieties of turmeric in Bangladesh (Tanvir et al. 2017). Flavonoids can have anticancer properties because of their ability to inhibit PKs, inhibit prooxidant enzyme, and modulate metabolism of carcinogen (Batra & Sharma 2013). The basic mechanism of this phenol is similar to that of tannin, namely, inducing apoptosis to the cells. This mechanism is induced by apoptotic proteins intrinsically and extrinsically, elevation of ROS, and induction of DNA damage (Abdotaleb et al. 2019).

Cytotoxicity of a compound was calculated to know the viability of the respective potential treatment to determine the concentration needed to make 50% of the cells die as a result of the treatment. Cancer treatment using natural resources needs to be checked in in-vitro conditions. Ethanolic extract of *Curcuma longa* from this study shows a different IC50 compared to previous studies, 41.95 µg/ml, even though it originated from the same Manoko Experimental Garden. This difference is possibly because the time of cultivation and method of extraction were different (Triyono et al. 2018).

4 CONCLUSIONS

Rhizomes of *Curcuma longa* originated from West Java have tannin and flavonoid content in the extract and provide a possible mechanism as an anticancer agent. The IC50 of the extract in cell line shows the potential use of the plant for breast cancer treatment.

REFERENCES

Abotaleb, M., Samuel, S. M., Varghese, E., Varghese, S., Kubatka, P., Liskova, A., & Büsselberg, D. 2019. Flavonoids in cancer and apoptosis. *Cancers*, 11(1): 28.

Bagchi, A. 2012. Extraction of curcumin. *Journal of Environmental Sciece, Toxicology and Food Technology*, 1(3): 1–16.

Cheng, Y.-Y., Hsieh, C.-H., & Tsai, T.-H. 2018. Concurrent administration of anticancer chemotherapy drug and herbal medicine on the perspective of pharmacokinetics. *Journal of Food and Drug Analysis*, 26(2): S88–S95.

González-Sarrías, A., Yuan, T., & Seeram, N. P. (2012. Cytotoxicity and structure activity relationship studies of maplexins A–I, gallotannins from red maple (*Acer rubrum*). *Food and Chemical Toxicology*, 50(5): 1369–1376.

International Agency for Research on Cancer. 2018. The Global Cancer Observatory: Indonesia.

Nie, F., Liang, Y., Jiang, B., Li, X., Xun, H., He, W., … & Ma, X. 2016. Apoptotic effect of tannic acid on fatty acid synthase over-expressed human breast cancer cells. *Tumor Biology*, 37(2): 2137–2143.

Oghenejobo, M. 2017. Antibacterial evaluation, phytochemical screening and ascorbic acid assay of turmeric (*Curcuma longa*). *MOJ Bioequiv Availab*, 4(2): 232–239.

Prasad, S., & Aggarwal, B. B. 2011. Turmeric, the golden spice. In W. Galor (ed.), *Herbal Medicine: Biomolecular and Clinical Aspects* (2nd ed.). CRC Press/Taylor & Francis.

Batra, P., & Sharma, A. K. 2013. Anti-cancer potential of flavonoids: Recent trends and future perspectives. *3 Biotech*, 3(6): 439–459.

Rajesh, H., Rao, S., Megha, R., Prathima, K., Rejeesh, E., & Chandrashekar, R. 2013. Phytochemical analysis of methanolic extract of *Curcuma longa* Linn rhizome. *International Journal of Universal Pharmacy and Bio Sciences*, 2(2): 39–45.

Ramli, M. 2015. Update breast cancer mangement, diagnostic, and treatment. *Majalah Kedokteran Andalas*, 28–53.

Sápi, J., & Kovács, L. 2015. Tumor volume estimation and quasi-continuous administration for most effective bevacizumab therapy. *PLoS ONE*, 10(11): 1–20.

Shareef, M. 2016. Natural cures for breast cancer treatment. *Saudi Pharm Journal*, 24(3): 233–240.

Subositi, D., & Wahyono, S. (2019). Study of the genus *Curcuma* in Indonesia used as traditional herbal medicine. *Biodiversitas*, 20(5): 1356–1361.

Tanvir, E. M., Hossen, M., Hossain, M., Afroz, R., Gan, S. H., Khalil, M., & Karim, N. 2017. Antioxidant properties of popular turmeric (*Curcuma longa*) varieties from Bangladesh. *Journal of Food Quality*, 2017.

Tiwari, P., Kumar, B., Kaur, M., Kaur, G., & Kaur, H. 2011. Phytochemical screening and extraction: A review. *Internationale Pharmaceutica Sciencia*, 1(1): 98–106.

Triyono, T., Chaerunisaa, A. Y., & Subarnas, A. 2018. Antioxidant activity of combination ethanol extract of turmeric rhizome (*Curcuma domestica Val*) and ethanol extract of trengguli bark (*Cassia Fistula L*) with DPPH Method. *Indonesian Journal of Pharmaceutical Science and Technology*, 5(2): 43–48.

Widsten, P. C., Cruz, G. C., & Fletcher, M. A. 2014. Tannins and extracts of fruit byproducts: Antibacterial activity against foodborne bacteria and antioxidant capacity. *J. Agric. Food Chem*, (62): 11146–11156.

Yildirim, I., & Kutlu, T. 2015. Anticancer agents: Saponin and tannin. *Int. J. Biol. Chem*, 9: 332–340.

Zhang, Q. Y., Wang, F. X., Jia, K. K., & Kong, L. D. 2018. Natural product interventions for chemotherapy and radiotherapy-induced side effects. *Frontiers in Pharmacology*, 9, 1253.

Zheng, J., Zhou, Y., Li, Y., Xu, D. P., Li, S., & Li, H. B. (2016). Spices for prevention and treatment of cancers. *Nutrients*, 8(8): 495.

Medical Technology and Environmental Health – Abdullah, Widiaty & Abdullah (eds)
© 2020 Taylor & Francis Group, London, ISBN 978-0-367-86053-0

Anticancer effect and co-chemotherapy of [1,2-epoxy-3 (3-(3,4-dimetoksifenil)-4h-1-benzopiran-4on)] propane with Doxorubicin in breast cancer cell line MCF7

P.N. Namira, R. Nilapsari, R.A. Indriyanti & A.F. Sobandi
Bandung Islamic University, Bandung, Indonesia

ABSTRACT: The cancer with the highest morbidity for women is breast cancer. Chemotherapy treatment still has negative points, such as low effectivity, high adverse effects, and resistance to therapy. This has encouraged a search for a natural alternative as a therapy or co-chemotherapy for cancer. This study evaluates the anticancer effect and combination of the compound [1,2-epoksi-3(3-(3,4-dimetoksifenil)-4h-1-benzopiran-4on)] propane (EPI) with doxorubicin in breast cancer culture MCF7. The study is experimental in vitro; the cytotoxicity is analyzed using (3-(4,5-dimetiltiazol-2-il)-2,5-difeniltetrazolium bromid) (MTT). The IC50 is analyzed using probit regression calculation using SPSS software. The synergistic effect of EPI and doxorubicin is analyzed based on the value of the Combination Index (CI) and the data is analyzed using Compusyn software 1.0. The result of this study shows that the IC50 for EPI is 50.77 μg/mL and for Doxorubicin is 26.80 μg/mL. CI for EPI and Doxorubicin (EPI-DOX) shows the average at value 0.1–0.7, which means it has a synergistic effect. The most synergistic effect at concentration $^1/_8$IC50 EPI with ½ IC50 doxorubicin which is 0.00302. In conclusion, the compound EPI shows anticancer effect toward cancer cell MCF7 and strong synergistic effect when combined with doxorubicin.

1 INTRODUCTION

Cancer is a collection of diseases that causes abnormal cell growth and has the ability to metastasize to other parts of the body. Cancer is the leading cause of mortality and morbidity worldwide, with around 14 million new cases in 2012, and caused about 8.8 million deaths in 2015. New cancer cases are expected to increase by 70% within the next two decades. The most common cancers are lung cancer (1.69 million), liver (788,000), colon (774,000), stomach (754,000) and breast (571,000) (WHO 2018). According to WHO, in 2014, in Indonesia, from 92,200 female deaths due to cancer, 21.4% were due to breast cancer (WHO 2014).

Treatment for all types of cancer, including breast cancer, is by performing surgery to remove tumors, then radiation therapy, chemotherapy, or hormone therapy. The latest is target therapy but the principle for all the therapies is the same: that is, to eliminate cancer cells (WHO 2018, Shah et al. 2014). Common cancer chemotherapy treatments include using doxorubicin drugs (NCI 2018). The mechanism of doxorubicin is by inhibition of DNA topoisomerase II, but doxorubicin has side effects such as cardiotoxicity and resistance (Thorn et al. 2011). Dangerous adverse effects and high resistance make scientists search for more effective therapies as anticancer drugs or in combination with chemotherapy drugs to produce higher synergistic effect.

One of the developments in an effort to find new anticancer drugs is to use flavonoids, potential active compounds that are often isolated from plants and are classified into isoflavones, bioflavones, and neoflavones. Isoflavones, a form of flavonoids, can be isolated from soy in the form of genistein (Linus Pauling Institute 2005). Epidemiological studies show that consuming genistein is associated with a significant reduction in the incidence of breast cancer, but due to high prices and the amount of soybean needed to isolate genistein from soybeans, researchers are looking for a cheaper, economical alternative for isolating isoflavone compounds other than soybean genistein (Xue et al. 2014).

The latest development shows clove leaf oil (eugonol) can isolate a type of compound, 1,2-epoxy-3 (3-(3,4-dimethoxyphenyl)-4H-1-benzopiran-4on) propane, also known as epoxy, which has a structure similar to genistein from soybeans. Epoxy can be isolated from clove leaf oil or made synthetically at a cheaper price than genistein (Yuniarti et al. 2018).

Based on the above background, researchers are interested in investigating the anticancer effect and the combination test of synthetic isoflavones [1,2-epoxy-3(3-(3,4-dimethoxyphenyl)-4H-1-benzopiran-4on)] propane] on cell culture MCF7 breast cancer.

2 METHOD

This research included in vitro experimental research, using the test method of the administration of compounds (1,2-epoxy-3(3-(3,4-dimethoxyphenyl)-4H-1-benzopiran-4on) propane (epoxy) and doxorubicin chemotherapy drugs against MCF7 breast cancer cell culture. There are two stages of testing, the determination of IC50 (concentration that can inhibit the growth of 50% of cancer cells) and testing the synergistic effect of a combination of compounds between epoxy and doxorubicin.

The epoxy synthesis compound was made at the Faculty of Mathematics and Natural Sciences, Gadjah Mada University, Indonesia, by Dr. Andi Hairil Alimuddin. MCF7 cells were cultured in the Parasitology Lab, Gadjah Mada University, Indonesia. Cells were cultured using DMEM and given fetal bovine serum/FBS 10%, fungison 0.5% penicillin and 1% streptomycin.

2.1 Cytotoxic test

In the determination of IC50 obtained by giving a single compound epoxy and doxorubicin against breast cancer MCF7 with 80% confluent culture conditions, MCF7 cells were distributed in 96 wells. Then different concentrations of epoxy and doxorubicin were given to each well: epoxy concentrations of 250 µg/ml, 125 µg/ml, 62.5 µg/ml, 31.25 µg/ml, 15.625 µg/ml, and 7.8125 µg/ml and doxorubicin concentrations of 100 µg/ml, 50 µg/ml, 25 µg/ml, 12, 5 µg/ml, 6,250 µg/ml, 3,125 µg/ml, and 1,563 µg/ml. Then the MTT test (tetrazolium method (3- (4,5-dimethyltiazol-2-il) -2.5 definil tetrazolium bromide)) was repeated four times, the absorbance results read using the Elisa reader. From the absorbance results of the test, cell control, and media control then get bioavaibility data at each test level with calculations.

$$Life\ cell\ percentage = \frac{(Absorbance\ of\ treatment\ -\ Absorbance\ of\ media\ control)}{(Absorbance\ of\ control\ solvent\ -\ Absorbance\ control\ media)} \times 100$$

After the compound test results were obtained, the calculation of the relationship between levels and bioavailability used probit regression calculations in SPSS software to obtain IC50 levels of epoxy and doxorubicin compounds on MCF7 breast cancer cells.

2.2 Determine CI (Combination Index)

The CI test was carried out by a combination of the CCRC test method with epoxy and doxorubicin concentration series treatments consisting of four concentrations: ½ IC50, $^3/_8$ IC50, ¼ IC50, and $^1/_8$ IC50 with three repetitions. CI calculation was performed using Compusyn software, which consists of calculating the effect of compound concentration for a combination of single effects, and the effect of the concentration of a combination compound.

3 RESULTS

This study consisted of two stages, namely the determination of IC50 EPI and doxorubicin and the determination of IC EPI and doxorubicin compounds against MCF7 breast cancer cells.

3.1 EPI and doxorubicin cytotoxic test

The test was carried out using the methods (3- (4,5-dimethyltiazol-2-il)-2,5-diphenyltetrazolium bromide) (MTT) to obtain the concentration of compounds that could inhibit the growth of 50% of the tumor cell population.

The cytotoxic test results of each compound are presented in Figure 1. Figure 1 shows the EPI compound has an IC50 of 50.77 µg/ml and doxorubicin 26.80 µg/ml.

3.2 Determine CI (Combination Index) of EPI and doxorubicin compounds against MCF7 cells

Based on the IC50 values of each test compound, the treatment was carried out with EPI and doxorubicin concentrations consisting of ½ IC50, ³/₈ IC50, ¼ IC50, and ¹/₈ IC50 with three replications.

The results of calculations in Table 1 showed a combination of EPI and doxorubicin compounds have a high CI that exceeds 1.0 with a value of ½ IC50 concentration with ¹/₈ IC50

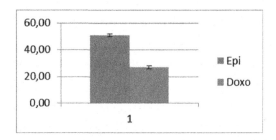

Figure 1.   Graphic of IC50 of EPI and doxorubicin.

Table 1.   Combination index of EPI and doxorubicin compounds on MCF7 cells.

Epoxy concentration (µg/ml)	Doxorubicin concentration (µg/ml)	Viability (%)	CI
25.39	13.4	0.49	0.65656
25.39	10.05	0.51	1.02755
25.39	6.7	0.48	0.52449
25.39	3.35	0.54	2.01452
19.04	13.4	0.33	0.01169
19.04	10.05	0.51	0.77079
19.04	6.7	0.48	0.39336
19.04	3.35	0.52	0.96373
12.69	13.4	0.51	0.51434
12.69	10.05	0.51	0.51403
12.69	6.7	0.45	0.13345
12.69	3.35	0.51	0.51342
6.35	13.4	0.32	0.00302
6.35	10.05	0.55	0.63571
6.35	6.7	0.55	0.63417
6.35	3.35	0.44	0.05323

doxorubicin with CI = 2.01452, which indicates the antagonistic effect (> 0.1), and the lowest CI at concentrations of $^1/_8$ IC50 epoxy with ½ IC50 doxorubicin with CI = 0.00302, indicating a high synergistic effect. Most show synergistic effects, with susceptibility 0.1–0.7.

Interpretation of CI Value

- < 0.1 synergistic effect is very strong
- 0.1–0.3 strong synergistic effect
- 0.3–0.7 synergistic effect
- 0.7–0.9 mild–moderate synergistic effect
- 0.9–1.1 approaches the additive effect
- 1,1–1.45 mild–moderate antagonistic effects
- 1.45–3.3 antagonistic effects
- > 3.3 strong antagonistic effect–very strong

4 DISCUSSION

Cancer is an abnormally proliferating cell and therapy for cancer in general is to inhibit or kill these excessive cells.

In this research, epoxy and doxorubicin were administered and MCF7 cancer cell death was assessed using the MTT method and then absorbance readings using an ELISA reader.

Epoxy (1,2-epoxy-3 (3-(3,4-dimethoxyphenyl)-4H-1-benzopiran-4on) propane) is an iso-flavone that can be isolated from clove leaves or made synthetically. Epoxy has a chemical structure resembling genistein and resembles 17β-Estradiol, which selectively binds to estrogen receptors (ER) with different affinity, and modulates the recruitment of co-receptors and co-activators to influence estrogen-receptor signals. Epoxy compounds have a structure similar to genistein and biological effects as anticancer like genistein (Yuniarti et al. 2018).

Genistein [4′,5,7-trihydroxyisoflavone or 5,7-dihydroxy-3-(4-hydroxyphenyl) chromen-4-one) (C15H10O5) is included in the multifunctional natural isoflavonoid flavonoid class with a 15 carbon skeleton. Similar to other plant constituents, such as lignans, which have estrogenic effects, genistein is a typical example of phytoestrogenic compounds. Phytoestro-gens in genistein can reduce the resistance of breast cancer cells to chemotherapy drugs if given in the right dose.

The results of this study show that in the first step, a single epoxy treatment of MCF7 cancer cells, shows IC50 (concentration that can inhibit 50% of cancer cells) of epoxy compounds is 50.77 μg/ml (± 0.58) and IC50 for doxorubicin which is a standard drug is 26.80 μg/ml (± 1.2) of MCF7 cells. Doxorubicin, which is a standard cancer therapy drug, requires lower concentra-tions to get IC50, indicating the effect of doxorubicin is higher than epoxy. These results indi-cate that, in addition to doxorubicin, epoxy also has an anticancer effect on MCF7 breast cancer cells but should be at a higher dose than doxorubicin.

In the second step, a combination of doxorubicin and epoxy was carried out. With combined doxorubicin and epoxy, the most synergistic anticancer effect is at a concentration of $^1/_8$ IC50 epoxy with ½ IC50 doxorubicin with IK = 0.00302, which indicates a high synergistic effect.

Thus epoxy synthesis (1,2-epoxy-3(3-(3,4-dimethoxyphenyl)-4H-1-benzopiran-4on) pro-pane) has a synergistic effect as a co-chemotherapy with doxorubicin against breast cancer cells MCF7 with an effect highest at concentrations of $^1/_8$ IC50 epoxy with ½ IC50 doxorubi-cin with CI = 0.00302.

5 CONCLUSIONS

It is concluded that compound 1,2,2-epoxy-3(3- (3,4-dimethoxyphenyl)-4H-1-benzopiran-4on) propane can be anticancer in MCF7 breast cancer cells and, when combined 1,2-epoxy-3

(3-(3,4-dimethoxyphenyl)-4H-1-benzopiran-4on) propane with doxorubicin shows synergistic results against MCF7 breast cancer cells.

REFERENCES

WHO Global., 2018. *Cancer Data Sheet.*

WHO. Int. 2014. *Cancer data country.*

Shah, R., Rosso, K., & Nathanson, S. D. 2014. Pathogenesis, prevention, diagnosis and treatment of breast cancer. *World Journal of Clinical Oncology*, 5(3): 283.

NCI Drug Dictionary. 2018. *Defnition of doxorubicin hydrochloride.*

Thorn, C. F., Oshiro, C., Marsh, S., Hernandez-Boussard, T., McLeod, H., Klein, T. E., & Altman, R. B. 2011. Doxorubicin pathways: Pharmacodynamics and adverse effects. *Pharmacogenetics and Genomics*, 21(7): 440.

Linus Pauling Institute., 2005 *Dietary Factors Phytochemicals Flavonoids.*

Xue, J. P., Wang, G., Zhao, Z. B., Wang, Q., & Shi, Y. 2014. Synergistic cytotoxic effect of genistein and doxorubicin on drug-resistant human breast cancer MCF-7/Adr cells. *Oncology Reports*, 32(4): 1647–1653.

Yuniarti, L., Mustofa, M., Aryandono, T., & Haryana, S. M. 2018. Synergistic Action of 1, 2-Epoxy-3 (3-(3, 4-dimethoxyphenyl)-4H-1-benzopiyran-4-on) propane with doxorubicin and cisplatin through increasing of p53, TIMP-3, and MicroRNA-34a in cervical cancer cell line (HeLa). *Asian Pacific Journal of Cancer Prevention: APJCP*, 19(10): 2955.

Infectious and non infectious diseases

Antiretroviral therapy (ART) substitution among HIV/AIDS patients visiting Sanjiwani hospital, Bali

S. Masyeni, I.W.A. Sudiarsana & I.D.G.W. Asmara
Universitas Warmadewa, Denpasar, Indonesia

ABSTRACT: Substitution of long-term ARTs provided to HIV/AIDS patients is common due to the ART adverse events. The study aims to describe the reason for ART substitution among HIV/AIDS patient at Sanjiwani Hospital Bali. A retrospective study was conducted of the medical records of HIV/AIDS patients at Sanjiwani hospital Bali, during 2006–2018. Clinical data were retrieved from the medical records and presented as descriptive data. Over 12 years, 1,112 HIV/AIDS patients were evaluated in the study. The ART regimens were zidovudine-based ART, tenofovir-based ART, and stavudine-based, at 12.2%, 87.3%, 0.5%, respectively. There was 2.2% switching of ART during the study period. The most common reason for switching was anemia (48%), followed by reduction of kidney function (28%), allergic reaction (16%), and 4% of nausea and suspected failure to ART clinically. We highlight that anemia is the main reason for ART substitution among HIV/AIDS patients.

1 INTRODUCTION

Antiretroviral therapy (ART) with highly active antiretroviral therapy (HAART) in HIV infection converted a fatal condition into a chronic and manageable illness. In resource-limited countries, the ART regimen mostly consists of a combination of two nucleoside analogue reverse transcriptase inhibitors (NRTIs) such as zidovudine (AZT), lamivudine (3TC), or tenofovir disoproxil fumarate (TDF), and one non-nucleoside reverse transcriptase inhibitor (NNRTI) such as nevirapine (NVP) or efavirenz (EFV) (Kementerian Kesehatan Republik Indonesia 2014). The mode of action of the NRTIs competes with the natural deoxynucleotides for incorporation into the growing viral DNA chain. Unlike the natural deoxynucleotides substrate, NRTIs lack a 3'-hydroxyl group on the deoxyribose moiety; hence, following incorporation of the NRTIs, the newly performed deoxynucleotide cannot form the next 5'-3' phosphodiester bond needed to extend the DNA chain (Tressler & Godfrey 2012). The NRTI's triphosphate inhibits the function of polymerase-γ, the enzyme responsible for mitochondrial DNA (mtDNA) replication; hence, the depletion of mtDNA is common among HIV-treated persons (Montaner et al. 2004, Wagner et al. 2013, Masyeni et al. 2018).

Even though HAART can fruitfully defeat viral replication in the long term, it is not without substantial toxicity, which can radically undermine treatment effectiveness. Central toxicity has been documented for more than a decade. The severity of the adverse events ranges from mild to life-threatening with short- and long-term effects in NRTI-related mitochondrial toxicity, which exhibits as severe side effects such as hepatic failure, cardiac dysfunction, skeletal myopathies, and lactic acidosis (Gudina 2017). Adverse events of ART are reported as high as 54% on AZT, in which the most ordinary adverse events were pain (30%) and skin rashes (18%) (Eluwa 2012). The general principle of ART toxicities depends on the severity of the adverse events. Mild toxicities do not require termination of therapy or drug substitution, and symptomatic management may give some relief (e.g., antihistamines for a mild rash). Moderate or severe toxicities may require substitution with a drug in the same ART class but with a different toxicity profile, or with a drug in a different class, but do not require discontinuation of all ART. Severe life-threatening toxicities need cessation of all ARV drugs, and the commencement of proper supportive therapy until the symptoms are alleviated. Substitution of long-term ARTs provided to HIV/AIDS patients is common due to ART adverse events (Eluwa 2012).

However, only a little information is known about ART adverse events in many HIV programs in the public health sector of developing countries. The study aim is to describe the reasons for ART substitution among HIV/AIDS patients at Sanjiwani Hospital Bali.

2 METHODS

The current study was a hospital-based, retrospective observational study conducted at HIV care clinics in Gianyar Bali from 2006–2018. The hospital has an HIV clinic, staffed with health professionals trained in ART treatment and adherence counseling services. Clinical data were retrieved from the medical records and presented as descriptive data. The reasons for substitution were retrieved from the medical record. A data-gathering format was used to collect data on the demographic settings, the starting and changing regimens, the period of the initial therapy, CD4 count, World Health Organization (WHO) stage of the disease, and reasons for regimen substitution. Adverse drug reactions (ADR) are defined as the occurrence of adverse events such as diarrhea, nausea, vomiting, anemia, rash, fatigue, peripheral neuropathy, lipodystrophy, metabolic disturbances, or any other effect related to HAART. Substitution is defined as single or triple drug changes due to side effects and initiating another drug of the same class and/or another category.

3 RESULTS

A total of 1,094 medical records of HIV-infected patients at Sanjiwani Hospital were assessed in the study. Female patients account for the minority, (29%) and 7 (0.6%) of the female participants were pregnant. The median age of the participants was 32.5 (IQR 13) years old. Total CD4+ $\leq$100 cell/mm^3, CD4+ 101–200 cell/mm^3, CD4+ 201–350 cell/mm^3, CD4+ 351–499 cell/mm^3, and CD4+ $\geq$ 500 cell/mm^3 accounting for 347 (31.7%), 139 (12.7%), 173 (15.8%), and 37 (93.4%) respectively.

The first-line original ART consists of stavudine-based, tenofovir-based, and zidovudine-based ART. The stavudine ART is in combination with lamivudine and nevirapine or efavirenz, as well as the combination of TDF and ZDV. Characteristics of the participants are presented in Table 1. A majority of the patients were on the combination of tenofovir+lamivudine+efavirenz, the fixed-dose combination ART (738; 67.5%).

Total switching of the ART was found in 26 (2.37%) patients. Ten out of 26 (38.46) cases of switching was ZDV-based as the original ART. Adverse events of ART were found in 22 (2.01%) patients. Types of adverse events included anemia 4 (18.18%), itching 10 (45.45%), a decrease of kidney function 7 (31.82), and gastrointestinal problems 1 (4.55%). The adverse effects of anemia and itching mostly relate to the combination of ZDV+3TC+NPV and another first-line ART, TDF+3TC+EFV is the most common ART of choice substitute for the previous ART. On the other hand, reduced kidney function may relate to TDF-based ART and change to ZDV-based ART. Some 50% of switching occurs in the first year of treatment (Figure 1).

The switching to second-line ART was found in 4 (0.36%) patients due to the suspicion of treatment failure, whether clinical failure (33.3%) or immunological failure (66.7%). The ART of choice for switching due to treatment failure is boosted lopinavir/ritonavir in combination with TDF+3TC.

Table 1. Characteristic of the participants (N = 1094).

Variable	Frequency	Percentage (%)
Gender		
Male	777	71.0
Age		
18–30	470	42.96
31–40	382	34.92
41–50	166	15.17
51–60	58	5.31
61–70	17	1.55
>70	1	0.09
Initial Body Weight		
Median (IQR)	55 (12)	
Initial CD4 (IQR)	136 (246)	
HIV stage (WHO)		
Stage 1	115	10.5
Stage 2	109	10.0
Stage 3	431	39.4
Stage 4	439	40.1

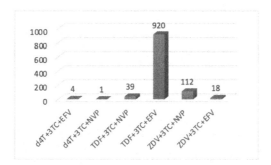

Figure 1. Profile of ART among the HIV-infected patients.

4 DISCUSSION

Acquired immunodeficiency syndrome (AIDS) caused by the human immunodeficiency virus (HIV) is a major global health problem (van Sighem et al. 2015). There are several studies of HIV infection in Bali due to subtype characteristics (Khairunisa et al. 2018), toxicity (Masyeni et al. 2018), adherence (Jiamsakul et al. 2014), and co-infection (Juliari 2018), which reflect the local epidemic of HIV infection. The main reasons for treatment conversion might be due to adverse events, poor adherence, a desire for pregnancy, or treatment failure (Haile & Berha 2019). The finding of high numbers of male patients who need ART substitution in the current study is supported by another study where they were found as high as 53.7%.

The median CD4 of the study participants is 136 cell/mm^3. This is in contrast with the finding of the previous study where they found the median CD4 of the patients was 201 cell/mm^3 (Zhang 2011). The discrepancy may associate with social demographics of the countries that affect the immune status of the patients. The study found toxicity (88%) as the most common reason to change the ART regimen. A concordance finding reported by other study found up to 72.73% ART changing due to toxicity (Assefa & Hussein 2014). This similar finding may be because the HIV sub-type is HIV-1, but we do not assess the genetic diversity of the patients.

The most common primary ART in the current study is a TDF-based regimen, which is in contrast with a previous study where the primary ART was zidovudine (ZDV) based (Sandeep 2014). In Indonesia, since 2014, the availability of TDF made it the ART of choice. Likewise, ZDV-adverse events such as anemia may cause a switch to another regimen in this study. Another study found that ART substitution is most commonly found due to NVP toxicities (Boulle 2007). This study found that toxicity due to NVP, although infrequent, may explain why the most common ART use at the hospital is TDF+3TC+EFV.

5 CONCLUSION

We highlight the most frequent ART substitution in the study was due to ART toxicities, instead of treatment failure. This finding may help physicians monitoring an ART-adverse event in improving the services for the patient's convenience.

ACKNOWLEDGEMENT

We thank the honorable doctors, nurses, and Faculty of Medicine and Health Sciences Universitas Warmadewa for the great support of this work.

CONFLICT OF INTEREST

The authors declare no conflict of interest.

REFERENCES

Assefa, D., & Hussein, N. 2014. Reasons for regimen change among HIV/AIDS patients initiated on first line highly active antiretroviral therapy in Fitche Hospital, Oromia, Ethiopia. *Adv Pharmacol Pharm*, 2(5): 77–83.

Boulle, A., Orrell, C., Kaplan, R., Van Cutsem, G., McNally, M., Hilderbrand, K., ... & Wood, R. 2007. Substitutions due to antiretroviral toxicity or contraindication in the first 3 years of antiretroviral therapy in a large South African cohort. *Antiviral Therapy* 12: 753–760.

Eluwa, G. I., Badru, T., & Akpoigbe, K. J. 2012. Adverse drug reactions to antiretroviral therapy (ARVs): Incidence, type and risk factors in Nigeria. *BMC Clinical Pharmacology*, 12(1): 7.

Gudina, E. K., Teklu, A. M., Berhan, A., Gebreegziabhier, A., Seyoum, T., Nega, A., ... & Assefa, Y. 2017. Magnitude of antiretroviral drug toxicity in adult HIV patients in Ethiopia: A cohort study at seven teaching hospitals. *Ethiopian Journal of Health Sciences*, 27(1): 39–52.

Haile, G. S., & Berha, A. B. 2019. Predictors of treatment failure, time to switch and reasons for switching to second line antiretroviral therapy in HIV infected children receiving first line anti-retroviral therapy at a Tertiary Care Hospital in Ethiopia. *BMC Pediatrics*, 19(1): 37.

Jiamsakul, A., Kumarasamy, N., Ditangco, R., Li, P. C., Phanuphak, P., Sirisanthana, T., ... & Merati, T. 2014. Factors associated with suboptimal adherence to antiretroviral therapy in Asia. *Journal of the International AIDS Society*, 17(1): 18911.

Juliari, I. G. A. M., & Susila, N. K. N. (2018, November). Ocular syphilis in HIV-positive male. In *IOP Conference Series: Materials Science and Engineering* (Vol. 434, No. 1, p. 012340). IOP Publishing.

Kementerian Kesehatan Republik Indonesia. 2014. Pedoman pengobatan antiretroviral *Peratur. Menteri Kesehat. Republik Indones. Nomor 87 Tahun 2014*, pp. 1–121.

Khairunisa, S. Q., Masyeni, S., Witaningrum, A. M., Budiyasa, D. G., & Nasronudin, M. K. 2018. Genotypic characterization of human immunodeficiency virus type 1 isolated in Bali, Indonesia in 2016. *HIV AIDS Rev.*, 17: 81–90.

Masyeni, S., Sintya, E., Megawati, D., Sukmawati, N. M. H., Budiyasa, D. G., Aryastuti, S. A., & Nasronudin, N. 2018. Evaluation of antiretroviral effect on mitochondrial DNA depletion among HIV-infected patients in Bali. *HIV/AIDS (Auckland, NZ)*, 10: 145.

Montaner, J. S., Côté, H. C., Harris, M., Hogg, R. S., Yip, B., Harrigan, P. R., & O'Shaughnessy, M. V. (2004). Nucleoside-related mitochondrial toxicity among HIV-infected patients receiving antiretroviral therapy: Insights from the evaluation of venous lactic acid and peripheral blood mitochondrial DNA. *Clinical Infectious Diseases*, 38(Supplement_2): S73–S79.

Sandeep, B., Chavan, V. R., Raghunandan, M., Arshad, M., & Sayana, S. B. 2014. Factors influencing the substitution of antiretroviral therapy in human immunodeficiency virus/acquired immunodeficiency syndrome patients on first line highly active antiretroviral therapy. *Asian J Pharm Clin Res.*, 7(5): 117–20.

Tressler, R., & Godfrey, C. 2012. NRTI Backbone in HIV Treatment. *Drugs*, 72(16): 2051–2062.

van Sighem, A., Nakagawa, F., De Angelis, D., Quinten, C., Bezemer, D., de Coul, E. O., ... & Phillips, A. (2015). Estimating HIV incidence, time to diagnosis, and the undiagnosed HIV epidemic using routine surveillance data. *Epidemiology (Cambridge, Mass.)*, 26(5): 653.

Wagner, T. A., Lin, C. H., Tobin, N. H., Côté, H. C., Sloan, D. D., Jerome, K. R., & Frenkel, L. M. 2013. Quantification of mitochondrial toxicity in HIV-infected individuals by quantitative PCR compared to flow cytometry. *Cytometry Part B: Clinical Cytometry*, 84(1): 55–58.

Zhang, F., Dou, Z., Ma, Y., Zhang, Y., Zhao, Y., Zhao, D., & Chen, R. Y. 2011. Effect of earlier initiation of antiretroviral treatment and increased treatment coverage on HIV-related mortality in China: A national observational cohort study. *The Lancet Infectious Diseases*, 11(7): 516–524.

Basic sanitation: Is it an important factor in dengue transmission?

T. Respati
Universitas Islam Bandung, Bandung, Jawa Barat, Indonesia

A. Raksanagara
Universitas Padjadjaran, Bandung, Jawa Barat, Indonesia

R. Wangsaputra
Institut Teknologi Bandung, Bandung, Jawa Barat, Indonesia

ABSTRACT: The unavailability of basic sanitation facilities, waste management, and clean water is one of the essential factors influencing dengue in developing countries. However, these important factors are mostly neglected by local government. This article describes the relationship between basic sanitation facilities, including clean water, toilet facilities, and sewage systems (liquid waste) or household solid waste, and the presence of mosquito-breeding places to dengue cases. This study conducted mixed-method research with data collected from the Health District Office and a survey conducted in 2,036 houses. Results showed that factors that influence the existence of a breeding place in the house is the condition of the house, trash and wastewater disposal facilities. Meanwhile, some other factors are related to breeding places outside the home are: latitude, two-story houses, and the availability of trash bins and wastewater facilities. In conclusion, dengue cases and breeding sites are influenced by land use, especially the formation of unplanned housing, and availability of basic sanitation facilities.

1 INTRODUCTION

The increasing population and the need for a place to live cause the growth of unplanned settlements that are not in a city's residential blueprint (WHO 2010). The unplanned settlements usually come with the unavailability of adequate basic sanitation. This problem occurs mainly in developing countries (Seng et al. 2009, Ohba et al. 2010). The unavailability of basic sanitation facilities, waste management, and clean water is one of the essential factors that rarely gets attention (Respati et al. 2017).

Poor areas in many countries have the fewest facilities, including clean water, garbage disposal, and basic sanitation (Respati et al. 2016b). Water supply is usually limited only to certain times so the community holds water in containers that do not meet health requirements (Ohba et al. 2010). These containers are vulnerable to contamination by several biological agents, including mosquito larvae (Heintze et al. 2007, Toledo et al. 2007). Researchers in some countries show that management of the use of water containers in households is an important determinant in the dynamic population of *Aedes aegypti* (Hales & Van Panhuis 2005, Rozhan et al. 2006, Heintze et al. 2007, Stahl et al. 2013).

The role of sanitation facilities is very important in dengue haemorrhagic fever (DHF) because most of the breeding sites occur due to the available sanitation facilities that are not optimal (Respati et al. 2016a). The problem of standing water, clean water sources, waste handling, and the availability of wastewater disposal facilities are some important factors to consider (Respati et al. 2016a). This article describes the relationship between basic sanitation facilities, including clean water, toilet facilities, and sewage systems (liquid waste), or household solid waste, and the presence of mosquito-breeding places to dengue cases.

2 METHOD

The research is part of a study entitled "Eco-health System Dynamics Model to Reduce Mosquito Breeding Places as an Effort to Cut Transmission of Dengue Haemorrhagic Fever." The study collected basic sanitation data from the Bandung District Health Office. Concurrently, the survey was conducted in 2,036 houses from 16 villages in 12 sub-districts. The number of containers inspected was 5,888, so the average household had around 3 containers to store water with a maximum number of 15 containers in a house. From 2,036 households participating, 1,951 (96.8%) samples were analysed. Data collected during survey were breeding sites and basic sanitation status including clean water sources, waste handling, and the availability of wastewater disposal facilities. We also observed any standing water in the study areas. Knowledge about the roles of family members in the prevention of the disease were also collected. Factors analysis was used for the data and was presented in table form using SPSS Ver. 17. The Health Research Ethics Committee of the Faculty of Medicine, Padjadjaran University approved the study with letter No. 464/UN6.c2.1.2/KEPK/PN/2014 on August 14th 2014.

3 RESULTS AND DISCUSSION

The results showed that most of the research areas were densely populated areas with occupancy between 2 people up to a maximum of 20 people in one house, with an average of 3.9 house occupants. The conditions of environmental sanitation vary with some areas having good sanitation, while other regions have poor sanitation.

In relation to dengue cases in Bandung City, Table 1 shows the correlation between cases and basic sanitation facilities.

The results showed that the correlations between dengue cases with SAB, Latrine, and SPAL variables are 0.640, 0.634, and 0.634, respectively, with a coefficient interval 0.60–0.799, which shows a unidirectional and robust relationship. The correlations between DHF case variables with healthy houses and trash Points are 0.589 and 0.585, respectively, with a coefficient interval 0.40–0.599, which means that the relationship between the two variables is moderate and unidirectional.

Table 2 showed factors that influence the existence of a breeding place in the house is the condition of the house, trash, and wastewater disposal facilities. The house with its environment greatly influences the formation of breeding places. The house subsystem for this model is the availability of basic sanitation and waste management, which affects mosquito-breeding

Table 1. Coefficient correlation of dengue cases and basic sanitation.

Basic Sanitation	Dengue Cases	
	Correlation Coefficient	Sig. (2-tailed)
Clean water	0.640	0.000*
Toilet	0.634	0.000*
Trash	0.589	0.000*
Sewage	0.634	0.000*
Healthy house	0.585	0.000*

Notes: *Significant $p < 0.005$

Table 2.   Indoor breeding sites and basic sanitation.

Effect	Univariabel			Model		
	Est. par.	SE	P Value	Est. par.	SE	P Value
Latitude (m)	−0.0027	0.0012	0.029	−0.006	0.0013	< 0.001
Two-story houses	0.1981	0.0948	0.037	0.321	0.0994	0.001
Healthy Houses	−0.5205	0.1034	< 0.001	−0.400	0.1172	0.001
	−0.4346	0.1080	< 0.001	−0.157	0.1250	0.209
Wastewater	−0.8119	0.1029	< 0.001	−0.915	0.1111	< 0.001

SE = standard error, * significant p < 0.005

Table 3.   Outdoor Breeding sites and basic sanitation.

Effect	Univariabel			Model		
	Est. par.	SE	P Value	Est. par.	SE	P Value
Latitude (m)	0.0019	0.0012	0.108	−0.0018	0.001	0.157
Two-story houses	0.1187	0.0947	0.210	0.1893	0.099	0.057
Healthy Houses	−0.3435	0.1045	0.001*	−0.2822	0.118	0.018
	−0.2801	0.1092	0.010	0.0208	0.126	0.869
Wastewater	−1.0199	0.0999	< 0.001*	−1.0530	0.107	< 0.001

SE = standard error, * significant p < 0.005

places. Houses with the availability of good basic sanitation become healthy homes that will affect mosquito-breeding places.

Table 3 showed factors related to breeding places outside the home are latitude, two-story houses, and the availability of trash bins and wastewater facilities.

A house that has a good basic sanitation facilities and is free from disease vectors including mosquito vectors is the definition of a healthy house. The increase in healthy homes is a response to the availability of basic sanitation (Respati et al. 2017).

Based on the model's behaviour, basic sanitation is the most significant factor influencing transmission. Universal basic sanitation access is set by the local government to be achieved in 2019. With the availability of basic sanitation, a healthy home, which is an indicator of the achievement of system performance, can be achieved.

It is known that providing basic sanitation facilities is the policy scenario that provides the best and fastest results to reduce transmission. To achieve universal access, the policy to add basic sanitation facilities, including the availability of clean water must be carried out with a minimum increase of 4% of the total increase in housing complexes so that the target can be achieved in 2019. The provision of this facility is closely related to the allocation of funds needed.

The policy scenario implemented in 2016 will only produce tangible results in 2019 if the policy to increase the provision of basic sanitation facilities applied at a high level (4%). It will be possibly achieved in 2026 if applied at the medium level (2.5%) and in 2046 if the policy is applied according to the existing conditions with an increase of 1% per year.

The city of Bandung has now become a destination for resettlement due to economic, educational, and tourism reasons that are facilitated by the availability of various forms of transportation. With the increasing number of migrants, there is also a need for a place to live so there will be settlements built without a plan. Many settlements are built in a hurry without considering land allotment or spatial planning, and without the availability of adequate basic facilities. The problem of the legality of a settlement will make the development of facilities and infrastructure programs such as basic sanitation and clean water more complicated. The environment that does not meet the rules of the spatial city of Bandung may be one of the causes of high incidence of cases.

Rapid population growth causes the use of land as a settlement to be increasingly widespread in the city of Bandung. More settlements in the form of horizontal settlements such as housing complexes or clusters cause land functions to change. Land use as a settlement has close links to DHF because it is related to the habitat of the *Aedes* spp. as a dengue vector. Dense settlements, inadequate basic sanitation facilities, and poorly maintained environmental hygiene increase the risk of dengue virus transmission.

4 CONCLUSIONS

The home environment that affects dengue cases and mosquito-breeding sites in Bandung is influenced by land use, especially the formation of unplanned housing; availability of basic sanitation facilities in the form of wastewater disposal facilities and waste management; and availability of clean water.

REFERENCES

Hales, S., & Van Panhuis, W. 2005. A new strategy for dengue control. *Lancet*, 365(9459): 551–552.

Heintze, C., Garrido, M. V., & Kroeger, A. 2007. What do community-based dengue control programmes achieve? A systematic review of published evaluations. *Transactions of the Royal Society of Tropical Medicine and Hygiene*, 101(4): 317–325.

Ohba, S.-Y., Kashima, S., Matsubara, H., Higa, Y., Piyaseeli, U. K. D., Yamamoto, H., & Nakasuji, F. 2010. Mosquito breeding sites and people's knowledge of mosquitoes and mosquito borne diseases: A comparison of temporary housing and non-damaged village areas in Sri Lanka after the tsunami strike in 2004. *Tropical Medicine and Health*, 38(2): 81–86.

Respati, T., Raksanagara, A., Djuhaeni, H., Sofyan, A., & Shandriasti, A. 2017. Ecohealth system dynamic model as a planning tool for the reduction of breeding sites. *IOP Conference Series: Materials Science and Engineering*, 180, 012108.

Respati, T., Piliang, B., Nurhayati, E., Yulianto, F. A., & Feriandi, Y. 2016a. Perbandingan Pengetahuan dengan Sikap dalam Pencegahan Demam Berdarah Dengue di Daerah Urban dan Rural. *Global Medical & Health Communication (GMHC)*, 4(1): 53.

Respati, T., Raksanegara, A., Djuhaeni, H., Sofyan, A., Agustian, D., Faridah, L., & Sukandar, H. 2016b. Berbagai Faktor yang Memengaruhi Kejadian Demam Berdarah Dengue di Kota Bandung. *Aspirator*, 9(November): 91–96.

Rozhan, S., Jamsiah, M., Rahimah, A., & Ang, K. T. 2006. The COMBI (Communication for Behavioural Impact) in the Prevention and Control of Dengue: The Hulu Langat Experience. *Journal of Community Health*, 12(1): 19–32.

Seng, C. M., Setha, T., Nealon, J., & Socheat, D. 2009. Pupal sampling for Aedes aegypti (L.) surveillance and potential stratification of dengue high-risk areas in Cambodia. *Tropical Medicine and International Health*, 14(10): 1233–1240.

Stahl, H. C., Butenschoen, V. M., Tran, H. T., Gozzer, E., Skewes, R., Mahendradhata, Y., ... Farlow, A. 2013. Cost of dengue outbreaks: Literature review and country case studies. *BMC Public Health*, 13(1).

Toledo, M. E., Vanlerberghe, V., Baly, A., Ceballos, E., Valdes, L., Searret, M., ... Van der Stuyft, P. 2007. Towards active community participation in dengue vector control: results from action research in Santiago de Cuba, Cuba. *Transactions of the Royal Society of Tropical Medicine and Hygiene*, 101(1): 56–63.

WHO. 2010. Developing guidance for health protection in the built environment: Mitigation and adaptation responses; Meeting report. *International Workshop on Housing, Health and Climate Change*, 1–28.

Relationship between body mass index and the degree of mitral valve stenosis: Supporting evidence for the obesity paradox phenomenon

A.N. Lestari & I.R. Alie
Universitas Islam Bandung, Bandung, Indonesia

M.R. Akbar
Universitas Padjadjaran, Bandung, Indonesia

ABSTRACT: The existence of an obesity paradox has been proposed lately, suggesting that obese individuals may have a survival benefit on acute cardiovascular decompensation. We look at mitral valve stenosis (MS) and a feature of rheumatic heart disease (RHD) to determine the relationship between body mass index (BMI) and degree of MS in RHD patients. An observational study was done, looking at 225 medical records from RHD patients at a heart service installation in Bandung. Anthropometric and echocardiography results were taken and analyzed using Kruskal–Wallace followed by Mann–Whitney for numeric data and chi-square for categorization. Underweight, normal, overweight, and obese percentages are 16%, 52%, 15%, and 17% respectively. Severe MS is distributed most in the normal BMI group. There is relationship between BMI and degree of MS. Median of mitral valve area (MVA) in the obese group shows the greatest among other groups (1.2 vs 0.8–0.9 cm^2) and significant. Statistically, this may support the evidence of an obesity paradox in RHD patients with MS.

1 INTRODUCTION

Rheumatic heart disease (RHD) is a burden for low- or middle-income countries, and for immigrants and the elderly in high-income countries (Rothenbühler et al. 2014). In Indonesia, the incidence of RHD is estimated to reach 1.18 million cases (Watkins et al. 2017). The feature of RHD pathogenesis is inflammation. Various studies relate the inflammation state with cytokine involvement produced from adipose tissue. High body mass index (BMI) can be an indicator for the level of visceral adipose tissue (Harrington et al. 2013). In contrast, there has been an ongoing controversion about the role of adipokines, a factor from adipose tissue, toward inflammation.

Positive correlation between BMI and inflammation was shown in a study conducted by Wisse (2004) indicating a role of adipose tissue cytokine on metabolic conditions, mainly in patients with obesity. However, contradictive results were shown by a retrospective analysis study by Wacharasint (2013) where patients with obesity and overweight has less IL-6 on initial state of inflammation and lower mortality numbers. In the context of acute rheumatic fever, a study in Turkey showed that adiponectine, one of the polypeptides produced by fat cells, was predicted to have an anti-inflammatory effect (Ozgen et al. 2015).

Based on the explanation above, we hypothesized that RHD, which started from upper respiratory tract infection, and its inflammatory feature on heart valves, is also related to BMI. Regarding the initial infection, a systematic review mentioned an increased risk of infection in underweight children and adolescents, especially in developing countries (Dobner & Kaser 2018). There is very little information regarding the association of body mass index and nutritional state with mitral stenosis. The only study mentioning this is a 1959 study stating that there is a trend of patients with mitral stenosis to develop cachexia as the disease progresses (Olesen et al. 1959). In this case, we'd like to know the relationship between body mass index and degree of mitral valve stenosis in RHD patients.

2 METHODS

We conducted an analytic observational study with a cross-sectional design at the Noninvasive Diagnostic Division of Cardiology, Dr. Hasan Sadikin General Hospital, Bandung, Indonesia. We took the data from medical records of RHD patients from 2016–2017 with consecutive sampling. Criteria of inclusion was RHD patients with echocardiography examination during January 2016–December 2017; aged ≥ 15 years old during examination; and showing mitral valve stenosis based on echocardiographic mitral valve area (MVA) examination with the planimetry method. Criteria of exclusion was incomplete medical records and RHD patients who previously had surgical valve repair and/or valve replacement such as double valve replacement (DVR), mitral valve replacement (MVR), and balloon mitral valvotomy (BMV).

Body mass index classification was based on WHO & Asia Pacific guidelines, consecutively underweight (<18.5 kg/m^2), normal (18.5–22.9 kg/m^2), overweight (23–24.9 kg/m^2), and obese (≥25 kg/m^2). Mitral valve area data were further classified by degree of mitral valve stenosis based on recommendation of European Association of Echocardiography/American Society of Echocardiography (EAE/ASE) 2009, consecutively: mild stenosis (>1.5 cm^2 and <4 cm^2), moderate stenosis (>1–1.5 cm^2), and severe stenosis (<1 cm^2). Data processing was done for categorical data using a chi-square test for normally distributed data and Kruskal–Wallace for abnormally distributed data. For abnormally distributed numeric data, the Mann–Whitney test was used. All statistical analysis was done with SPSS for Windows version 24 with 95% confidence interval and p value ≤0.05.

3 RESULTS

From 391 medical records, 225 samples were found to fit the inclusion criteria. There were only 217 data that completely included the patients' age. There were some patients who had two or three follow-up sessions per year or had double input on medical records that we called "specific group": for the flow diagram of this research, see Figure 1.

Table 1 illustrates that the most populous age group was 35–44 years, consisting of 84 people (38.7%). The majority of RHD patients in Dr. Hasan Sadikin General Hospital in 2016–2017 were women (69.8%). Table 1 shows that, in general, in all age groups, patients mostly have a severe degree of mitral stenosis with the number of patients at other degrees of mitral stenosis varying. This difference is statistically significant (p = 0.001). In distribution, there is an increase in all groups of mitral stenosis severity from the age range of 15–24 years to 35–44 years, then a decrease to the age range of 65–74, forming a U-shaped pattern. Female dominates in all degrees of mitral stenosis. Chi-square test shows this is statistically significant (p = 0.02).

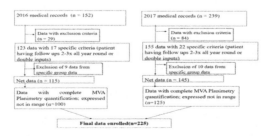

Figure 1.   Flow diagram of this research.

275

Table 1. Comparison of age and sex according to BMI of rheumatic heart disease patients.

| Variable | Mitral Stenosis Degree | | | Total | p-value |
	Mild	Moderate	Severe		
Age (year)					
15–24	3 (1.4%)	1 (0.5%)	4 (1.8%)	8 (3.7%)	<0.01[a]** (0.001)
25–34	3 (1.4%)	7 (3.2%)	20 (9.2%)	30 (13.8%)	
35–44	10 (4.6%)	18 (8.3%)	56 (25.8%)	84 (38.7%)	
45–54	5 (2.3%)	14 (6.5%)	28 (12.9%)	47 (21.7%)	
55–64	13 (6.0%)	18 (8.3%)	11 (5.1%)	42 (19.4%)	
≥65	2 (0.9%)	3 (1.4%)	1 (0.5%)	6 (2.8%)	
Total	36	61	120	217	
Sex (n = 225)					
Male	10 (4.4%)	12 (5.3%)	46 (20.4%)	68 (30.2)	0.02[b]**
Female	26 (11.6%)	53 (23.6%)	78 (34.7%)	157 (69.8%)	
Total	36	65	124	225	

Note: [a] shows statistical quantification using Kruskall–Wallace and sign [b] is quantification using chi-square; * shows unsignificant p-value; ** shows significant p value. Total data for age variable is 217 data and sex variable is 225 data. BMI criteria based on WHO for Asia Pacific: underweight (<18.5 kg/m²), normal (18.5–22.9 kg/m²), overweight (23–24.9 kg/m²) and obese (>25 kg/m²)

Table 2 shows that the combination of mitral stenosis, mitral regurgitation, and tricuspid regurgitation is the most common valve abnormality, 45 patients (20%). The relationships of BMI and the degree of mitral valve stenosis in RHD patients are numerically shown in Table 3. In general, planimetric mitral valve area measurements tend to be stable except for the obese group, which has a median value that is quite different from other BMI groups. The p value is calculated by the Kruskall–Wallace test and shows significant results (p = 0.008). After the Kruskall–Wallace test, a follow-up test was carried out (post-hoc test) and the Mann–Whitney test used for abnormally distributed data. The relationship of BMI and the degree of mitral valve stenosis in RHD patients are also categorically shown in Tables 3 and 4. Based on Tables 3 and 4, it is found that most patients have normal BMI in all mitral stenosis (MS) degrees. Most patients have severe MS conditions with normal BMI (31.1%).

For underweight, normal, and overweight groups, the distribution of cases of mitral stenosis from the most to the least common is severe, moderate, and mild stenosis. In contrast, for the obese BMI group, the order of distribution of cases of mitral stenosis from the most to the least common is moderate, severe, and mild stenosis. Statistical test results showed that there are differences between the degrees of MS and BMI (p = 0.006). Boxplot of mitral valve area in rheumatic heart disease patient according to body mass index based on mitral valve area planimetry method can see in Figure 2.

4 DISCUSSION

Table 1 showed that the majority of rheumatic heart disease patients in this study were in the age group of 35–44 years and the majority of patients were female (157 people; 69.8%). Ratio of female patients to males (68 people; 30.1%) was 2:3. This was in line with the findings in 2014–2016 in central Nepal, where the majority of 235 RHD patients were in the 30–44 years age group (28.78%) and the ratio of female and male patients was 2.1:1 (Laudari & Subramanyam 2017).

Several studies have found an association between sex differences and immune responses. One of it was a study in 2016, which explained that regardless of a woman's age, women have

Table 2. The description of valve lesions of rheumatic heart disease patient.

Valve lesion *	n	%
Single valve lesion		
Pure mitral stenosis (MS)	13	5.8
Combination valve lesion		
MS + MR	14	6.2
MS + TR	27	12.0
MS + AR	7	3.1
MS + MR + TR	45	20.0
MS + TR + AR	27	12.0
MS + MR +AR	16	7.1
MS + MR + AS	1	0.4
MS + MR +PR	1	0.4
MS + TR + AS	1	0.4
MS + TR + PR	5	2.2
MS + AR + AS	3	1.3
MS + MR + AR + TR	36	16.0
MS + TR + AR + TS	3	1.3
MS + MR + AR + AS + TR	7	3.1
MS + MR + AR + AS + TR + PR	1	0.4
MS + MR + TR + PR	1	0.4
MS + MR + AS + TR	5	2.2
MS + AR + AS + PR	2	0.9
MS + MR + AS + PR	6	2.7
MS + MR + AR + AS + PR	4	1.8
Total	225	100

MS = mitral stenosis; MR = mitral regurgitation; AS = aortic stenosis; AR = aortic regurgitation; TS = tricuspid regurgitation

Table 3. Comparison of median and mean rank of mitral valve area in rheumatic heart disease patients according to BMI based on MVA planimetry method.

Variable	Body Mass Index (BMI) (kg/m^2)					
	Underweight (<18.5) (n=36)	Normal (18.5–22.9) (n= 117)	Overweight (23–24.9) (n= 33)	Obese (≥25) (n= 39)	n	p value
Mitral Valve Area (cm^2)						
(Median; Min– Max)	0.9 (0.3 – 2.8)	0.8 (0.3 – 2.5)	0.9 (0.4 – 3.7)	1.2 (0.4 – 4)	0.9 (0.3 – 4)	<0.05* (0.008)
Mean rank	111.31	103.56	110.77	144.76	–	

Note : *p-value is calculated using Kruskal–Wallace test.

more activated CD4 + and CD8+ T cells and proliferated T cells in peripheral blood compared to men (Klein & Flanagan 2016).

The distribution of valve lesions in Table 2 showed that the mitral stenosis (MS)-mitral regurgitation (MR)-tricuspid regurgitation (TR) combination was found in the most patients (45; 20%). Research at a heart health center in Bangalore also showed that multiple valve lesions showed in a greater percentage of cases than single lesions. The most common combination found in the study was the MS-MR combination (46.6%) then followed by a combination of MS and aortic valve regurgitation (AR) (26.5%). These multiple valve

Table 4. Comparison of mitral valve stenosis severity based on 2009 EAE/ASE classification* in rheumatic heart disease patients according to BMI.

Variable	Body Mass Index (BMI) (kg/m²)					
	Underweight (<18.5) (n=36)	Normal (18.5–22.9) (n= 117)	Overweight (23–24.9) (n= 33)	Obese (≥25) (n= 39)	n = 225	p value
Mitral Valve Area (cm²)						
Mild	6 (2.7%)	16 (7.1%)	2 (0.9%)	12 (5.3%)	36 (16%)	<0.05** (0.006)
Moderate	10 (4.4%)	31 (13.8%)	10 (4.4%)	14 (6.2%)	65 (28.9%)	
Severe	20 (8.9%)	70 (31.1%)	21 (9.3%)	13 (5.8%)	124 (55.1%)	

* European Association of Echocardiography/American Society of Echocardiography (EAE/ASE) mitral valve area classification based on planimetry is: mild stenosis (>1.5 cm²), moderate (1–1.5 cm²), and severe (<1 cm²). **p-value was calculated using chi-square test.

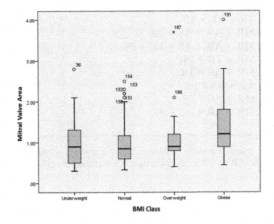

Figure 2. Boxplot of mitral valve area in rheumatic heart disease patient according to body mass index based on mitral valve area planimetry method.

lesions are also more often found in women than men with a 1.2:1 ratio (Manjunath et al. 2014).

In a guideline of echocardiographic examination of valve stenosis, Baumgartner et al. (2009) stated that there are other valve abnormalities that can be caused by mitral valve stenosis due to its rheumatic nature. Tricuspid valve disease, which is often associated with this case, is functional tricuspid regurgitation. Pulmonary valve stenosis can be very rare, although it is related to the rheumatic process. In mitral valve regurgitation due to rheumatism, the main mechanism is the restriction of movement of the valve leaflets (Baumgartner et al. 2009).

Further, another study also mentioned the mechanism of mitral valve regurgitation in rheumatic heart disease. When there is a molecular mimicry process, antibodies produced by streptococcal bacteria recognize myosin in the heart muscle, valve endothelium via laminin protein, and collagen. This causes an inflammatory response and exposure to collagen. Endothelial cells targeted by these antibodies will express vascular cell adhesion molecule-1 (VCAM-1), which will increase the inflammatory reaction and also the release of cytokines into the valve, causing valve fibrosis and interstitial neovascularization. This results in the elongation of the chorda tendineae, causing a prolapse of the anterior valve leaflet and mitral regurgitation (Dal-Bianco et al. 2014).

The statistical tests in Table 3 using the Kruskal–Wallace test at a 95% confidence interval showed that there was a significant relationship between valve size and BMI in RHD patients with a value of p = 0.008 (p < 0.05). In general, it is found that planimetric MVA size tends to be stable at a median of 0.8 and 0.9 except in the obese group which has a median that is quite different, at 1.2. With this significant result, a follow-up test in the form of a post-hoc test using the Mann–Whitney method was done and showed that there were significant differences between the data in the underweight, normal, and overweight group compared with the obese group.

The significance of the results is also strengthened by the results of statistically significant tests (p = 0.006) in Table 4, showing the relationship between BMI and the degree of categorical mitral stenosis in patients with RHD.

This result can be explained by the phenomenon of "the obesity paradox," a term that explains that although obesity is a major risk factor in the development of cardiovascular disease and peripheral vascular disease, when there is acute cardiovascular decompensation, such as myocardial infarction or congestive heart failure, some obese individuals can have a survival benefit (Amundson et al. 2010). A study by Parto et al. (2016) also mentioned that the obesity paradox applies in the context of heart failure. In the context of heart failure, overweight patients had the best prognosis for heart failure compared to obese or underweight patients, which is then followed by patients with normal BMI (Parto et al. 2016).

Although the phenomenon of the obesity paradox is still widely studied throughout the world, many global studies have shown this. A meta-analysis found that the obese and overweight groups had lower risk of cardiovascular mortality compared to the normal BMI group. Shah and colleagues in Parto et al. (2016), 2014 evaluated 6,142 patients with decompensated acute heart failure on four continents and found that for every 5 kg/m^2 increase in BMI, there was an 11% reduction in 30-day mortality rate and also a 9% decrease in annual mortality.

Adipokines, which are often found in overweight patients, have roles in the inflammatory process. One example of adipokines is adiponectin, which circulates in the blood. It is a protein residue produced by white adipose tissue (WAT) and has a structural homology with collagen VIII and X and C1q complement factors. Adiponectin has anti-inflammatory effects at the cardiovascular level where it will inhibit interleukin-6 (IL-6) and TNF-$\propto$ (Conde et al. 2011). Adiponectin effects various cardiovascular components, such as myocytes, endothelial cells, vascular smooth muscle cells, and macrophages. It also converts the macrophage phenotype from M1, which is proinflammatory, to M2, which is M2 anti-inflammatory (Shibata et al. 2017).

Discussing an inflammatory state of a disease will be related with discussion of cachexia. In our case, discussion of cachexia is important as Olesen et al. (1959) previously examined patients with mitral stenosis in 1933–1949 at Copenhagen and stated that the weight trend in patients with mitral stenosis was underweight. It was evidenced by 30 patients who experienced an average loss of 3.9 kilograms within 10 years (Olesen et al. 1959). This phenomenon can certainly be explained through cardiac cachexia, which is a decrease of body mass found in heart failure settings. In addition, cardiac cachexia is important to discuss because there are many studies on reverse epidemiology of obesity, a term that also becomes an umbrella term for the obesity paradox. The term "reverse epidemiology" is used to indicate the role of cardiovascular risk and metabolic syndromes such as obesity, hypercholesterolemia, and hypertension paradoxically associated with higher survival in individuals with chronic disease status (Kalantar-Zadeh 2007).

The definition of cardiac cachexia is a loss of >5% of body mass for 12 months in the presence of heart failure. This entity affects about 5–15% of patients with heart failure and is generally in the New York Heart's Association (NYHA) III or IV functional class. Some of the things that underlie the process of cardiac cachexia include the imbalance between catabolic and anabolic processes, neurohornonal alteration, immunological activation, reduction of food intake, and gastrointestinal abnormalities in patients with heart failure (Okoshi et al. 2014).

Intracellular degradation of skeletal muscle protein occurs in the catabolic process of heart failure. One thing that causes it is the activation of the ubiquitin-protease system. During this process, the damaged cytosol protein then connects with the ubiquitin molecule and into the pro-teasome, where they will split into short peptides and also amino acids. This catabolic process is also enhanced by a process of inflammation and neurohormonal alteration (Okoshi et al. 2014).

Heart-failing patients with cachexia experience increased plasma concentrations of nor-adrenaline, epinephrine, cortisol, and aldosterone compared with healthy patients or patients with cardiac failure without cachexia. Adrenergic stimulation can increase resting-phase energy requirements and induce vasoconstriction, and interfere with intestinal perfusion and also bacterial translocation. An experimental study has shown that angiotensin II can contrib-ute to muscle wasting by increasing protein enumeration, decreasing protein synthesis, and inhibiting the process of muscle regeneration. In contrast to adrenergic hormones, it is also found that heart-failure patients have a condition of growth hormone resistance. As a consequence, insulin growth factor-1, which stimulates protein synthesis, myoblast differen-tiation, and muscle growth, also decreases (Okoshi et al. 2014).

So far, there has not been a specific reference that includes the relationship between BMI and mitral stenosis. The new phenomenon is called the obesity paradox, and is under the umbrella of reverse epidemiology terms; also, research on cardiac cachexia explains BMI and its relationship with heart failure.

With various limitations to the existing research, this study only shows that there is a significant relationship of BMI and the degree of mitral stenosis, where patients with obese BMI have valve median sizes larger than the other BMI categories. This study has not been able to show a causal relationship – such as, the more underweight a person is, the more severe the degree of mitral stenosis – similar to Olesen's research in 1959 on weight trend in RHD patients or vice versa. Even so, the authors posit that the obesity paradox can be con-sidered to explain the prognosis of mitral stenosis patients. A limitation of this study is that the amount of data in each group, especially the BMI group, is not the same.

5   CONCLUSION

We have observed echocardiography results of rheumatic heart disease (RHD) patients with mitral stenosis. There is no significant relationship between BMI and degree of mitral valve stenosis in RHD patients. Greater median of mitral valve area (MVA) in obese groups is also found. Statistically, this may support the evidence of obesity paradox in RHD patients with mitral stenosis, especially regarding the effect of cardiac cachexia and the cardioprotective role of adipokines.

ACKNOWLEDGEMENTS

We would like to thank all boards and staffs of the Faculty of Medicine, Universitas Islam Bandung (UNISBA), Indonesia, and Dr. Hasan Sadikin General Hospital Bandung, Indo-nesia, for the support of this research.

ETHICAL ISSUES

Ethics approval was obtained from the Health Research Ethics Committee, Faculty of Medi-cine, Universitas Islam Bandung and the Research Ethics Committee Universitas Padjadjaran Bandung. Data collection proposal also approved by Director of Education and Human Resources of Dr. Hasan Sadikin General Hospital Bandung.

REFERENCES

Amundson, D. E., Djurkovic, S., & Matwiyoff, G. N. 2010. The obesity paradox. *Critical Care Clinics* 26(4): 583–596.

Baumgartner, H., Hung, J., Bermejo, J., Chambers, J. B., Evangelista, A., Griffin, B. P., ... & Quiñones, M. 2009. Echocardiographic assessment of valve stenosis: EAE/ASE recommendations for clinical practice. *Journal of the American Society of Echocardiography* 22(1): 1–23.

Conde, J., Scotece, M., Gómez, R., López, V., Gómez-Reino, J. J., Lago, F., & Gualillo, O. 2011. Adipokines: biofactors from white adipose tissue. A complex hub among inflammation, metabolism, and immunity. *Biofactors* 37(6): 413–420.

Dal-Bianco, J. P., Beaudoin, J., Handschumacher, M. D., & Levine, R. A. 2014. Basic mechanisms of mitral regurgitation. *Canadian Journal of Cardiology* 30(9): 971–981.

Dobner, J., & Kaser, S. 2018. Body mass index and the risk of infection-from underweight to obesity. *Clinical Microbiology and Infection* 24(1): 24–28.

Harrington, D. M., Staiano, A. E., Broyles, S. T., Gupta, A. K., & Katzmarzyk, P. T. 2013. BMI percentiles for the identification of abdominal obesity and metabolic risk in children and adolescents: evidence in support of the CDC 95th percentile. *European journal of clinical nutrition* 67(2): 218–222.

Kalantar-Zadeh, K. 2007. Cardiovascular and Survival Paradoxes in Dialysis Patients: What Is So Bad about Reverse Epidemiology Anyway?. *Seminars in dialysis* 20(6): 593–601.

Klein, S. L. & Flanagan, K. L. 2016. Sex differences in immune responses. *Nature Reviews Immunology* 16(10): 630–634.

Laudari, S. & Subramanyam, G. 2017. A study of spectrum of rheumatic heart disease in a tertiary care hospital in Central Nepal. *IJC Heart and Vasculature* 15(2017): 27–28.

Manjunath, C. N., Srinivas, P., Ravindranath, K. S., & Dhanalakshmi, C. 2014. Incidence and patterns of valvular heart disease in a tertiary care high-volume cardiac center: a single center experience. *Indian heart journal* 66(3): 320–326.

Okoshi, M. P., Romeiro, F. G., Martinez, P. F., Oliveira Junior, S. A. D., Polegato, B. F., & Okoshi, K. 2014. Cardiac cachexia and muscle wasting: definition, physiopathology, and clinical consequences. *Research Reports in Clinical Cardiology* 319–326.

Olesen, K. H. 1959. Body Weight in Mitral Stenosis: Documentation and Body Compositional Interpretation of the Underweight in Progressive Mitral Valvular Disease. *Acta Medica Scandinavica* 165(2): 137–145.

Ozgen, H., Ucar, B., Yildirim, A., Colak, O., Bal, C., & Kilic, Z. 2015. Plasma adiponectin levels and relations with cytokines in children with acute rheumatic fever. *Cardiology in the Young* 25(5): 879–892.

Parto, P., Lavie, C. J., Arena, R., Bond, S., Popovic, D., & Ventura, H. O. 2016. Body habitus in heart failure: understanding the mechanisms and clinical significance of the obesity paradox. *Future cardiology* 12(6): 639–653.

Rothenbühler, M., O'Sullivan, C. J., Stortecky, S., Stefanini, G. G., Spitzer, E., Estill, J., ... & Pilgrim, T. 2014. Active surveillance for rheumatic heart disease in endemic regions: a systematic review and meta-analysis of prevalence among children and adolescents. *The Lancet Global health* 2 (12): e717–e726.

Wacharasint, P., Boyd, J. H., Russell, J. A., & Walley, K. R. 2013. One size does not fit all in severe infection: obesity alters outcome, susceptibility, treatment, and inflammatory response. *Critical care* 17(3): R122.

Watkins, D. A., Johnson, C. O., Colquhoun, S. M., Karthikeyan, G., Beaton, A., Bukhman, G., ... & Nascimento, B. R. 2017. Global, regional, and national burden of rheumatic heart disease, 1990–2015. *New England Journal of Medicine* 377(8): 713–722.

Wisse, B. E. 2004. The inflammatory syndrome: the role of adipose tissue cytokines in metabolic disorders linked to obesity. *Journal of the American society of nephrology* 15(11): 2792–2800.

Medical Technology and Environmental Health – Abdullah, Widiaty & Abdullah (eds)
© 2020 Taylor & Francis Group, London, ISBN 978-0-367-86053-0

Rett syndrome with all its problems in Indonesia: A review of case reports

D. Santosa
Universitas Islam Bandung, Bandung, Jawa Barat, Indonesia

D.A. Gurnida & A. Subarnas
Universitas Padjadjaran, Bandung, Jawa Barat, Indonesia

ABSTRACT: Rett syndrome is a progressive neurodegenerative disorder accompanied by autistic behavior accompanied by stereotypic hand movements, dementia, ataxia, epilepsy, growth disturbance, mental retardation, and often occurs in girls. Rett syndrome is caused by a mutation in the MeCP2 gene. The purpose of this case report is to review a case, because many health experts may not be familiar with Rett syndrome. A 13-year-old girl came to a hospital in Bandung, Indonesia, with a diagnosis of dengue fever, pulmonary tuberculosis, epilepsy, and Rett syndrome. The patient was treated for two weeks, with complications of respiratory failure, UTI, gastrointestinal bleeding, anemia, stomatitis, and feeding problems. Patients are known to suffer from Rett syndrome from the age of three. At this time, the patient has reached the fourth stage of Rett syndrome (late motor decline), with symptoms of scoliosis, muscle spaticity, severe physical disability, convulsions, dystonia, and bradykinesia. The management of Rett syndrome has so far been aimed at improving and slowing progression, as well as symptomatic treatment, thus requiring a medical team. The prognosis of Rett syndrome is unknown. Some patients die during childhood due to airway complications, arrhythmias, nutritional disorders, but many reach middle age.

1 INTRODUCTION

Rett syndrome is a neurodegenerative disorder that inhibits the development of the central nervous system and can affect all races in the world. Rett syndrome was firstly introduced by Andreas Rett in 1966. Rett syndrome is characterized by normal psychomotor development in the first month of life, then is followed by loss of psychomotor abilities, the onset of stereotypic hand movements, dementia, ataxia, epilepsy, growth disorders, and mental retardation (Macini 2004). Rett syndrome is caused by a mutation in the MeCP2 gene (Methyl CpG binding Protein 2), which is an essential gene for nerve cell maturity, and is located on the X chromosome. Rett syndrome often occurs in girls with a prevalence of 1:10,000 to 1:23,000 live births. Boys with a genetic history of Rett Syndrome usually do not survive after birth (Percy & Lane 2004, Weaving et al. 2005, Samaco & Neul 2011). Rett syndrome is often diagnosed as autism, cerebral palsy, or non-specific developmental delay (Samaco & Neul 2011). The diagnosis is made based on history and clinical symptoms, and there are no biological markers for this disease, except for specific chromosome examinations (Macini 2004, Percy & Lane 2004, Weaving et al. 2005, Samaco & Neul 2011). In Indonesia, discussion about Rett syndrome is rare, so information is still difficult to obtain by the public. It is very important to recognize the causes, symptoms, characteristics, and even treatment of this disorder (Herini et al. 2004). This case review may be helpful because many health experts may not be familiar with Rett syndrome, especially in Indonesia.

2 CASE REPORT

A 13-year-old girl came to Al Islam Bandung Hospital, Indonesia, on November 29, 2017, with a diagnosis of pneumonia + pulmonary tuberculosis + epilepsy + Rett syndrome. The patient had been treated for two weeks in the PICU room with indications of respiratory failure + UTI + gastrointestinal bleeding, anemia, stomatitis, and feeding problems. The patient then was returned to the clinic of pediatric for recovery and management of tuberculosis therapy and epilepsy therapy, and medical rehabilitation for physiotherapy for muscle spasticity. On July 8, 2018, the patient was hospitalized and treated for one week with an indication of status epilepticus + dengue fever. The patient then was returned home for recovery. On February 10, 2019, the patient was again treated for two weeks with a diagnosis of pneumonia + pulmonary tuberculosis. Patient was admitted to the PICU room with complications of respiratory failure. The patient was then returned home for recovery.

The patient was first diagnosed with Rett syndrome when she was three years old, after Rett symptoms began to appear. The patient is the first child born to parents who have no history of consanguinity. Patient's brother is a healthy and normal child. There is no history of congenital defects, neurological disorders, or seizures in the patient's family. The patient's mother was 25 years old and the patient's father was 28 years old when the patient was born. Patient was born without a history of asphyxia, with a birth weight of 3,000 grams, a body length of 49 cm, and a head circumference of 33 cm. From birth until the age of 2.5 years, the patient's growth and development was still within normal limits (Figure 1). After this age, she became more silent, often daydreaming, interacting less; it appears both hands often appeared to be wringing and clapping, rarely making a sound. According to information from the patient's parents, the size of the head circumference begins to slow down at ± 2.5 years of age (<–2SD Z-Score). At 3 years old, she often put her finger into her mouth. At the age of 8 years old, the patient was unconscious, and was admitted to the hospital. At the age of 11 years, symptoms of seizures (epilepsy) occurred frequently, 4–15 times a day. Symptoms of epilepsy were first discovered at the age of 2 years old, in the form of clonic tonics. At 8 years old, the symptoms of epilepsy underwent a change in the form of focal to date. Due to the frequent seizures, she became weaker and her legs stiffened, and she often experienced respiratory infections, especially coughing colds, and weight loss and difficulty gaining weight. The last data availabel was from September 2019; the patient's weight was 32 kg (<–3 SD Z score) and the patient's height was 141 cm (<–3 SD Z score).

The patient still had the habit of putting fingers into her mouth. She was not able to walk, so used a wheelchair (Figure 2). Symptoms of spasticity showed in her both legs. Head CT scans in 2015 showed mild cerebral atrophy (Figure 3). The BERA test showed the patient's hearing was within normal limits. EEG examination results in 2006 showed cortical dysfunction in the right and left central temporal regions, with epileptogenic focus in the independent right and left central temporall region. The results of the re-examination of EEG in 2015 showed the presence of epileptogenic waves in the right central region with cortical dysfunction in both hemispheres. The results of repeated EEG examinations in 2018 showed cortical dysfunction in the left prefrontal, left temporo-parieto-occipital, and right temporal regions. Antiepileptic drugs were still given in the form of valproic acid and topiramate. Echocardiography results in 2015 showed mild mitral regurgitation and tricuspid regurgitation. Spine X-ray showed thoracic vertebrae scoliosis with an angle of >20 degrees, the apex in the sixth thoracic vertebra (Figure 4).

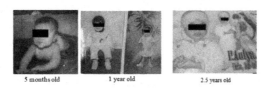

5 months old   1 year old   2.5 years old

Figure 1.   Patients at the age of 5 months, 1 year, and 2.5 years still appear normal.

Figure 2. The patient is dependent on a wheelchair, and she is seen frequently inserting her fingers in her mouth.

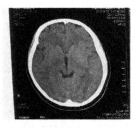

Figure 3. CT scan of the patient's head showing mild atrophy.

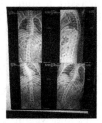

Figure 4. X-ray examination of a patient's vertebral bone showing thoracic vertebra scoliosis with an angle > 20 degrees, apex in the sixth thoracic vertebra.

3 DISCUSSION

The diagnosis of Rett syndrome in patients is made based on DSM IV-TR (Table 1), Trevathan E and Naidu S criteria (Table 2), and the clinical stage of Rett syndrome (Table 3).

Symptoms that exist in this patient meet almost all criteria of diagnosis of Rett syndrome. From birth to the age of 2.5 years, both growth and development were still in accordance with milestones. Head circumference became microcephal after 2.5 years of age, when the symptoms of Rett syndrome began to appear. At this time the patient has reached the fourth stage of Rett Syndrome (late motor decline), with symptoms of scoliosis, muscle spaticity, severe physical disability, convulsions, dystonia and bradykinesia (Clarke 1996).

Hand wringing is a specific sign of Rett syndrome. All Rett syndrome case reports show this sign, including the habit of clapping and inserting the fingers into the mouth. Patients begin to show these symptoms after approximately 2.5 years of age (Siew & Rani 1991, Kalra & Sud 1994, Herini et al. 2004, Ahuja et al. 2005, Sitholey et al. 2005, Pereira et al. 2017).

CT-scan images of the head show mild atrophy. Oldfors et al. found cerebellar changes in Rett syndrome begin in childhood and develop for years (Oldfors et al. 1990).

The results of the patient's first EEG examination and subsequent repetitions showed different results. Symptoms of epilepsy were first discovered at the age of 2 years, in the form of

Table 1. Diagnostic criteria for Rett syndrome based on DSM IV-TR (Sadock & Sadock 2011).

A. Found

1. Normal prenatal and perinatal development
2. Normal psychomotor development during the first five months after birth
3. Normal head circumference at birth

B. Symptoms that arise after a period of normal development

1. Slowing head growth between the ages of 5 and 48 months
2. Loss of purposeful hand skill which was previously achieved between the ages of 5 and 30 months followed by the development of stereotypic hand movements
3. Loss of social relations (although social interactions often occur later)
4. Lack of coordination when walking or body movements
5. Severe disturbances in expressing feelings and new language development with severe psychomotor setbacks

Table 2. Diagnostic criteria of Rett syndrome (Trevathan & Naidu 1988).

Criteria	Patient
Necessary criteria	
1 Apparently normal prenatal and perinatal period	+
2 Apparently normal psychomotor development through the first 6 months up to 18 months	+
3 Normal head circumference at birth	+
4 Deceleration of head circumference at birth, deceleration of head growth between the age of 5 months and 4 years	+
5 Loss of acquired purposeful hand skills between the age of 6 and 30 months, that is temporarily associated with communication dysfunction and social withdrawal	+
6 Development of severely impaired expressive and receptive language and the presence of apparent, severe psychomotor retardation	+
7 Stereotypic hand movements such as hand wringing/squeezing, clasping/tapping, mouthing and "washing"/rubbing automatisms appearing after purposeful hand skills are lost	+
8 Appearance of gait apraxia and truncal apraxia/ataxia between 1 and 4 years	+
9 Diagnosis tentative until 2 to 5 years of age	+
Supportive Criteria	
1 Breathing dysfunction	-
2 EEG abnormalities	+
3 Seizures	+
4 Spasticity, often associated with muscle wasting and dystonia	+
5 Peripheral vasomotor disturbances	-
6 Scoliosis	+
7 Growth retardation	+
8 Hypotrophic, small feet	+

clonic tonics. By 8 years of age, the symptoms of epilepsy underwent changes in the form of focal seizures. Herini et al. found changes in the evolution of epilepsy according to age (Herini et al. 2005).

The patient has been hospitalized several times, especially with respiratory tract infections. Yang T et al. found the MeCP2 gene abnormality causes immunodeficiency (Yang et al. 2012).

The management of Rett syndrome has so far been aimed at improving and slowing progression, as well as symptomatic treatment, thus requiring a medical team. The prognosis of Rett syndrome is unknown. Some patients die during childhood due to airway complications, arrhythmias, nutritional disorders, but many reach middle age (Lotan 2006).

Table 3. Clinical stages of Rett syndrome (Clarke 1996).

Stage I Early onset of stagnation

- Onset age 5–18 months
- Progress delayed but not very significant
- Postural delay and hypotonia
- Often diagnosed retrospectively
- Duration: week or month

Stage II Developmental regressions that occur rapidly

- Onset age 1–4 years, occurs suddenly
- Loss of skills achieved (using hands, voice, communication, active play)
- Coarse motor function may be relatively maintained
- Temperament can change, and sometimes it feels like stress
- Autistic symptoms appear, including stereotypical ones
- Eye contact is often maintained
- Developmental delay is significant and dementia is apparent
- Irregular breathing and seizures occur
- Duration: from weeks to months, can reach one year.

Stage III Pseudostationary period

- Onset age 2–10 years
- Some restoration of communication
- Ambulation is maintained, slowing down slowly from the motor nerve
- Prominent apraxia/ dyspraxia hands
- Seizures often appear
- Duration: from years to decades

Stage IV Final Motor Setback

- This onset starts when ambulation stops at stage III, often in adolescence, but can also start at 5 years of age
- Severe physical disability, convulsions, dystonia and bradykinesia, distal distortion
- Full dependence on wheelchairs
- Duration: decade

4 CONCLUSION

Rett syndrome is a developmental disorder in the form of regression and delays in brain development in the first year of life that was previously relatively normal, and often occurs in girls. Characteristic symptoms include hand movements such as hand wringing that is stereotypic, hyperventilation, convulsions, and loss of ability to control motor movements of the hand. The etiology of Rett syndrome is not yet known with certainty, and the diagnosis of Rett syndrome is based on clinical symptoms and neurological examination, and confirmed by examination of the MECP2 gene mutation. The treatment of Rett syndrome is not specific, it is only intended to treat symptomatic symptoms. Pharmacological and physiotherapy approaches are needed in Rett syndrome to overcome clinical symptoms that may arise. The prognosis of Rett syndrome is unknown. Some patients die during childhood due to airway complications, arrhythmias, nutritional disorders, but many reach middle age.

REFERENCES

Ahuja, S. R., Karande, S., Kulkarni, H. V., & Kulkarni, M. 2005. Rett syndrome: A case report and overview. *Journal of the Indian Medical Association*, 103(10): 533–535.
Clarke, A. 1996. Rett syndrome. *Journal of Medical Genetics*, 33(8): 693.
Herini, E. S., Mangunatmadja, I., Solek, P., & Pusponegoro, H. D. 2004. Rett syndrome in childhood: The clinical characteristics. *Paediatrica Indonesiana*, 44(4): 160–164.

Herini, E. S., Sunartini, H., Mangunatmadja, I., Purboyo, S., & Pusponegoro, H. D. 2005. The epilepsies of Rett syndrome in Indonesia. *Paediatrica Indonesiana*, 45(5): 203–206.

Kalra, V., & Sud, D. T. 1994. Rett syndrome. *Indian Pediatrics* 31(6): 711–715.

Lotan, M. 2006. Management of Rett syndrome in the controlled multisensory (Snoezelen) environment: A review with three case stories. *The Scientific World Journal*, 6: 791–807.

Macini, J. 2004. *Rett syndrome*. [Online]. http://www.orpha.net/data/ patho/GB/uk-rett.pdf.

Oldfors, A., Sourander, P., Armstrong, D. L., Percy, A. K., Witt-Engerström, I., & Hagberg, B. A. 1990. Rett syndrome: Cerebellar pathology. *Pediatric Neurology* 6(5): 310–314.

Percy, A. K., & Lane, J. B. 2004. Rett syndrome: Clinical and molecular update. *Current Opinion in Pediatrics* 16(6): 670–677.

Pereira, K. G., Alves, G. R., Jóias, R. M., Josgrilberg, E. B., & Jóias, R. P. 2017. General and oral characteristics of Rett syndrome: Case report. *Brazilian Dental Science*, 20(3): 142–150.

Sadock, B. J., & Sadock, V. A. 2011. *Kaplan and Sadock's synopsis of psychiatry: Behavioral sciences/clinical psychiatry*. Lippincott Williams & Wilkins.

Samaco, R. C., & Neul, J. L. 2011. Complexities of Rett syndrome and MeCP2. *Journal of Neuroscience* 31(22): 7951–7959.

Siew, H. F., & Rani, J. M. 1991. Rett syndrome: Two case reports. *The Medical Journal of Malaysia* 46(2): 192–198.

Sitholey, P., Agarwal, V., & Srivastava, R. 2005. Rett syndrome. *Indian Journal of Psychiatry* 47(2): 116.

Trevathan, E., & Naidu, S. 1988. The clinical recognition and differential diagnosis of Rett syndrome. *Journal of Child Neurology* 3(1 suppl): S6–S16.

Weaving, L. S., Ellaway, C. J., Gecz, J., & Christodoulou, J. 2005. Rett syndrome: Clinical review and genetic update. *Journal of Medical Genetics* 42(1): 1–7.

Yang, T., Ramocki, M. B., Neul, J. L., Lu, W., Roberts, L., Knight, J., ... & Corry, D. B. 2012. Overexpression of methyl-CpG binding protein 2 impairs TH1 responses. *Science Translational Medicine* 4(163): 163ra158–163ra158.

Can rose apple leaf be developed for antileucorrhoea and antidandruff?

S. Suwendar, F. Lestari, S.P. Fitrianingsih, D. Mardliyani & N. Fitriani
Universitas Islam Bandung, Bandung, Jawa Barat, Indonesia

ABSTRACT: Fungal infections are very prevalent in Indonesia. Rose apple [*Eugenia aqueum* (Burm. F) Alston] leaf is one kind of natural agent in Indonesia that has been suggested as a natural antifungal. The objectives of this research were to scientifically prove the potential antifungal activity of rose apple leaf and to obtain the value of minimum inhibitory concentration (MIC). Evaluation of antifungal activity was carried out by the agar diffusion method using well techniques. Tests were carried out to test efficacy against *Candida albicans* and *Pityrosporum ovale*. The results showed that the ethanol extract of rose apple leaf showed antifungal activity against both *Candida albicans* and *Pityrosporum ovale*. The MIC of ethanol extract was 1% both to for *Candida albicans* and *Pityrosporum ovale*. Rose apple leaf has the potential to be developed as an antifungal agent, especially for antileucorrhoea and antidandruff.

1 INTRODUCTION

Indonesia's tropical climate allows the growth of various fungi, including pathogenic fungi, so that fungal infections have a high rate of prevalence (Ministry of Health of the Republic of Indonesia 2011), for example, leucorrhoea and dandruff (Mayer et al. 2013). Leucorrhoea and dandruff are prevalent fungal infections (Zoya et al. 2016, Chen et al. 2017). Most women (between 70–75%) have had leucorrhoea (Bubakar & Amiruddin 2012). No population in any geographic area is free of being affected by dandruff (Ranganathan & Mukhopadhyay 2010). Currently, antifungal drugs have been widely used. However, based on safety evaluations, these medicines cause adverse side effects (Castinetti et al. 2014). In addition, there have been many cases of fungi resistance to these medicines (Ministry of Health of the Republic of Indonesia 2011, Sony et al. 2018). Further, it is known that those problems are caused by antifungals, such as ketoconazole (Castinetti et al. 2014).

One of the efforts made to overcome these problems is to develop drugs from natural ingredients. Rose apple leaf has the potential to be developed as an antimicrobial. It contains flavonoids and tannins with antifungal activity (Monisha et al. 2018). So far, research on rose apple leaf activity on bacteria has been widely carried out. Research on the activity of fungi is relatively rare (Mapatac & Mamaoag 2019). Tests of antifungal effects of various plants have been conducted, however, so far none of these plants is simultaneously effective on both fungi. For example, there are 142 plants that are proven to have anticandidal properties but there are no data on their activity on *Pityrosporum ovale* (Zida et al. 2017). As another example, a plant that had been proven to have antidandruff properties was urang aring (*Eclipta prostate*) (Sukandar & Ekawati 2006), however, it had not been proven to have an effect on leucorrhoea. According to these data, the purpose of this study was to evaluate the activity and minimum inhibitory concentration (MIC) of ethanol extract of rose apple leaf on the fungi that cause leucorrhoea and dandruff.

2 METHOD

The research phase included the collection of rose apple leaves, making simplicia, making ethanol extracts, making extract fractions, and testing antifungal activity. The study was conducted at the Laboratory of Pharmacy Unit D, Department of Pharmacy, Faculty of Mathematics and Natural Science, Universitas Islam Bandung (UNISBA), Indonesia.

Simplicia was made by drying without direct sunlight (Purnomo & Indarti 2017). Extract was made by maceration (Potluri et al. 2013). Antifungal activity tests were carried out on *Candida albicans* and *Pityrosporum ovale* in vitro. All stages were carried out with aseptic techniques. The activity test was carried out using the agar diffusion method in order to use well techniques using a series of test concentrations (in % w/v) Antifungal activity was stated by MIC. Determination of MIC was carried out by the agar dilution method. The concentration of the test on the determination of the MIC was based on the results of the activity test (Sharma et al. 2011, Kandimalla et al. 2016). Fluconazol was used as a comparison. Dimethylsulfoxide (DMSO) was used as a control.

3 RESULTS AND DISCUSSION

Antifungal activity of rose apple leaf extract is caused by tannins (Monisha et al. 2018). Tannins shrink the cell wall so that they interfere with cell permeability. Disruption of cell permeability causes these cells to be unable to perform their living activities so that their growth is stunted or the cell dies (Ajizah 2004). Antifungal activity is also caused by flavonoids. Flavonoids inhibit growth and decrease the cell's biofilm production (Serpa et al. 2012).

Based on Table 1, it can be seen that the lowest concentration of ethanol extract of rose apple leaves that still shows activity on *Candida albicans* is 1% w/v. At a concentration of 1% w/v, the zone of growth inhibition of *Candida albicans* was formed with an average diameter of 1.14±0.02 cm after three observations were made. At a test concentration lower than 1% w/v, no inhibition zone was observed. Thus the MIC of ethanol extract of rose apple leaves in *Candida albicans* is 1% w/v. When compared with the results of research on guava (*Psidium guajava*) leaf conducted in Cameroon, activity on *Candida albicans* from guava leaf turned out to be stronger. The MIC of guava leaves in *Candida albicans* based on the study in Cameroon was > 4% w/v (Elisabeth et al., 2016).

Based on Table 2, it can be seen that the lowest concentration that still shows activity in *Pityrosporum ovale* is 1% w/v. At a concentration of 1% w/v, the zone of growth inhibition of *Pityrosporum ovale* was formed with an average diameter of 1.21±0.02 cm after three

Table 1. Activity test of ethanol extract of rose apple leaf results on *Candida albicans.*

Concentration (% w/v)	Average inhibition zone diameter (mm)
32	1.47 ± 0.03
16	1.19 ± 0.04
8	1.12 ± 0.02
4	1.20 ± 0.01
2	1.21 ± 0.02
1	1.14 ± 0.02
0.5	–
0.25	–
0.125	–
0.625	–
Control (DMSO)	–
flukonazol	1.10 ± 0.10

Table 2. Activity test of ethanol extract of rose apple leaves results on *Pityrosporum ovale*.

Concentration (%)	Average inhibition zone diameter (cm)
32	1.93 ± 0.06
16	1.84 ± 0.06
8	1.64 ± 0.05
4	1.49 ± 0.09
2	1.28 ± 0.11
1	1.21 ± 0.02
0.5	–
0.25	–
0.125	–
0.625	–
Control (DMSO)	–
flukonazol	2.29 ± 0.05

observations were made. At a test concentration lower than 1% w/v, no inhibition zone was observed. Thus the MIC of ethanol extract of water guava leaves in *Pityrosporum ovale* is 1% w/v.

The activity on *Pityrosporum ovale* from rose apple leaves was stronger than urang aring (*Eclipta prostata*). MIC value of urang aring leaves on *Pityrosporum ovale* was 5% w/v. Urang aring leaves are commonly used in shampoo preparations (Sukandar & Ekawati 2006). The results of research on the leaves of pacar air (*Impatiens balsamina* L.) at a concentration of 5% w/v showed that the diameter of the inhibitory zone in *Pityrosporum ovale* was 1.388 cm (Malonda et al. 2017), which is lower than the diameter of the inhibitory zone from rose apple leaves.

Results of another study showed that the effect on *Pityrosporum ovale* of rose apple leaves was lower than singalawang (*Petiveria allicea* L.) roots. The MIC of singalawang root (ethyl acetate fraction) was 3.2 mg/100 mL or 0.03% w/v (Indriyanti et al. 2013).

The growth inhibition zone diameter to *Candida albicans* and *Pityroporum ovale* from rose apple leaves extract was more than 6 mm. According to the category of antimicrobial inhibition based on the diameter of the inhibitory zone (Table 3), the extract of the rose apple leaf was strongly active both on *Candida albicans* and on *Pityrosporum ovale* (Pan et al. 2009). Thus, ethanol extracts of rose apple leaf have good potential to be developed as antileucorrhoea and antidandruff preparations.

Both the observation of activity on *C. albicans* (Table 1) and on *P. ovale* (Table 2) showed that the control group, DMSO (Arafath et al. 2019), did not show any diameter of inhibition. This shows that the effect occuring in the form of an inhibition zone diameter in the test system in the form of ethanol extract of rose apple leaf is really the activity of the test material and not due to other factors.

In addition, it can be seen that in Tables 1 and 2, the comparison system, contain fluconazole (Mashaly & Shrief 2019), shows a zone of growth inhibition in both *C. albicans* and

Table 3. Antimicrobial inhibition categories based on inhibition zone diameter (Pan et al. 2009).

Diameter (mm)	Plant-inhibited response
0–3	Low
3–6	Medium
More than 6	High

P. ovale This shows that the method used is valid and work procedures have been implemented correctly. Thus the results shown are not false positive results.

4 CONCLUSIONS

The ethanol extract of rose apple [*Eugenia aqueum* (Burm. F) Alston] leaf has the property of inhibiting the growth of *Candida albicans* and *Pityrosporum ovale*. The MIC of ethanol extract of apple rose was 1% both to *Candida albicans* and *Pityrosporum ovale*. Therefore, rose apple leaf has the potential to be developed for antileucorrhoeal and antidandruff properties.

REFERENCES

Ajizah, A. 2004. Sensitivitas *Salmonella thypimurium* Terhadap Ekstrak Daun *Psidium guajava* L *Bioscientiae*, 1(1): 31–38.

Arafath, M. A., Kwong, H. C., & Adam, F. 2019. The crystal structure of ((cyclohexylamino) {(Z)-2-[(E)-5-methoxy-3-nitro-2-oxidobenzylidene-jO]-hydrazin-1-ylidene-jN2} methanethiolato-jS)-(dimethyl sulfoxide-jS) platinum(II): A supramolecular two-dimensional network. *Acta Crystallographic Communications*, 75: 1486–1489.

Bubakar, A. R., & Amiruddin, M. D. 2012. Clinical aspects fluor albus of female and treatment. *Indonesian J Dermatol Venerol*, 1(1): 19–29.

Castinetti, F., Guignat, L., Giraud, P., Muller, M., Kamenicky, P., Drui, D., . . . & Bihan, H. 2014. Ketoconazole in Cushing's disease: Is it worth a try?. *The Journal of Clinical Endocrinology & Metabolism*, 99(5), 1623–1630.

Chen, Y., Bruning, E., Rubino, J., & Eder, S.E. 2017. Role of female intimate hygiene in vulvovaginal health: Global hygiene practices and product usage. *Women's Health*, 13(3): 58–67.

Elisabeth, Z. M., Lopez, P. C., Soto, S. M., & Fabrice, F. B. 2016. Anti-*Candida* biofim properties of cameroonian plant extract. *Journal of Medicinal Plant Research*, 10(35): 603–611.

Indriyanti, N., Adnyana, I. K., & Sukandar, E. Y. 2013. Aktivitas Ekstrak Etanol dan Fraksi Akar Singawalang (*Petiveria alliacea* L.) terhadap Jamur Penyebab Ketombe dengan Metode Broth Microdilution. *Journal of Tropical Pharmaceutical Chemistry*, 2(2): 113–117.

Kandimalla, R., Kalita, S., Choudhury, B., Dash, S., Kalita, K., & Kotoky, J. 2016. Chemical composition and anti-candidiasis mediated wound healing property of *Cymbopogon nardus* essentials oil on chronic diabetes wounds. *Frontiers in Pharmacology*, 7: 2–3.

Malonda, T. C., Yamlean, P. V. Y., & Citraningtyas, G. 2017. Formulasi Sediaan Sampo Antiketombe Ekstrak Daun Pacar Air (*Impatiens balsamina* L.) dan Uji Aktivitasnya terhadap Jamur *Candida albicans* ATCC 10231 secara in vitro. *Pharmacon*, 6(4): 97–109.

Mapatac, L. C., & Mamaoag, N. R. 2014. Efficacy of three varieties of *Syzygium aqueum* (Tambis) as antimicrobial agent and its bioactive component. *Int J Sci Clin Lab*, 9: 1.

Mashaly, G., & Shrief, R. 2019. *Candida glabrata* complex from patients with healthcare-associated infections in Mansoura University Hospitals, Egypt: Distribution, antifungal susceptibility and effect of fluconazole and polymyxin B combination. *Germs*, 9(3): 125–132.

Mayer, F.L., Wilson, D., & Hube, B. 2013. *Candida albicans* pathogenicity mechanisms. *Virulence*, 4(2): 119–128.

Ministry of Health of the Republic of Indonesia. 2011. *Peraturan Menteri Kesehatan Republik Indonesia Nomor 2406/Menkes/Per/XII/2011 Tentang Pedoman Penggunaan Antibiotik*. Jakarta. 4–5.

Monisha, P., Shabna, E., Subhashri, S. H. R., Sridevi, R., & Kavimani, S. 2018. Phytochemistry and pharmacology of *Syzygium aqueum*: A critical review. *European Journal of Biomedical and Pharmaceutical Sciences*, 5(6): 271–276.

Pan, X., Chen, F., Wu, T., Tang, H., & Zhao, Z. 2009. The acid, bile tolerance and antimicrobial property of *Lactobacillus acidophilus* NIT. *Journal of Food Contro,l* 20(6): 598–602.

Potluri, A., Shaheda, A. S. K., Rallapally, N., Durrivel, S., & Harish, G. 2013. A review on herbs used in anti-dandruff shampoo and its evaluation parameters. *Indo American Journal of Pharmaceutical Research*, 3(4): 3266–3278.

Purnomo, C.W. & Indarti, S. 2018. Modification of indirect solar dryer for simplicia production. *IOP Conference Series: Earth and Enviromental Sciences* 120: 1–6.

Ranganathan, S., & Mukhopadhyay, T. 2010. Dandruff: The most commercially exploited skin disease. *Indian Journal of Dermatology*, 5(2): 130–134.

Serpa, R., França, E. J., Furlaneto-Maia, L., Andrade, C. G., Diniz, A., & Furlaneto, M. C. 2012. In vitro antifungal activity of the flavonoid baicalein against Candida species. *Journal of Medical Microbiology*, 61(12): 1704–1708.

Sharma, K. K., Saika, R., Kotoky, J., Kalita, J. C., & Das, J. 2011. Evaluation of antidermatophytic activity of *Piper betle, Allamanda cathertica*, and their combination: An in vitro and in vivo study. *International Journal of PharmTech Research*, 3(2): 645–646.

Sony, P., Kalyani, M., Jeyakumari, D., Kannan, I., & Sukumar, R. G. 2018. In vitro antifungal activity of *Cassia fistula* extracts againts fluconazole resistant strains of Candida species from HIV patients. *Journal de Mycologie Medicale*, 28: 193–200.

Sukandar, E. Y., & Ekawati, E. 2006. Activity of ethanol extracts of seledri (*Apium graveolens*) herbs and urang aring (*Eclipta prostata* (L.) L.) leaves against *Pityrosporum ovale*. *Indonesian Journal of Pharmacy*, 17(1): 7–12.

Zida, A., Bamba, S., Yacouba, A., Ouedraoqo-Traore, R., & Guiquemde, R.T. 2017. Anti-*Candida albicans* natural products, sources of new antifungal drugs: A review. *Journal de Mycoogiel Medicale*, 27(1): 1–19.

Zoya, M., Bhikhu, M., & Gaurav, S. 2016. Anti-dandruff activity of synthetic and herbal shampoos on dandruff causing isolate: Malassezia. *International Journal of Applied Research*, 2(7): 80–86.

Medical Technology and Environmental Health – Abdullah, Widiaty & Abdullah (eds)
© 2020 Taylor & Francis Group, London, ISBN 978-0-367-86053-0

Intra-cytoplasmic Cytokine Staining (ICS): Optimizing antigen stimulation for measuring *M. tuberculosis*-specific T cell response

H. Muflihah
Centenary Institute, The University of Sydney, NSW, Australia
Faculty of Medicine, Universitas Islam Bandung, West Java, Indonesia

W.J. Britton
Centenary Institute, The University of Sydney, NSW, Australia
Sydney Medical School, The University of Sydney, NSW, Australia

ABSTRACT: Protective immunity induced by new tuberculosis (TB) vaccines was assessed by an immunology assay measuring cytokines. Intra-cytoplasmic cytokine staining (ICS) requires optimization as an easy-to use package kit is not available. This preliminary study aims to optimize the period of *Mycobacterium tuberculosis* (*Mtb*) antigen stimulation and inhibition of cytokine secretion. Three designs on the stimulation of peripheral blood mononuclear cells (PBMC) with *Mtb* culture filtrate protein (CFP) and additional protein transport inhibitor Brefeldin A were compared. The incubation period is 1 day or 5 days of CFP stimulation with Brefeldin A at the last 4 or 18 hours. The production of Th1 cytokines was analysed by flow cytometry. The 18-hour Brefeldin A resulted in increased detection of frequency of CD4$^+$ T cells producing IFN-γ, IL-2, and TNF and combined cytokine producers. Five-day antigen stimulation improved the detection of IL-2, but not IFN-gamma or TNF as compared to the one day. In conclusion, detection of *Mtb*-specific T cell response using ICS of human PBMC is optimum using 1 day of antigen stimulation with additional Brefeldin A at the last 18 hours. The duration of antigen stimulation and protein transport inhibition affects the frequency and the functionality of cytokines analysed using flow cytometry.

1 INTRODUCTION

Tuberculosis (TB) remains a devastating infectious disease globally. Indonesia was the third highest TB burdened country in 2017 (WHO 2018). The current *Mycobacterium bovis* Bacille Calmette-Guérin (BCG) vaccine protects against severe TB in childhood (Mangtani et al. 2014). However, the protection of BCG against pulmonary TB ranges from no protection to 99% in 12 cohort studies (Abubakar et al. 2013). Therefore, long-term control of TB depends on development of new tools, including a TB vaccine more effective than BCG.

The development of TB vaccine candidates has progressed rapidly in the last decade, with 12 new vaccines entering clinical trials (WHO 2018). Protective immunity of TB vaccine candidates is evaluated by immunological assays. Detection of IFN-γ, TNF, and IL-2 that are specific to *M. tuberculosis* has been evaluated as protective immunity as well as a biomarker for diagnostic TB infection in human (Qiu et al. 2012, Kaufmann et al. 2017) and pre-clinical TB vaccine studies (Muflihah et al. 2018). Methods for detection of IFN-γ on human blood that have been widely used are ELISA, interferon-γ enzyme-linked immunosorbent spot (ELISpot) and intra-cytoplasmic cytokine staining (ICS) flow cytometry.

Unlike ELISA and ELISpot, ICS can simultaneously detect responder cells and phenotype markers of cells in addition to secretion of cytokines (Flórido et al. 2018).

Commercial package kits that is easy to use are available for ELISA and ELISpot, but not for ICS. The protocol for ICS need to be optimized to adjust assays intention and laboratory settings.

This is a preliminary study that aims to optimize the antigen stimulation step on ICS in measuring *M. tuberculosis*-specific response in humans. The focus of optimization is the duration of antigen stimulation and the duration of cytokine secretion inhibition. These are particularly important when using fresh blood samples in the setting where the availability of an incubator with CO_2 is limited.

2 METHODS

2.1 Ethical approval

The use of human blood samples in this study was approved by University of Sydney Human Research Ethics Committee (HREC) Project no. 2015/346.

2.2 Isolation of human peripheral blood mononuclear cells (PBMC)

A total volume of 15 ml of venous blood was collected from two healthy persons who were known to have a positive protein-purified derivative (PPD) skin test. The PBMC was isolated by standard density-gradient centrifugation technique using Ficoll-Plaque Plus (GE Healthcare, Amersham, UK). Briefly, heparinized whole blood was diluted 1:3 in phosphate-buffered saline (PBS). Diluted blood was gently layered into tubes containing Ficoll (10 ml for 3 ml diluted blood) to have the blood layered on top of the Ficoll. The tube was then centrifuged at 1800 rpm for 10 minutes at room temperature with the brake off. The PBMC layer was collected, washed three times with PBS, and resuspended in 2 ml of pre-warmed (room temperature) complete RPMI tissue culture media consisting of RPMI 1640 (Gibco, Australia), 10% foetal calf serum (FCS), penicillin and streptomycin (100 µg/ml, Gibco). The number of viable PBMC was counted in a hemocytometer. Approximately 1 million PBMC were isolated per 1 ml of whole blood.

2.3 Stimulation of human PBMC with M. tuberculosis antigen

Fresh PBMC ($2x10^6$ cells) in the complete RPMI media were incubated with culture filtrate protein (CFP) *M. tuberculosis* H37Rv (BEI Resources NIAID, NIH, Manassas, VA) at concentration 10 µg/ml at 37°C in 5% CO_2. The cells were stimulated with T cell activation and expansion beads (TAE, Miltenyi Biotech) as positive controls or with media alone as negative controls. An inhibitor of protein transport Brefeldin A (10 µg/ml, Sigma) was added to retain cytokines within T cells. The duration of antigen stimulation was either 1 day or 5 days with the presence of Brefeldin A in the last 4 hours or 18 hours.

2.4 Intra-cytoplasmic cytokine staining flowcytometry

PBMC were stained with fluorochrome-conjugated monoclonal antibodies against surface markers and cytokines following antigen stimulation. Cells were incubated with BV605-conjugated anti-CD3, APC-conjugated anti-CD4 (Biolegend), and live/dead fixable blue staining (Life Technologies) in FACS wash for 30 minutes on ice. The cells were washed, fixed and permeabilized with Cytofix (BD Biosciences) for 20 minutes on ice. The cells were washed with PERM wash buffer (BD Biosciences) prior to staining with FITC-conjugated anti-IFN-γ, AF700-conjugated anti-TNF AF700, and PE-conjugated anti-IL-2 in PERM wash buffer for 30 minutes on ice. After washing with PERM wash, cells were resuspended with FACS wash and 10% neutral buffered formalin. Stained cells were acquired using LSR Fortessa flow cytometer (BD) and the data was analyzed with FlowJo (Tree Star) and FlowJo Boolean gating tool.

3   RESULTS

3.1   *Prolonged duration of secretion inhibitor increased the detection of cytokines*

Production of cytokine by CD4⁺ T cells following stimulation with *M. tuberculosis* CFP antigen and different duration of additional protein transport inhibitor Brefeldine A is shown in Figure 1. The frequency of CD4⁺ T cells producing IFN-γ, TNF, and IL2 at 4 hours of Brefeldin A were 0.02 %, 0.04% and 0.1%, respectively. These were increased up to 0.14% following 18-hour Brefeldin A for all cytokines. The frequency of TNF producers had the highest increase following prolonged Brefeldin A, whereas IL-2 producers had the lowest one.

3.2   *CD4⁺ T cells producing combined cytokines increased by prolonged inhibition of cytokine transport*

The combination of IFN-γ, TNF, and IL2 produced by CD4+ T cells was compared in the presence of 4 hours and 18 hours of Brefeldin A within the same type and duration of antigen stimulation. The majority of CD4+ T cells produced single cytokine at the 4 hour of Brefeldin A (Figure 2A). Interestingly, the frequency of double and triple cytokine producers was increased at the longer period of cytokine transport inhibition (Figure 2B).

3.3   *5-day stimulation of M. tuberculosis antigen decreased IFN-γ and TNF, but not IL-2*

Secretion of Th1 cytokines was measured following different total duration of *M. tuberculosis* CFP antigen stimulation (Figure 3). There was 0.14% of CD4 T cells produced IFN-γ or TNF

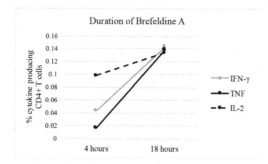

Figure 1.   Cytokine-producing CD4⁺ T cells following stimulation with *M. tuberculosis* CFP antigen and Brefeldine A. Human PBMC was stimulated with *M. tuberculosis* CFP antigen for 1 day (24 hours) in the presence of Brefeldine A at the last 4 or 18 hours. The production of IFN-γ, TNF, and IL2 was measured by ICS flow cytometry.

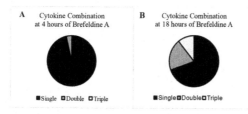

Figure 2.   The proportion of CD4⁺ T cells producing combination of IFN-γ, TNF, and IL2 in different duration of cytokine transport inhibition. (A) The proportion of single cytokine producers was dominant at 4-hour inhibition of cytokines with Brefeldin A. (B) The proportion of single, double and triple cytokine producers detected following 18 hours of Brefeldin A.

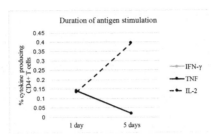

Figure 3. The effect of duration of antigen stimulation on cytokine secretion of cytokine by CD4[+] T cells following 1-day or 5-day stimulation. Stimulation of human PBMC with one day or five days of *M. tuberculosis* CFP antigen. The secretion cytokine inhibitor was added at the last 18 hours of stimulation.

or IL-2 at 1-day antigen stimulation. The frequency of IFN-γ or TNF was decreased following 5-day antigen stimulation. In contrast, the proportion of IL-2 producers increased up to 0.4 %.

4 DISCUSSION

Optimization on ICS flow cytometry methods is necessary for the standardization of assays in human TB vaccine studies. We have shown that the duration of antigen stimulation and inhibition of cytokine secretion has an effect on detection of cytokine produced by CD4[+] T cells.

This study used fresh PBMC instead of frozen PBMC to have the optimal result on cytokine detection. Fresh PBMC has shown to have higher frequencies of cytokine-expressing cells than the frozen PBMC (Smith et al. 2015). A possible reason is the loses of activated, effector T cells and antigen-presenting cells that are required for response to antigen stimulation on recall assyas (Smith et al. 2015). However, the frozen PBMC would be beneficial for future batch analysis in settings where the facilities for cell culture and flow cytometry are limited. Future direction is to optimize the use of prestimulated frozen PBMC. The PBMC was stored in –80°C following incubation with antigen and cytokine secretion inhibitors.

A cytokine secretion inhibitor is often used in most protocols of ICS to trap the cytokine within the cells. The proposed mechanism of Brefeldin A is inhibition some of proteins that activate ADP-ribosilation factors (Arf), which is critical for vesicular transport between endoplasmic reticulum and the Golgi (Chardin & McCormick 1999). Cytokine secretion inhibitors seem to be time and cytokine dependent. For example, another cytokine inhibitor, namely Monesin, in ICS decreased detection of IL-4 within cells (Smith et al. 2015), but the shorter incubation of Monesin improved the detection of IL-10 (Muris et al. 2012). The incubation of Brefeldin A for 3.5 hours (Muflihah et al. 2018) or less than 12 hours is for practicality (Miguel et al. 2012) and avoiding cell toxicity. Our work showed that prolonged incubation of Brefeldin A up to 18 hours improved the frequency (Figure 1) and functionality of T cells secreting IFN-γ, TNF, and IL2 (Figure 2B).

Prolonged incubation of antigen may allow slower responder T cells to proliferate and to provide more time for "resting" or central memory T cells. We have found that prolonged incubation of PBMC with *M. tuberculosis* CFP improved detection of IL-2, but not IFN-γ and TNF. This is in contrast to a previous study that showed 5-day incubation with PPD improved the detection of all these cytokines (Smith et al. 2015). The difference of our result is probably related to the antigen used and responding cytokines. For weaker immunogen and low-rate expression of cytokine, the prolonged antigen stimulation might be a good option to improve detection. TNF is an inflammatory cytokine during *M. tuberculosis* infection and marking effector T cells (Pollock et al. 2013, Dorhoi & Kaufmann 2016). This cytokine might be released rapidly following antigen stimulation but wanes by prolonged antigen stimulation. We showed that for this cytokine, shorter antigen stimulation with longer inhibition of secretion improved detection. In contrast, prolonged antigen stimulation decreased detection as the cytokine released at the early phase of antigen stimulation. Prolonged incubation for ICS may also need an evaluation for the viability of cells. The frequency of CD4$^+$ T cells producing cytokines might be high in the low rate of viable cells. Thus, this study suggests a total duration of 1-day antigen for ICS flow cytometry using fresh PBMC.

5 CONCLUSION

Detection of cytokine in ICS depends on the incubation period of antigen and cytokine secretion inhibitors. Prolonged duration treatment of cytokine secretion inhibitor Brefeldin A increases the frequency and functionality of CD4$^+$ T cells producing IFN-γ, TNF, and IL2. In contrast, prolonged duration of antigen stimulation decreased the frequency of IFN-γ and TNF producers, but increased the IL-2 producers. In our study, the optimal antigen stimulation is 1-day incubation of *M. tuberculosis* CFP with Brefeldin A in the last 18 hours.

ACKNOWLEDGEMENT

We would like to thank Dr. Umaimainthan Palendira for providing monoclonal antibodies used in this study.

REFERENCES

Abubakar, I., Pimpin, L., Ariti, C., Beynon, R., Mangtani, P., Sterne, J. A. C., ... & Watson, J. M. 2013. Systematic review and meta-analysis of the current evidence on the duration of protection by bacillus Calmette-Guérin vaccination against tuberculosis. *Health Technology Assessment (Winchester, England)* 17(37): 1.

Chardin, P., & McCormick, F. 1999. Brefeldin A: The Advantage of Being Uncompetitive. *Cell*, 97(2): 153–155.

Dorhoi, A., & Kaufmann, S. H. E. 2016. Pathology and immune reactivity: understanding multidimensionality in pulmonary tuberculosis. *Seminars in Immunopathology*, 38(2): 153–166.

Flórido, M., Muflihah, H., Lin, L. C., Xia, Y., Sierro, F., Palendira, M., ... & Britton, W. J. 2018. Pulmonary immunization with a recombinant influenza A virus vaccine induces lung-resident CD4+ memory T cells that are associated with protection against tuberculosis. *Mucosal Immunology*, 1(6): 1743–1752.

Kaufmann, S. H., Dockrell, H. M., Drager, N., Ho, M. M., McShane, H., Neyrolles, O., ... & Stenger, S. 2017. TBVAC2020: Advancing tuberculosis vaccines from discovery to clinical development. *Frontiers in Immunology*, 8: 1203.

Mangtani, P., Abubakar, I., Ariti, C., Beynon, R., Pimpin, L., Fine, P. E., ... & Sterne, J. A. 2014. Protection by BCG vaccine against tuberculosis: A systematic review of randomized controlled trials. *Clinical Infectious Diseases*, 58(4): 470–480.

Miguel, R. D. V., Maryak, S. A., & Cherpes, T. L. 2012. Brefeldin A, but not monensin, enables flow cytometric detection of interleukin-4 within peripheral T cells responding to ex vivo stimulation with Chlamydia trachomatis. *Journal of Immunological Methods*, 384(1-2): 191–195.

Muflihah, H., Flórido, M., Lin, L. C. W., Xia, Y., Triccas, J. A., Stambas, J., & Britton, W. J. 2018. Sequential pulmonary immunization with heterologous recombinant influenza A virus tuberculosis vaccines protects against murine *M. tuberculosis* infection. *Vaccine*, 36(18): 2462–2470.

Muris, A.H., Damoiseaux, J., Smolders, J., Cohen Tervaert, J. W., Hupperts, R., & Thewissen, M. 2012. Intracellular IL-10 detection in T cells by flowcytometry: The use of protein transport inhibitors revisited. *Journal of Immunological Methods*, 381(1): 59–65.

Pollock, K. M., Whitworth, H. S., Montamat-Sicotte, D. J., Grass, L., Cooke, G. S., Kapembwa, M. S., ... & Lalvani, A. 2013. T-cell immunophenotyping distinguishes active from latent tuberculosis. *The Journal of Infectious Diseases*, 208(6): 952–968.

Qiu, Z., Zhang, M., Zhu, Y., Zheng, F., Lu, P., Liu, H., ... & Chen, X. 2012. Multifunctional CD4 T cell responses in patients with active tuberculosis. *Scientific Reports*, 2: 216.

Smith, S. G., Smits, K., Joosten, S. A., van Meijgaarden, K. E., Satti, I., Fletcher, H. A., ... & Dockrell, H. M. 2015. Intracellular cytokine staining and flow cytometry: Considerations for application in clinical trials of novel tuberculosis vaccines. *PloS one*, 10(9).

WHO 2018. *Global Tuberculosis Report 2018*. [Online]. https://apps.who.int/medicinedocs/en/m/abstract/Js23553en/.

Determinants of tuberculosis-preventive behavior in rural areas of Indonesia

H.S. Rathomi & N. Romadhona
Universitas Islam Bandung, Bandung, Indonesia

ABSTRACT: Sustainable development goals mandate all countries struggle to reduce tuberculosis (TB) incidence by up to 80%. One of the essential efforts for eliminating TB is practicing preventive behaviors. This study aims to explore TB-preventive behavior and associated factors. This is a cross-sectional study using a multistage cluster sampling method. We involved 210 residents in one of the villages with the highest TB prevalence in Bandung, Indonesia. Data regarding behaviors and related determinants were obtained through a validated questionnaire. We analysed the data using STATA 13. Most of the respondents had good TB-preventive behavior. While the majority of the respondents implemented specific behavior such as closing their mouth when coughing, general behavior, such as regular exercise, was only performed by a limited number of respondents. Based on the chi-square test, the level of education and knowledge was the only factors that significantly related to preventive behavior, while gender, occupation, income, and experience getting TB education did not have a meaningful relationship. It can be concluded that more than half of respondents have practised adequate TB prevention, and there is a significant relationship between the level of education and respondents' knowledge about TB and TB-preventive behavior.

1 INTRODUCTION

Tuberculosis is a global problem and one of the top 10 causes of death. In 2017, WHO stated that 10 million people suffer from the illness, of whom 900,000 also have HIV. Indonesia itself is currently ranked as having the 3rd highest incidence rate and accounts for 8% of the total number of sufferers after India and China. The incidence of tuberculosis in Indonesia is very high, reaching 316 cases per 100,000 population (WHO 2019).

The magnitude of the burden of tuberculosis makes this problem a global priority. In the target of sustainable development goals, it is targeted to reduce mortality due to TB by 90% and decrease incidence by up to 80% by 2030 (WHO 2018). If this target is not achieved, TB will increasingly be a considerable burden, not only on the healthcare systems but also on the economy. The cost burden for TB treatment is due not only to long-term treatment, but also the existence of catastrophic events that may occur in TB patients, at an incidence rate of 36% in all patients (Fuady et al. 2018).

Among the various efforts to reduce the burden of tuberculosis, prevention is the essential thing. Various efforts have been made, such as increasing the coverage of BCG immunization, and also developing vaccines for adults through intrapulmonary methods (Muflihah et al. 2018). Fundamental efforts such as getting the community used to preventing transmission are also an urgent priority.

Improving preventive behavior, especially in rural areas, is a challenge because there are many determinant factors. Lack of access to health facilities in rural areas, lower levels of knowledge in the community, and the presence of a thicker sociocultural element can make efforts to change behavior in rural communities more difficult (Rathomi & Nurhayati 2019). In terms of TB prevention, several studies have confirmed disparities in levels of knowledge and community behavior between rural and urban areas (Mushtaq et al. 2011). Various studies have also shown differences in what factors can influence TB-preventive behavior, both in urban and rural areas (Mushtaq et al. 2011, Bati et al. 2013, Duru et al. 2016, Rahman et al. 2017). This study aims to explore TB-preventive behavior among rural residents in Rancakasumba village and associated factors.

2 METHODS

This research is a cross-sectional study with a sample selected by multistage cluster sampling. Respondents in this study were 210 residents of Rancakasumba Village, Solokan Jeruk Sub-District, Bandung Regency, Indonesia. We choose this location due to its high prevalence of tuberculosis. Data was collected during May 2019 and was carried out by a door-to-door process.

Data was retrieved using a validated instrument containing inquiry about demographic data, level of knowledge, TB-preventive behavior, and housing conditions. The knowledge level was assessed by six components, the behavior assessed by seven components, and house conditions by five aspects – the answer to each item valued and accumulated after that. Respondents with knowledge and behavior scores >80% were stated to have good knowledge and behavior, while respondents with score >50% and <50% were categorised as medium and bad, respectively. The good category for housing conditions is given to respondents with a value >50%. Data was analysed using STATA 13 software with a chi-square test to see the relationship between factors, namely demographics, access, ownership of health insurance, knowledge, and home conditions, with TB-preventive behavior. A significant relationship was stated if the test results p values <0.05.

3 RESULTS

Table 1 shows the distribution of respondents' characteristics in Rancakasumba village. The median age of the respondents was 43, where most of the respondents were female and housewives. This is because the survey is conducted by visiting each house directly during business hours. The level of education and economy is relatively low, as indicated by the majority having primary and secondary education levels, and 72% of respondents are families with income less than the minimum wage district of Bandung, which is Rp. 2,890,000.

Respondents' access to health services in this study was also identified by covering geographical, financial, and health education access. From Table 1, it can be seen that the majority of respondents do not have access within a distance of 5km (68.6%), 53.5% have not received education about TB in the past year, and there are 34.3% have not become BPJS participants.

The level of knowledge, preventive behavior, and the situation of the environment/housing is shown in Table 2. Most of the respondents have a good level of knowledge, preventive behavior, and housing conditions. Only a small proportion still has a level of knowledge and behavior that is classified as poor. The distribution of TB-preventive behavior in respondents can be seen in the following Figure 1.

Figure 1 Shows the distribution of respondents' behavior, which is divided into seven types of behavior. A specific preventive behavior, such as closing the mouth when coughing, has been consistently carried out by 64.3% of respondents. However, doing regular exercise, which is a more general behavior, as well as the routine action of drying new mattresses, is done routinely by only a small proportion of respondents (12.4% and 19% respectively).

Table 3 shows the results of the chi-square test to find the relationship between determinants AND TB-preventive behavior. There are only two determinants that have a significant relationship with TB-preventive behavior in the Rancakasumba village community, namely the level of education and level of knowledge. Respondents with middle to higher education levels and who have better knowledge tend to have better preventive practices.

Table 1. Respondent characteristics.

Characteristics	Frequency (n=210)	%
Age		
Median (min-max)	43 (15–89)	
Sex		
Men	67	31.9
Women	143	68.1
Education		
Not completed	31	14.8
Elementary	59	28.1
Junior high school	60	28.6
Senior high school	57	27.1
Bachelor	3	1.4
Work Status		
Does not work	152	72.4
Work	58	27.6
Income		
< Regional minimum wage	164	78.1
≥ Regional minimum wage	46	21.9
Health insurance		
BPJS subsidized	85	40.5
BPJS nonsubsidized	53	25.2
Non-BPJS member	72	34.3
Healthcare access		
< 5 km	66	31.4
≥ 5 km	144	68.6
Get TB education in last year		
Yes	98	46.7
No	112	53.3
Has a family member with TB		
Yes	20	9.5
No	190	90.5

Table 2. Knowledge, practice, and housing conditions of TB prevention.

	Good	Fair	Poor	Total
Knowledge	126 (60%)	51 (24%)	33 (16%)	210
Preventive practice	107 (51%)	82 (39%)	21 (10%)	210
Housing conditions	158 (75%)		52 (25%)	210

Figure 1. Distribution of preventive behavior among respondents.

300

Table 3. Factors associated with TB-preventive behavior.

Determinants	Behavior level (%)			P-value
	Poor	Fair	Good	
Sex				
Men	13.43	38.81	47.76	p = 0.507
Women	8.39	39.16	52.45	
Education				
Low	11.11	55.56	33.33	p = 0.000*
Middle to high	9.17	26.67	64.17	
Work status				
Does not work	7.24	38.16	54.61	p = 0.506
Work	17.24	41.38	41.38	
Income				
< Minimum wage	10.98	39.63	49.39	p = 0.566
> Minimum wage	6.52	36.96	56.52	
Health insurance				
BPJS subsidized	10.59	37.65	51.76	p = 0.764
BPJS nonsubsidized	5.66	43.4	50.94	
Non BPJS member	12.5	37.5	50	
Get TB education in last year				
Yes	13.27	38.78	47.96	p = 0.319
No	7.14	39.29	53.57	
Knowledge				
Poor	15.15	66.67	18.18	p = 0.000*
Fair	11.76	52.94	35.29	
Good	7.94	26.19	65.87	
Total	10	39.05	50.95	

4 DISCUSSION

TB-preventive behavior in rural areas (Rancakasumba Village) in this study was quite good. This can be seen, particularly in the behaviour of closing the mouth when coughing. However, in behaviors, the proportion of respondents who consistently make preventive efforts is still low. The habit of not spitting carelessly, drying the mattress, and opening the window is practised routinely by less than half the respondents. More general prevention efforts such as exercise and the consumption of nutritious foods are even lower.

The level of knowledge, behavior, and housing conditions in the community in this study was not much different from research in Demak and Banjar Districts (Rahman et al. 2017, Setiyadi & Adi 2019). Both studies are also located in rural areas so they show consistent results. Research in other developing countries, such as Ethiopia, Nigeria, and Timor Leste, shows more varied results. The level of TB-preventive behavior also varies depending on profession. This can be seen from Gebrehiwot's research that states that TB-preventive behavior in bus drivers is still low (Gebrehiwot & Tesfamichael 2017) while on the passenger bus as researched by Malede, it is even better (Malede et al. 2019).

Although quite good, the level of knowledge and behavior of TB prevention is still lower than in urban areas. This is as stated by Fuady, who examined the level of public knowledge in the Jakarta area with a good level of knowledge of 88.7% (Fuady et al. 2014).

Health behavior, including the prevention of tuberculosis, is theoretically influenced by many factors. In the theory of behavior-change wheel model, behavior is formed by capability, motivation, and opportunity factors (Michie et al. 2011). However, in this study, only aspects of knowledge and education level have a significant relationship. Higher levels of education provide the ability to reason more precisely and produce more rational behavior. Increased levels of knowledge also result in a better level of behavior. The importance of this level of knowledge not only has an impact on preventive behavior but also in the long term, that is, healing therapy if the person is suffering from TB. Nurkomarasi et al. (2014) proved a significant relationship between low levels of knowledge and the risk of treatment dropouts in TB patients.

The relationship of TB-preventive behavior with level of knowledge and education has been confirmed by other studies, both in Indonesia and other developing countries (Fuady et al. 2014, Duru et al. 2016, Rahman et al. 2017, Pengpid & Peltzer, 2019). However, some research also states that there are other determinants that are statistically significant, such as gender (Bati et al. 2013), marital status, religion, work, and age (Duru et al. 2016), and housing factors (Malede et al. 2019). Explicitly, Mushtaq et al. (2011) also confirmed the effect of residence in TB-prevention practices, especially the existence of significant differences between rural and urban areas. In addition to lower levels of preventative behavior, Musthaq et al. (2011) also stated that people in rural areas are more dependent on health workers as a source of information about preventing tuberculosis.

Differences regarding what factors are related to TB-preventive behavior are influenced by various things, such as the location and culture of the community where the study was conducted. Various studies conducted in African countries, such as Nigeria and Ethiopia, show that primary demographic factors such as gender and religion are included in the determinants of behavior (Bati et al. 2013, Gebrehiwot & Tesfamichael 2017). This is understandable because the level of gender equality and religion is still very different between the situation in Indonesia and in these countries.

The demographics of respondents in this study show the characteristics of rural communities which have low income and education levels. BPJS ownership is still low, even though this research was conducted in a period targeted for all regions to reach universal health coverage. Rathomi et al.'s (2018) research showed that in the third year of the JKN program, almost all community users of hospital services in Bandung Regency used BPJS for treatment.

Access to health facilities, both geographically and financially, as well as health education in this study was not significantly related. However, access should be an essential determinant in shaping one's knowledge and health behavior. This paradox is an input for health facilities, particularly Puskesmas, to increase the effectiveness of their programs in changing community behavior. The efforts to prevent and educate about tuberculosis require a sociocultural approach to be more effective and make an impact (Respati & Sufrie 2014).

5 CONCLUSION

Most people in rural areas have a good level of TB-preventive behavior, especially closing their mouths when coughing. However, general preventive behavior and actions to produce a healthier environment are still not widely practised. There is a significant relationship between knowledge level and education level with TB-preventive behavior, while other socioeconomic and financial factors, as well as geographical access to health services, are not significantly related. Puskesmas, which is the territorial authority area responsible, is expected to be able to increase the effectiveness of education and health promotion efforts to improve community knowledge and behavior, especially regarding tuberculosis prevention.

ACKNOWLEDGEMENT

We thank all medical students in Puskesmas Solokan Jeruk, Bandung District, between May and June 2019 who have helped with data collection.

REFERENCES

Bati, J., Legesse, M., & Medhin, G. 2013. Community's knowledge, attitudes and practices about tuberculosis in Itang Special District, Gambella Region, South Western Ethiopia. *BMC Public Health*, 13(1): 1.

Duru, C. B., Iwu, A. C., Uwakwe, K. A., Diwe, K. C., Merenu, I. A., Chineke, H. N., & Adaeze, C. A. 2016. Pulmonary tuberculosis preventive practices and its determinants among residents in communities in Orlu, Imo State, Nigeria. *Int J Sci Healthc Res*, 1(2): 57–69.

Fuady, A., Houweling, T. A., Mansyur, M., & Richardus, J. H. 2018. Catastrophic total costs in tuberculosis-affected households and their determinants since Indonesia's implementation of universal health coverage. *Infectious Diseases of Poverty*, 7(1): 3.

Fuady, A., Pakasi, T. A., & Mansyur, M. 2014. The social determinants of knowledge and perception on pulmonary tuberculosis among females in Jakarta, Indonesia. *Medical Journal of Indonesia*, 23(2): 93–105.

Gebrehiwot, T. T., & Tesfamichael, F. A. 2017. Knowledge, risk perception and practice regarding tuberculosis transmission among long distance bus drivers in Addis Ababa, Ethiopia: A cross sectional study. *Ethiopian Journal of Health Sciences*, 27(6): 601–612.

Malede, A., Taye, B., Legesse, M., Debie, A., & Shibabaw, A. 2019. Pulmonary tuberculosis preventive practices among Anibessa bus users at Addis Ababa, Ethiopia: A cross-sectional study. *BMC Research Notes*, 12(1): 104.

Michie, S., Van Stralen, M. M., & West, R. 2011. The behavior change wheel: A new method for characterising and designing behavior change interventions. *Implementation Science*, 6(1): 42.

Muflihah, H., Flórido, M., Lin, L. C. W., Xia, Y., Triccas, J. A., Stambas, J., & Britton, W. J. 2018. Sequential pulmonary immunization with heterologous recombinant influenza A virus tuberculosis vaccines protects against murine *M. tuberculosis* infection. *Vaccine*, 36(18): 2462–2470.

Mushtaq, M. U., Shahid, U., Abdullah, H. M., Saeed, A., Omer, F., Shad, M. A., ... & Akram, J. 2011. Urban-rural inequities in knowledge, attitudes and practices regarding tuberculosis in two districts of Pakistan's Punjab province. *International Journal for Equity in Health*, 10(1): 8.

Nurkomarasi, N., Respati, T., & Budiman, B. 2014. Karakteristik penderita drop out pengeobatan tuberkulosis paru di Garut. *Global Medical & Health Communication*, 2(1): 21–26.

Pengpid, S., & Peltzer, K. 2019. Knowledge, attitudes, and practices regarding tuberculosis in Timor-Leste: Results from the demographic and health Survey 2016. *Journal of Preventive Medicine and Public Health*, 52(2): 115–122.

Rahman, F., Adenan, A., Yulidasari, F., Laily, N., Rosadi, D., & Azmi, A. N. 2017. Pengetahuan Dan Sikap Masyarakat Tentang Upaya Pencegahan Tuberkulosis. *Media Kesehatan Masyarakat Indonesia Universitas Hasanuddin*, 13(2): 183–189.

Rathomi, H. S., & Nurhayati, E. 2019. Hambatan dalam Mewujudkan open defecation free. *Jurnal Integrasi Kesehatan & Sains*, 1(1): 68–73.

Rathomi, H. S., Yulianto, F. A., & Romadhona, N. 2018. Dampak Program Jaminan Kesehatan Nasional Terhadap Utilisasi Layanan Kesehatan Pasien Kanker Serviks. *Jurnal Kebijakan Kesehatan Indonesia: JKKI*, 7(3): 126–133.

Respati, T., & Sufrie, A. 2014. Socio cultural factors in the treatment of pulmonary tuberculosis: A case of Pare-Pare Municipality South Sulawesi. *Global Medical & Health Communication*, 2(2): 60–65.

Setiyadi, D., & Adi, M. S. 2019. Pengetahuan, Praktik Pencegahan Dan Kondisi Rumah Pada Kontak Serumah Dengan Penderita Tb Paru Di Kabupaten Demak. *Visikes Jurnal Kesh Masyarakat*, 18(1): 36–45.

WHO. 2018. *The End TB Strategy Report*. [Online]. https://www.who.int/tb/post2015_TBstrategy.pdf?ua=1.

WHO. 2019. *Global Tuberculosis Report*.

Medical Technology and Environmental Health – Abdullah, Widiaty & Abdullah (eds)
© *2020 Taylor & Francis Group, London, ISBN 978-0-367-86053-0*

Antituberculosis induced drug reaction with eosinophilia and systemic symptoms in a pediatric latent tuberculosis infection overdiagnosed as tuberculosis disease

W. Setiowulan
Rumah Sakit Muhammadiyah, Bandung, Indonesia
Universitas Islam Bandung, Bandung, Indonesia

R. Rulandani
Rumah Sakit Angkatan Udara dr. M. Salamun, Bandung, Indonesia

H.S. Rachman
Rumah Sakit Umum Daerah Al Ihsan, Bandung, Indonesia
Universitas Islam Bandung, Bandung, Indonesia

ABSTRACT: Pediatric drug reaction with eosinophilia and systemic symptoms (DRESS) is a life-threatening uncommon disease that can be difficult to diagnose. We present a case of a 10-year old girl with ten days of high fever during her fifth week of ATT (consisting of iso-niazid, rifampicin, and pyrazinamide), followed by appearance of pruritic morbiliform rash and severe liver injury. The fever decreased after discontinuation of the ATT and oral steroid administration, followed by resolution of other symptoms. The diagnosis of TB disease in this patient was made in primary health care using pediatric TB scoring system. Despite the total score of 6 (positive tuberculin test and TB exposure), she has normal chest X-ray and no symptoms of TB. Therefore, the patient has latent tuberculosis infection (LTBI) but overdiag-nosed as TB disease. Due to the negative HIV status of the patient, TB prophylaxis is not indicated. Therefore, reintroduction of ATT was contraindicated, which led to lack identifica-tion of specific drug causing DRESS in this case.

1 INTRODUCTION

Drug reaction with eosinophilia and systemic symptoms (DRESS), also known as drug induced hypersensitivity syndrome (DIHS), is a rare but severe drug reaction with mortality rate up to 10% (Singh et al. 2019). The features consist of fever, rash, abnormal hematologic findings (eosinophilia, leukocytosis, atypical lymphocytes), lymphadenopathy, and visceral involvement (hepatitis, pneumonitis, myocarditis, pericarditis, nephritis, and colitis) (Chen et al. 2013, Singh et al. 2016). The symptoms may occur anywhere from 2 to 8 weeks after initiating the offending drug (Choudhary et al. 2013). The estimated incidence of DRESS ranges from 1 in 1000 to 1 in 10,000 drug exposures and more prevalent in adult than pediatric patients (Castellazzi et al. 2018).

Many drugs have been implicated to induce DRESS in children, such as antibiotics (tri-methoprim-sulfamethoxazole, vancomycin, amoxicillin-clavulanic acid), antiepileptic medica-tions (valproic acid, lamotrigine, carbamazepine, phenytoin), ibuprofen, nevirapine, acetylsalicylic acid, griseofulvin, fluoxetine, and sulthiamine (Oberlin et al. 2019). Although antituberculosis drugs such as isoniazid, rifampicin, streptomycin, and pyrazinamide are known to cause DRESS among adults, there is only one report of this occurrence in pediatric patients (Allouchery et al. 2017, Cheng et al. 2013).

Tuberculosis among pediatric patients is prevalent in Indonesia with first line treatment consist of 6 months of isoniazid + rifampicin and two months pyrazinamide during intensive phase. Establishing TB diagnosis in children is a challenge due to the low rate of bacterial confirmed case. This problem has led to developmental of a Pediatric TB Scoring System by the Indonesia Pediatric Society; which a total score ≥ 6 is diagnosed as clinical TB. However, the sensitivity and specificity of this scoring system is low (47% and 68% respectively) which may cause a lot of under or overdiagnosed cases (Triasih & Graham 2011).

2 CASE PRESENTATION

A ten-year old girl was admitted to our hospital with 7 days of remittent high fever, mostly during the nights, accompanied with headache, nausea, mild cough, and constipation. She appeared weak and lost 1 kg during the course of illness. The patient was on her fifth week of antituberculosis treatment (ATT) consisting of isoniazid, rifampicin, and pyrazinamide in a primary health care where the diagnosis was made during households contact investigation from a tuberculosis case using TB scoring system (score 6). Her mother had sputum smear positive tuberculosis (score 3) and tuberculin skin test (TST) of the patient was positive (score 3). The patient showed no TB signs or symptoms and her chest X ray was normal. Her HIV status was negative.

On admission, she was moderately ill, alert, BP 100/70 mmHg, pulse rate 64x/min, RR 24x/ min, and temperature 38.7°C. Her body weight was 26 kg, her height 138 cm, BMI 13.1 kg/m2 (moderately malnourished). She had typhoid tongue, liver enlargement 3 cm below processus xyphoideus, and other physical examination was within normal limit. The CBC was normal (hemoglobin 12.6 g/dL, hematocrit 36%, leucocyte 5,200/mm3, thrombocyte 203,000/mm3) and ALT/AST was 29/68 U/L. Although the Widal test was negative (S. typhi O and H 1/80, S. paratyphi B O and H 1/40, S. paratyphi C H 1/40), due to the classical symptoms and high endemicity of this disease, the patient was diagnosed with typhoid fever and treated with ceftriaxone. Her ATT was continued.

After four days of ceftriaxone the fever persisted and the patient developed generalized morbiliform rash that began on the trunk and symmetrical on upper extremities. It was mildly pruritic initially but became more severe on the next day. The eruption also became more apparent and progressed caudally. Drug fever due to ceftriaxone was suspected. The medication was discontinued and changed into azithromycin. Not only did the fever persist, the generalized rash on her body showed no improvement with cetirizine administration and another rash appeared on the face mimicking the butterfly rash from systemic lupus erythematosus (SLE). Other signs of SLE were not present.

To confirm the diagnosis of typhoid fever, a five-day apart serial Widal test was taken and showed no increased results (S typhi O -, H 1/40, S paratyphi A O 1/40, S paratyphi C O 1/ 40). Investigations revealed increased CRP (40 mg/dL) and ESR (35/60 mm/hour). A blood stream nosocomial infection was suspected and amikacin was added. Urinalysis, electrocardiography, and chest X ray results were normal.

On the seventh day of hospitalization, additional information was obtained from the father that her mother had similar symptoms before she was deceased. She was also on the fifth week of ATT when a week of high grade fever developed, followed by pruritic rash. The mother was later jaundiced, unconscious, and died within three days. Severe drug reactions in the patient and her mother were suspected.

Although the patient was not jaundiced and only vomited once a day after meal, she appeared more lethargic. A repeated ALT and AST level was elevated to 830 and 1178 U/L. The eosinophil counts slightly increased to 9%. A diagnosis of DRESS was established. The ATT was stopped and 0.6 mg/kg intravenous dexamethasone was started in conjunction with ethambutol and streptomycin. Ursodeoxycholic acid was added to treat the liver injury. The rash disappeared in one day and the fever resolved in two days. Intravenous dexamethasone was changed into 1 mg/kg oral prednisone She appeared well and gained a good appetite on the following days. After negative results from ANA -

anti ds DNA test were received, oral prednisone, ethambutol, and streptomycin was discontinued after 7 days. She was discharged with serial liver function tests had already decreased by twofold. Her liver function tests normalized within two weeks. During the last visit, she had gained 1.5 kg and returned to her normal nutritional status. A note was given to the primary health care prescribing her the ATT that she had latent tuberculosis infection (LTBI) and not TB disease. Therefore, the ATT should be discontinued and there was no indication for prophylaxis treatment.

3 DISCUSSION

The diagnosis of DRESS was not considered on admission because the patient was presented with classical symptoms of typhoid, such as remittent fever with peak temperature during the nights, headache, nausea, constipation, malaise, relative bradycardia, and hepatomegaly (Crump et al. 2015). Widal test has a low sensitivity (73.5%) (Keddy et al. 2011), therefore the result in this patient was initially considered as false negative and antibiotic treatment for typhoid was started. Defervescence of fever in typhoid treated with ceftriaxone occur within 4-6 days (Crump et al. 2015, Nair et al. 2017), which cause a longer period to wait for the response of therapy. Blood culture and Tubex test were not performed in this patient due to financial reason (the patient wasn't covered by The Indonesian National Health Insurance System).

DRESS during initial phase (before appearance of rash) is easily mistaken as typhoid fever because the patient also has relative bradycardia, a condition that occurs when the heart rate does not increase to the extent that typically accompanies the temperature elevation. This is widely known to be a patognomonic sign of typhoid fever. However, unlike typhoid, various patterns of fever may occur in patients with drug fever, including: continuous fevers; remittent fevers, in which temperatures vary but are consistently elevated from normal; intermittent fevers, which are interrupted by daily normal temperatures; and hectic fevers, which manifest as a combination of intermittent and remittent fever patterns. Hectic fever is the most common pattern of drug fever (Patel & Gallagher 2010).

On admission, the complete blood count in this patient was normal. Unfortunately, the leucocyte differential count was not performed. Most patients with typhoid fever have relative lymphocytosis, eosinopenia, moderate anemia, and thrombocytopenia (Ifeanyi 2014). While in DRESS, eosinophilia and leukocytosis are frequently found (Castellazzi et al. 2018).

Persistence of fever after 4 days of antibiotic treatment followed by pruritic morbiliform rash prompt a suspicion of drug reaction. Ceftriaxone was considered as the causative agent and was replaced with azithromycin, which is proven to be equivalent or superior to ceftriaxone for the management of uncomplicated typhoid fever (Crump et al. 2015). However, no decline of fever and worsening symptoms of rash despite antihistamine therapy concluded that drug reaction due to ceftriaxone was unlikely.

Evaluation whether typhoid was the correct diagnosis in this patient was also problematic. A serial Widal test is supposedly to be taken 10 days apart and a positive result is determined by a 4-fold increase in antibody titer, which in practice is very difficult to do. On the other hand, discrepancies of two Widal test results with 5 day interval in this patient proved why this test has been associated with some controversies, including the inherent variabilities and lack of reproducibility (Olopoenia & King 2000).

Vigorous effort to find any source of infection as the cause of fever in this patient should be performed, such as blood, urine, and stool culture (Choudhary et al. 2013). Increased result of C-reactive protein (CRP) and blood sedimentation rate (BSE) test could not differentiate between bacterial infection or DRESS, because CRP and BSE may also be elevated in DRESS. Taegtmeyer et al. (2014) reported that the CRP lever in DRESS ranged from 28 – 420 mg/L.

The RegiSCAR (multinational registry of severe cutaneous adverse reactions) study group has developed a diagnostic score which classify potential cases of DRESS as definite, possible, and no case.

Based on this score, the diagnosis of the patient was possible case as shown in Table 1 (Choudhary et al. 2013).

The pathogenesis of DRESS syndrome is not well understood. It is hypothesized to be a combination of delayed T-cell-dependent hypersensitivity reaction to drug, genetic susceptibility, and virus-drug interaction associated with viral reactivation (e.g. Epstein-Barr virus, HHV-6, HHV-7, and cytomegalovirus) (Choudhary et al. 2013). The presence of DRESS syndrome in both mother and daughter shows that genetic susceptibility is very likely to play an important role. Human leukocyte antigen (HLA) has been associated with several drug hypersensitivities, such as HLA-B*1502 with carbamazepine induced Steven Johnson Syndrome (SJS)/toxic epidermal necrolysis (TEN) and HLA-B*1508 with allopurinol induced SJS/TEN. An association between HLA Cw*0401 and antituberculosis induced DRESS was observed (Kim et al. 2012). Virus-drug interaction could not be proven in this case because no viral serology test was performed.

It is still not clear whether the first sign of fever in this patient was a part of DRESS syndrome or due to other cause, which is clinical typhoid fever. Median onset of fever in DRESS syndrome is one day before to three days after the onset of skin rash. However, it can occur as early as thirteen days before the onset of rash (Kardaun et al. 2013). Therefore, as observed in this patient, a long duration between fever and onset of rash may cause a prolonged delay in diagnosis of DRESS.

Appearance of facial rash mimicking Lupus Erythematosus prompt a possible diagnosis of Drug Induced Lupus Erythematosus (DILE). Suspicion of DILE is made by finding of these conditions: one or more clinical symptoms of SLE (eg. arthralgias, butterfly rash, serositis), presence of antinuclear antibodies, no previous history of SLE, symptoms appear within 3 weeks to 2 years after drug administration, and rapid clinical improvement following discontinuation of the drug (Pramatarov 1998). Because ANA and anti-DNA test in this patient are negative, diagnosis of DILE may be excluded.

The most important treatment of DRESS syndrome is immediate removal of the offending drug. In some cases, this may be sufficient to achieve the resolution of clinical and laboratory abnormalities (Allouchery et al. 2017, Choudhary et al. 2013). The pharmacological approach to treating this syndrome is not completely defined as such treatments have not yet been evaluated in clinical trials. Intravenous corticosteroids, administered alone or followed by oral

Table. 1.   The RegiSCAR-Group diagnosis score for DRESS.

	NO	YES	UNKNOWN	RESULT	SCORE
Fever (≥ 38.5 °C)	-1	0	-1	YES	0
Enlarge lymph nodes (≥ 2 sites, > 1 cm)	0	1	0	NO	0
Atypical lymphocytes	0	1	0	UNKNOWN	0
Eosinophilia	0		0	NO	0
700-1499 or 10-19.9		1			
≥ 1500 or ≥ 20%		2			
Skin rash	0		0		
extent > 50%	0	1	0	YES	1
at least 2 of: edema, infiltration, purpura, scaling	-1	1	0	YES	1
biopsy suggesting DRESS	-1	0	0	UNKNOWN	0
Internal organ involved	0		0		
One		1		YES	1
2 or more		2			
Resolution in > 15 days	-1	0	-1	NO	-1
At least 3 biological investigations done and negative to exclude alternative diagnosis	0	1	0	UNKNOWN	0

Total Score

Final score: < 2 no case; 2-3 possible case; 4-5 probable case; > 5 definite case steroid therapy, have been shown to be an effective treatment for DRESS syndrome. However, there is no consensus regarding the dose and route of administration (Oberlin et al. 2019).

We used dexamethasone 0.6 mg/kg followed by oral prednison 1 mg/kg, in conjunction with ethambutol and streptomycin to prevent activation of TB disease which may be caused by long term systemic corticosteroid without prophylaxis treatment in a patient with LTBI (Kemenkes 2016). Due to the rapid clinical and laboratory improvement, we decided to give short term corticosteroid treatment in this patient.

The pivotal issue in this case is the implementation of the Indonesian scoring system (Table 2) in the context of contact screening. All primary health care facilities (puskesmas) in Indonesia must adhere to government TB DOTS program, which apply a score of ≥ 6 to diagnose pediatric TB. The scoring system gives a score of 3 for a history of close contact with such patient and a score of 3 for a positive TST result. Therefore, children with household contact but have no symptoms, normal CXR, and positive TST will have a total score of 6 and may receive unnecessary antituberculosis treatment (Triasih & Graham 2011, Kemenkes 2016).

A new revised technical guideline for diagnosing pediatric tuberculosis was published in 2016 but has not been widely disseminated among primary health care facilities in Indonesia. This guideline advocates symptom based screening to household contacts of smear positive pulmonary tuberculosis cases. TST and chest X ray are not required in asymptomatic children. Prophylaxis treatment is only indicated in children below 5 years or HIV positive (Kemenkes 2016).

All antituberculosis drugs pose a risk of DRESS, with rifampicin as the most suspected, followed by isoniazid, ethambutol, and finally by pyrazinamid. A reintroduction of suspected drugs is contraindicated after a diagnosis of DRESS. However, because the nature of a tuberculosis disease, the lack of adequate therapeutic alternatives, and the risk/benefit balance, a rechallange could be justified to identify the culprit drug. Rechallange test should be done in hospital at least two weeks after discontinuation of the ATT (Allouchery et al. 2017).

There is no indication for antituberculosis treatment or prophylaxis in this patient. Rechallenge test to identify specific drug that cause DRESS pose a greater risk than benefit and must be avoided. (Allouchery et al. 2017)

Table 2.   The Indonesian pediatric TB scoring system.

Variable	0	1	2	3
Household contact	Unknown		Contact with smear negative TB patient or unknown sputum smear result	Contact with smear positive TB patient
TST	Negative			Positive (>10 mm, or in immunocompromised children >5 mm)
Nutritional state		BW/age < 80%	Severe malnutrition (BW/age < 80%)	
Fever unknown origin ≥ 2 weeks		Present		
Cough ≥ 3 weeks		Present		
Lymph node (cervical, axillary, inguinal)		Multiple, non-tender, diameter ≥ 1 cm		
Joint swelling (knee, phalanges)		Present		
Chest X ray	Normal	Suggestive TB		

4 CONCLUSION

We should suspect a case of DRESS in a patient with fever and rash appearing within 2 to 8 weeks after initiating a specific drug. Laboratory tests to confirm DRESS and exclude other infectious causes should be carried out. Immediate discontinuation of drug and administration of systemic corticosteroid are important measures to prevent further complications.

REFERENCES

Allouchery, M. Logerot, S. Cottin, J. Pralong, P. Villier, C. & Saïd, B.B. 2017. Antituberculosis Drug-Associated DRESS: A Case Series. *The Journal of Allergy and Clinical Immunology: In Practice* 6(4): 1373–1380.

Castellazzi, M. L. Esposito, S. Claut, L.E. Daccò, V. & Colombo, C. 2018. Drug reaction with eosinophilia and systemic symptoms (DRESS) syndrome in two young children: the importance of an early diagnosis. *Italian Journal of Pediatrics* 44(1): 93.

Chen, Y. C. Cho, Y. T. Chang, C. Y. & Chu, C. Y. 2013. Drug reaction with eosinophilia and systemic symptoms: A drug-induced hypersensitivity syndrome with variable clinical features. *Dermatologica Sinica* 31(4): 196–204.

Cheng, J. Rawal, S. Roberts, A. & Guttman, O. R. 2013. Drug reaction with eosinophilia and systemic symptoms syndrome associated with antituberculosis medications. *The Pediatric Infectious Disease Journal* 32(12): 1388–1390.

Choudhary, S. McLeod, M. Torchia, D. & Romanelli, P. 2013. Drug Reaction with Eosinophilia and Systemic Symptoms (DRESS) Syndrome. *The Journal of clinical and aesthetic dermatology* 6(6): 31–37.

Crump, J. A. Sjölund-Karlsson, M. Gordon, M. A. & Parry, C. M. 2015. Epidemiology, Clinical Presentation, Laboratory Diagnosis, Antimicrobial Resistance, and Antimicrobial Management of Invasive Salmonella Infections. *Clinical microbiology reviews* 28(4): 901–937.

Ifeanyi, O.E. 2014. Changes in some Haematological Parameters in Typhoid Patients Attending University Health Services Departement of Michael Okpara University of Agriculture, Nigeria. *International Journal of Current Microbiology and Applied Sciences* 3(1): 670–674.

Kardaun, S. H., Sekula, P., Valeyrie-Allanore, L., Liss, Y., Chu, C. Y., Creamer, D., & RegiSCAR Study Group. 2013. Drug reaction with eosinophilia and systemic symptoms (DRESS): an original multisystem adverse drug reaction. Results from the prospective R egi SCAR study. *British Journal of Dermatology* 169(5): 1071–1080.

Keddy, K. H. Sooka, A. Letsoalo, M. E. Hoyland, G. Chaignat, C. L. Morrissey, A. B. & Crump, J. A. 2011. Sensitivity and specificity of typhoid fever rapid antibody tests for laboratory diagnosis at two sub-Saharan African sites. *Bulletin of the World Health Organization* 89(9): 640–647.

Kim, S.H. Lee, S.K. Kim, S.H. Park, H.W. Chang, Y.S. Lee, K.W. & Jee, Y.K. 2013. Antituberculosis drug-induced hypersensitivity syndrome and its association with human leukocyte antigen. *Tuberculosis* 93(2): 270–274.

Nair, B.T. Simalti, A.K. Sharma, S. 2017. Study comparing ceftriaxone with azithromycin for the treatment of uncomplicated typhoid fever in children of India. *Annals of Tropical Medicine and Public Health* 10(1): 205–210.

Oberlin, K.E. Rahnama-Moghadam, S. Alomari, A.K. & Haggstrom, A. N. 2019. Drug reaction with eosinophilia and systemic symptoms: Pediatric case series and literature review. *Pediatric Dermatology* 1–6.

Olopoenia, L.A. & King, A.L. 2000. Widal agglutination test - 100 years later: still plagued by controversy. *Postgrad Med J* 76: 80–84.

Patel, R.A. & Gallagher, C.G. 2010. Drug Fever *Pharmacotherapy* 30(1): 57–69.

Pramatarov K.D. 1998. Drug-Induced Lupus Erythematosus *Clinics in Dermatology* 16: 367–377.

Kemenkes, R. I. 2016. *Petunjuk teknis manajemen dan tatalaksana TB anak.* Jakarta: Kementerian Kesehatan RI.

Singh, J. Dinkar, A. Atam, V. Gupta, K. K. & Sahani, K. K. 2016. Drug reaction with eosinophilia and systemic symptoms syndrome associated with Nitrofurantoin. *Journal of research in pharmacy practice* 5(1): 70–73.

Singh, T. Niazi, M. Karri, K. Rudikoff, D. & Gonzalez, E. 2019. A Rare Case of DRESS (Drug Reaction with Eosinophilia and Systemic Symptoms) Syndrome with Cholecystitis in a Patient on Levetiracetam. *Cureus* 11(3): e4245.

Taegtmeyer, A. Bravo, A. R. Zimmermanns, B. Liakoni, E. Kraehenbuehl, S. & Haschke, M. 2014. C-reactive protein and procalcitonin in patients with DRESS syndrome. *Clinical and Translational Allergy* 4(3): 12.

Triasih, R. & Graham, S. M. 2011. Limitations of the Indonesian Pediatric Tuberculosis Scoring System in the context of child contact investigation. *Paediatrica Indonesiana* 51(6): 332–337.

Medical Technology and Environmental Health – Abdullah, Widiaty & Abdullah (eds)
© 2020 Taylor & Francis Group, London, ISBN 978-0-367-86053-0

Description of mild cognitive impairment for stroke patients in the department of neurology at Jakarta Islamic Hospital, September–November 2015

F.S. Farhan & M. Ramadhani
University of Muhammadiyah Jakarta, Jakarta, Indonesia

ABSTRACT: Stroke is the third most common disease after heart disease and cancer. Post-stroke disability can include motoric, sensory, autonomic, and cognitive disorders. Cognitive impairment caused by stroke can cause interference, such as language, memory, visuospatial, attention, orientation, cognition, and emotional disorders. The aim of this research is to evaluate cognitive impairment in post-stroke patients in Islamic Hospital in Jakarta, Cempaka Putih, Indonesia, in September–November 2015, using MoCA-INA (Montreal Cognitive Assessment Version Indonesia). In 83 stroke patients, ischemic stroke was the most common with 75 respondents (90.4%) and for a total of 76 respondents (91.6%) there was positive impaired cognitive function. The age range 55–64 years was the most represented with 29 respondents (34.9%). Under the category of gender, men represented 47 respondents (56.6%). Elementary school level is the highest level of education represented with a total of 22 respondents (26.5%). In terms of results, among stroke patients in Poli Neural Jakarta Islamic Hospital Cempaka Putih, 76 respondents (91.6%) have positive impaired cognitive function.

1   INTRODUCTION

Stroke is a health problem that needs special attention and can attack anyone and at any time, regardless of race, gender, or age. Stroke is the third most common disease after heart disease and cancer, and is the highest cause of disability in the world. According to the American Heart Association (AHA), the mortality rate of stroke patients in America is 50–100 of 100,000 sufferers every year (Dinata et al. 2013).

In ASEAN countries, stroke is also a major health problem that causes death. From the South East Asian Medical Information Center (SEAMIC) data it is known that the largest stroke death rate occurred in Indonesia, followed sequentially by the Philippines, Singapore, Brunei, Malaysia, and Thailand. In Indonesia, ischemic stroke is the most suffered type, amounting to 52.9%, followed sequentially by intracerebral hemorrhage, embolism, and subarachnoid hemorrhage with incidence of 38.5%, 7.2%, and 1.4%, respective;y (Dinata et al. 2013).

Based on the latest data and the results of the 2013 Basic Health Research (Riskesdas), stroke is the leading cause of death in Indonesia (Departemen Kesehatan 2013). The prevalence of stroke in Indonesia based on the diagnosis of health is 7.0% and based on the diagnosis of health or symptoms is 12.1%. Stroke tends to be higher in people with low education, both those diagnosed by health (16.5%) and diagnosed by health or symptoms (32.8%); and those who do not work, both health (11.4%) and health or symptoms (18%) (Departemen Kesehatan 2013).

Stroke falls into two categorizes, namely hemorrhagic stroke (SH) and non-hemorrhagic stroke (SNH). SH is a cerebral or possibly subarachnoid hemorrhage caused by rupture of blood vessels in certain areas of the brain. SNH is due to thromboembolic blockage of blood vessels causing an ischemic blockage. Cognitive impairment and memory dysfunction following stroke diagnosis are common symptoms that significantly affect the survivor's quality of life. Stroke patients have a high potential to develop dementia within the first year of stroke onset. Currently, efforts are being exerted to assess stroke effects on the brain, particularly in the early stages. Numerous neuropsychological assessments are being used to evaluate and differentiate cognitive impairment and dementia following stroke (Al-Qazzaz et al. 2014). Both ischemic and hemorrhagic strokes can cause damage, including death of brain cells. Progressive neurological deficits and impaired cognitive function can result from multiple small cerebral infarctions in the temporal region. Clinical disorders can arise as a result of damage to brain cells in certain parts of the brain and this can end in disability after a stroke (Panentu & Irfan 2013).

Post-stroke records can show motor, sensory, autonomic, or cognitive impairment. Cognitive disorders caused by stroke can result in disorders such as language, memory, visuospatial, attention, cognitive, and emotional orientation (Panentu & Irfan 2013).

Cognitive function is the whole process by which an individual receives, records, stores, and uses information and can be defined as all mental processes, including perception, memory, imagination creation, and thinking, that shape awareness and alertness and the process of making decisions (Panentu & Irfan 2013). Specific risk factors for cerebrovascular disease are associated with cognitive dysfunction. It can be concluded that stroke can cause from very mild to severe cognitive dysfunction (Trinita et al. 2014).

2 METHODS

This research study is a quantitative descriptive cross-sectional study to illustrate cognitive disorders in stroke patients. In this study, the technique used in data collection was primary data consisting of interviews and questionnaires. Data collection was conducted in September–December 2015, at the Neural Poly in the Central Jakarta Islamic Hospital. The respondents of this study were all post-stroke patients at the Jakarta Islamic Hospital in the September–November 2015 period. To assess cognitive function, the Indonesian version of Montreal Cognitive Assessment (MoCA-InA) was used, with a maximum value of 30. A final total score of 26 or more was considered normal. A total value of ≤25 indicates cognitive impairment (Friedman 2012).

3 RESULTS

Research respondents were found by type of stroke (ischemic or non-haemorrhagic), with a total of 75 respondents (90.4%). See Table 1.

3.1 Characteristics of respondents

Research respondents by age group found the highest number in the 55–64 year group with a total of 29 respondents (34.9%). See Table 2.

Research respondents by sex group found the majority were male with a total of 47 respondents (56.6%) and the remaining 36 respondents (43.4%) were female. See Table 3.

Table 1. Distribution of stroke patients.

Stroke type	(n)	Percent (%)
Haemorrhagic	8	9.6
Ischemic	75	90.4
Total	83	100.0

Table 2. Age of respondents.

Age	(n)	Percent (%)
35–44	4	4.8
45–54	20	24.1
55–64	29	34.9
65–74	24	28.9
>75	6	7.2
Total	83	100.0

Table 3. Gender of respondents.

Gender	(n)	Percent (%)
male	47	56.6
female	36	43.4
Total	83	100.0

Table 4. Education level of respondents.

Level education	(n)	Percent (%)
No education	7	8.4
Primary	22	26.5
Junior high	17	20.5
Senior high	21	25.3
University	16	19.3
Total	83	100.0

Table 5. Cognitive impairment of stroke patients.

Cognitive function	(n)	Percent (%)
Impairment	76	91.6
Normal	7	8.4
Total	83	100.0

Research respondents based on education level found the highest number, a total of 22 respondents (26.5%), had an elementary education. See Table 4.

Based on cognitive impairment in stroke patients, a total of 76 respondents (91.6%) had significant impairment, and the remaining 7 respondents (8.4%) did not experience disorders or were normal. See Table 5.

4 DISCUSSION

Based on the results of this study, of the 83 stroke patients studied at the Poly Neurology at the Jakarta Cempaka Putih Islamic Hospital, ischemic or non-haemorrhagic strokes were the most suffered compared to haemorrhagic strokes, with a total of 75 respondents (90.4%) suffering from ischemic stroke and the remaining 8 respondents (9.6%) suffering from haemorrhagic stroke.

This research study is in line with research done by Wibowo et al., in the neurology clinic of BLU PROF General Hospital. DR. R. D. Kandou Manado Period November–

December 2014. That study obtained data for 34 patients (97.1%) with ischemic stroke from a total of 35 respondents (Wibowo & Karema 2015).

Another study conducted by Ahangar et al., at Yahyanejad Hospital Babol, northern Iran from April 2001–April 2003 found more ischemic stroke cases than hemorrhagic stroke with a percentage of 67.2% and 32.8%, respectively (Ahangar et al. 2005).

This is because one of the risk factors for stroke is atherosclerosis, which means that there is an artery blockage that inhibits blood flow to the brain or other parts of the body, and this makes a person more at risk for one type of ischemic stroke, namely thrombotic stroke.

Based on the results of this study, of the 83 stroke patients studied at the Poly Neurology at Jakarta Cempaka Putih Islamic Hospital, > 55 years was the most common age group, broken down into the 55–64 years age group with a total of 29 respondents (34.9%), then followed by the 65–74 years age group with a total of 24 respondents (28.9%), and 6 respondents (7.2%) in the > 75 years age group. For respondents aged 45–54 years there were 20 respondents (24.1%) and 4 respondents (4.8%) in the 35–44 years age group.

The results of this study are in line with those in previous studies conducted by Caesaria Trinita et al, in the neurology clinic of General Hospital. DR. R. D. Kandou Manado, Period October–December 2013. That study found of 20 respondents (48.8%) out of a total of 41 respondents in the 55–64 years group (Trinita et al. 2014).

This proves that old age is a risk factor for increased incidence of stroke. Age factors double the risk of stroke after the age of 55 years., because the elasticity of blood vessel walls decreases at that age so that blood vessels become stiff and accelerate the formation of atherosclerosis, which ultimately disrupts blood supply to the target tissue (Gund et al. 2013).

Based on the results of this study, it was found that from 83 samples of stroke patients in the Poly Neurology of Jakarta Cempaka Putih Islamic Hospital, 76 (91.6%) of them had cognitive impairment and only 7 (8.4%) had no cognitive impairment.

The results of this study are in line with those in the research conducted by Wibowo et al. using the same research instrument, namely MoCA-INA, showing 34 people experienced cognitive impairment with the most common age > 55 years.

Indha Wardani et al.'s research in 2014 also stated the same thing, namely that out of 40 stroke patients, 25 people (67.5%) experienced impaired cognitive function, while the remaining 15 (40.5%) were declared normal. The same study was also conducted in 2014 with a percentage of 67.5% experiencing cognitive impairment and only 32.4% declared normal (Hasra et al. 2014).

In the MoCA-INA examination on this study, many stroke patients experienced memory disorders. This can occur if, during a stroke, the media cerebral artery is blocked in the left side of the brain causing patients to tend to suffer from dementia. The influence of stroke on cognitive impairment is usually abundant damage in small blood vessels (small vessel disease) in the brain, due to the presence of multiple small cerebral infarctions in the temporal area that clog arteries (Hasra et al. 2014). Such damage can result in neurological deficits that will cause several things, one of which is impaired cognitive function in the form of momentary memory disorders, attention disorders, visuofacial disorders, language disorders and abstraction disorders (Panentu & Irfan 2013). Based on the medical corporation center, cerebrovascular diseases such as stroke are the second cause of cognitive impairment. Based on the available literature, 20–30% of patients who have a stroke will experience cognitive impairment. The smallest number of infarcts and lesions in a stroke that can cause impaired cognitive function is more than 10 ml and less than 50 ml, which is 1–4% of brain volume (Hasra et al. 2014).

5 CONCLUSION

Based on the results of research conducted at the Cempaka Putih Hospital RSIJ, the MoCA-INA examination showed 91.6% of respondents (n = 76), experienced impaired cognitive function. Stroke patients in Cempaka Putih Hospital RSIJ have more ischemic strokes, with a total of 75 respondents (90.4%). Stroke patients in Cempaka Putih RSIJ

Neurology Poly have more cognitive impairment in the age range > 55 years, which in this study were further devided into the age group 55–64 years with 34.9% (n = 29), and the age group 65–74 years with 28.9% (n = 24). Male stroke patients suffered more cognitive impairment at 56.6% (n = 47) than women at 43.4% (n = 36) respondents. In terms of education level, stroke sufferers with an elementary school education had the highest risk of cognitive impairment at 26.5% (n = 22) compared to other respondents with higher education.

REFERENCES

Ahangar, A. A., Vaghefi, S. B. A., & Ramaezani, M. (2005). Epidemiological evaluation of stroke in Babol, northern Iran (2001–2003). *European Neurology*, 54(2): 93–97.

Al-Qazzaz, N. K., Ali, S. H., Ahmad, S. A., Islam, S., & Mohamad, K. (2014). Cognitive impairment and memory dysfunction after a stroke diagnosis: A post-stroke memory assessment. *Neuropsychiatric Disease and Treatment*, 10: 1677.

Dinata, C. A., Safrita, Y. S., & Sastri, S. (2013). Gambaran faktor risiko dan tipe stroke pada pasien rawat inap di bagian penyakit dalam RSUD Kabupaten Solok Selatan periode 1 Januari 2010–31 Juni 2012. *Jurnal Kesehatan Andalas*, 2(2): 57–61.

Departemen Kesehatan 2013. *Hasil Riset Kesehatan Dasar 2013*. [Online]. http://www.depkes.go.id/resources/download/general/Hasil%20Riskesdas%202013.pdf.

Friedman, L. (2012). Evaluating the Montreal Cognitive Assessment (MoCA) and the Mini Mental State Exam (MMSE) for cognitive impairment post stroke: A validation study against the Cognistat. Ontario: Western Libraries.

Gund, B. M., Jagtap, P. N., Ingale, V. B., & Patil, R. Y. (2013). Stroke: A brain attack. *IOSR Journal of Pharmacy*, 3(8): 1–23.

Hasra, I. W. P., Munayang, H., & Kandou, J. (2014). Prevalensi Gangguan Fungsi Kognitif Dan Depresi Pada Pasien Stroke Di Irina F Blu Rsup Prof. Dr. RD Kandou Manado. *e-CliniC*, 2(1).

Panentu, D., & Irfan, M. 2013 Uji validitas dan reliabilitas butir pemeriksaan dengan montreal cognitive assesment versi Indonesia (MOCA-Ina) Pada Insan Pasca Stroke Fase Recovery. *Jurnal fisioterapi*, 13(1): 55–67.

Trinita, C., Mahama, C. N., & Tumewah, R. (2014). Penurunan Fungsi Kognitif Pada Pasien Stroke Di Poliklinik Neurologi Blu Rsup Prof. Dr. RD Kandou Manado Periode Oktober–Desember 2013. *e-CliniC*, 2(2).

Wibowo, M. M., & Karema, W. (2015). Gambaran Fungsi Kognitif Dengan INA-MoCA Dan MMSE Pada Penderita Post-Stroke Di Poliklinik Saraf BLU RSUP Kandou Manado November-Desember. *e-CliniC*, 3(3).

Medical Technology and Environmental Health – Abdullah, Widiaty & Abdullah (eds)
© *2020 Taylor & Francis Group, London, ISBN 978-0-367-86053-0*

Pediculosis capitis at Islamic boarding schools

R.D.I. Astuti & T. Respati
Universitas Islam Bandung, Bandung, Indonesia

ABSTRACT: Pediculosis capitis is a common disease in Islamic boarding schools in Indonesia. This study aims to describe pediculosis capitis at an Islamic boarding school and the effort to eradicate it. This research is a cross-sectional study that was conducted at a Salafi Islamic boarding school in Cililin Kabupaten Bandung Barat, Indonesia, in October 2018. Respondents of this study were female students with pediculosis capitis. The diagnosis of pediculosis capitis established by the discovery of adults and or viable eggs lice on each case. The number of respondents involved was 45 of 50 female students. The results showed that 44.44% of female students were infected by pediculosis capitis at the Islamic boarding school. As many as 64.44% female students also had a family with pediculosis capitis at their homes. All the female students had made efforts to treat pediculosis capitis. The fine-tooth combs were preferred by female students (97.78%) rather than pediculocides (42.22%). However, appropriate use of fine-tooth combs that can treat pediculosis capitis is only 44.44%. Pediculosis capitis eradication requires simultaneous appropriate treatment both at the Islamic boarding school and at home.

1 INTRODUCTION

Pediculosis capitis is a disease caused by *Pediculus humanus capitis* infestation on human hair and scalp. This disease causes itching on the scalp that is disruptive (Madke & Khopkar 2012, Rassami & Soonwera 2012). The itching induces scratching and can lead to secondary infection and glomerulonephritis (Ash & Philips 2016, Bragg & Simon 2019). Besides causing itching of the scalp, pediculosis capitis causes negative stigma in society, especially in people in developed countries. Social pressure due to pediculosis capitis, including taking the blame for being a carrier of the parasite and shame, can be worse than the physical impact (Mumcuoglu et al. 2009, Madeira et al. 2015, Alla et al. 2016).

Many options are available to treat pediculosis, such as manual removal, combing, chemical and pediculocides, but the pediculosis is still a problem today in many countries (Sangaré et al. 2016). Pediculosis capitis occurs mostly in densely populated environment such as boarding schools or dormitories (Falagas et al. 2008). Several Islamic boarding schools in Indonesia are reported to have a quite high incidence of pediculosis capitis, 48.7–74.6%. The behavior of students who often share clothing and other personal things also causes a high incidence of pediculosis in Islamic boarding schools (Anggraini et al. 2018, Lukman et al. 2018, Arsinta et al. 2019).

The high incidence of pediculosis can also be because of drug resistance, ineffective treatment, and reinfection (Mumcuoglu et al. 2009, Canyon et al. 2014a). The resistance of pediculocides is currently reported; it is no longer effective in killing all lice. The other treatment that can be used is fine-tooth combs. The fine-tooth combs must be used at least every 3–4 days on wet hair to be effective. However, the use of fine-tooth combs is often uncomfortable and often not routinely used. Patients who have recovered can be infected again from patients who have not recovered because of improper treatment (Mumcuoglu et al. 2009, Madke & Khopkar 2012, Alla et al. 2016).

This study aims to describe pediculosis capitis in an Islamic boarding school. It includes the infestation sources, the impact, and the treatment efforts.

Table 1. Female student characteristics with pediculosis capitis.

Characteristics	N	%
Source of infection		
Exposed to the disease in Islamic boarding school	20	44.44 %
Having a family member with pediculosis	29	64.44 %
Impact		
Annoying itch	45	100 %
Shame	44	97.78 %
Treatment effort		
Using pediculocides	19	42.22 %
Using fine-tooth combs	44	97.78 %
Appropriate use of fine-tooth combs	20	44.44%

2 METHOD

This is a cross-sectional study conducted in October 2018 at one of the Salafi Islamic boarding schools in Cililin, West Bandung Regency. Criteria for inclusion of respondents were female student with pediculosis capitis. The diagnosis was with the discovery of adult lice or viable eggs on physical examination using a fine-tooth comb. Respondents who met the inclusion criteria were as many as 45 out of a total of 50 female students. Respondents in this study were 14–17 years old. A self-filled questionnaire used to collect the characteristics of respondents regarding pediculosis. The results are displayed as a percentage in Table 1.

3 RESULTS

Physical examination showed that the incidence of pediculosis capitis in this Islamic boarding school is 90%. The results of processing the questionnaire describe the pediculosis capitis at the Islamic boarding school (Table 1). The source of infection from pediculosis capitis came from the Islamic boarding school and home. Most of the female students got pediculosis capitis before entering the Islamic boarding school, but the proportion was almost equal between female students who were infested by pediculosis capitis at Islamic boarding school and home.

Female students with family members with pediculosis capitis was 64.44%. This family member can be a new source of infection for the students at home whenever the female student is already having treatment at the Islamic boarding school.

The impact of capitis pediculosis on female pediculosis capitis patients is also evident. All female students with pediculosis capitis suffer from disrupting itch on the scalp, and almost all female students feel ashamed of having pediculosis capitis.

The female students also made efforts to treat pediculosis capitis. The treatment that is often chosen by female students is the fine-tooth comb (97.78%), but only 44.44% of them do it in the right way.

4 DISCUSSION

Almost all female students who contracted pediculosis capitis become infestated when entering the Islamic boarding school. This is due to a large number of other female students who have already got pediculosis capitis when entering the Islamic boarding school. Islamic boarding school is a densely populated place where the habits of female students facilitated the transmission of pediculosis capitis by sharing personal things such as clothes, towels, and hairbrushes (Anggraini et al. 2018, Lukman et al. 2018, Arsinta et al. 2019). Transmission of pediculosis capitis can occur due to the migration of *Pediculus humanus capitis* from person to person through physical contact. *Pediculus humanus capitis* cannot jump, it can only crawl quickly. The transmission of pediculosis capitis can also occur through shared items such as combs, clothes, hats, and beds (Rassami & Soonwera 2012, Anggraini et al. 2018, Lukman et al. 2018, Arsinta et al. 2019).

The impact of pediculosis capitis is clearly illustrated by the respondents. Severe itchiness is caused by the saliva of *Pediculus humanus capitis* lice that enters the skin while sucking blood on the scalp and can cause an allergic reaction (Vázquez-Herrera et al. 2018). The female students can be embarrassed about having a disease that often occurs in people with poor hygiene and there is a negative stigma in society (Mumcuoglu et al. 2009, Canyon et al. 2014a, Madeira et al. 2015, Alla et al. 2016). The obvious impact made all female students with pediculosis capitis undertake treatment. Almost all the female students choose a fine-tooth comb in their treatment. Treatment using a fine-tooth comb is easy, inexpensive, and there are no side effects, although its use is often uncomfortable, especially in thick and curly hair (Mumcuoglu et al. 2009, Madke & Khopkar 2012). Fine-tooth combs are effective in treating pediculosis capitis if used on wet hair at least twice a week. Fine-tooth combs remove adult lice and the eggs (Madke & Khopkar 2012). The use of fine-tooth combs is more effective than pediculocides in the treatment of pediculosis capitis. In addition to treatment, fine-tooth combs are also effective for diagnosis and prevention (Sangaré et al. 2016).

Treatment with pediculocides was less popular than fine-tooth combs. Reinfection events require patients to treat their pediculosis capitis repeatedly and this makes treatment using pediculocides more expensive (Canyon et al. 2014b) Repeated use of these pediculocides can also cause resistance and becomes ineffective. Currently, there are many reports of pediculocides resistance all over the world (Mumcuoglu et al. 2009). Skin irritation by pediculocides can also be an obstacle to pediculocide use (Mac-Mary et al. 2012).

However, treatment efforts by female students were not effective. Nearly half of the female students with pediculosis capitis have not taken proper treatment and become a source of reinfection for other students who have recovered. This led to a high incidence of pediculosis in Islamic boarding schools. Education on pediculosis is needed to increase control over pediculosis capitis (Madeira et al. 2015). In addition to female students who have not been appropriately treated, other family members at their homes having pediculosis capitis can be a source of infection for students when returning home (Canyon et al. 2014b).

5 CONCLUSION

The results of the above study indicate that the treatment of pediculosis capitis must be done at Islamic boarding school and home appropriately, simultaneously, and continuously until all female students and family members get rid of pediculosis capitis.

REFERENCES

Alla, N. A., Al Megrin, W. A., & Alkeridis, L.A. 2016. Faye Abdellah model to banishing social stigma of head lice among school students. *Science Journal of Clinical Medicine*, 5(1): 1–11.

Anggraini, A., Anum, Q., & Masri, M. 2018. Hubungan Tingkat Pengetahuan dan Personal Hygiene terhadap Kejadian Pedikulosis Kapitis pada Anak Asuh di Panti Asuhan Liga Dakwah Sumatera Barat. *Jurnal kesehatan Andalas*, 7(1): 131–136.

Arsinta, D., Anwar, C., & Ramdja, M. 2019. Association of sharing materials with pediculosis capitis in students of Pondok Islamic boarding school Tahfidzil Qur'an Yayasan Tijarotal Lan Tabur Palembang. *Majalah Kedokteran Sriwijaya*, 51(3): 155–163.

Ash, M. M., & Philips, C. M. 2016. Parasitic diseases with cutaneus manifestation. *NCMJ*, 7(5): 350–354.

Bragg. B. N., & Simon, L. V. 2019. *Pediculosis Humanis (lice, Capitis, Pubis)*. Treasure Island, FL: Stat-Pearls Publishing.

Canyon, D. V., Canyon, C., & Milani, S. 2014a. Parental and child attitudes towards pediculosis are a major of reinfection. *The Open Dermatology Journal*, 8: 24–28.

Canyon, D. V. Canyon, C., & Milani, S. 2014b. Characterizing the nature of human carriers of head lice. *The Open Dermatology Journal*, 8: 29–31.

Falagas, M. E., Matthaiou, D. K., Rafailidis, P. I., Panos, G., & Pappas, G. 2008. Worldwide prevalence of head lice. *Emerg Infect Dis.*, 14(9): 1493–1494.

Lukman, N., Armiyanti, Y., & Agustina, D. 2018. Hubungan Faktor-Faktor Risiko Pediculosis capitis terhadap Kejadiannya pada Santri di Pondok Islamic boarding school Miftahul Ulum Kabupaten Jember. *Journal of Agromedicine and Medical Sciences*, 4(2):102–109.

Mac-Mary, S., Messikh, R., Jeudy, A., Lihoreau, T., Sainthillier, J. M., Gabard, B., & Humbert, P. (2012). Assessment of the efficacy and safety of a new treatment for head lice. *ISRN Dermatology*, 2012.

Madeira, N. G., Tomé de Souza, P. A., & da Silva Diniz, R. E. 2015. Perception and action of teachers and head lice in school. *Revista Electronica de las Ciencias*, 14(2): 119–130.

Madke, B., & Khopkar, U. 2012. Pediculosis capitis: An update. *Indian Journal of Dermatology Venerology and Leprology*, 78(4): 429–438.

Mumcuoglu, K. Y., Gilead, I., & Ingber, A. 2009. New insight pediculosis and scabies. *Expert Review of Dermatology*, 4(3): 285–302.

Rassami, W., & Soonwera, M. 2012. Epidemiology of pediculosis capitis among school children in the eastern area of Bangkok, Thailand. *Asian Pacific Journal of Tropical Biomedicine*, 2(11): 901–904.

Sangaré, A. K., Doumbo, O. K., & Raoult, D. 2016. Management and treatment of human lice. *BioMed Research International Article*, ID 8962685.

Vázquez-Herrera, N. E., Sharma, D., Aleid, N. M., & Tosti, A. 2018. Scalp itch: A systematic review. *Skin Appendage Disord*, 4: 187–199.

Identification of cardiovascular risk factors among Hajj pilgrims from Bali in 2018

N.W. Widhidewi, S. Masyeni & A.E. Pratiwi
Universitas Warmadewa, Denpasar, Indonesia

ABSTRACT: Almost 2 million Muslims from more than 183 countries conduct Hajj to Mecca, Saudi Arabia, every year. This event carries significant public-health challenges – not only infectious disease transmission but also exacerbation of noncommunicable diseases. Although the prevalence of cardiovascular disease (CVD) was not high, it confers high mortality rate. The aim of this study is to identify major risk factors of cardiovascular events among Hajj pilgrims from Bali in 2018. A cross-sectional study was conducted in Western Denpasar Community Health Center, where Hajj preparations take place. Demographical data, history of preexisting medical illness, physical examination, laboratory, electrocardiogram (ECG) and thorax x-ray results were collected. CVD risk defined as the finding of at least one of the parameters (hypertension, diabetes mellitus, dyslipidemia, hyperuricemia, ECG or x-ray abnormalities). A total of 99 participants were enrolled in the study. Among them, 97 (98%) participants had at least one CVD risk factor. Significant findings of Hajj pilgrimage with CVD risk indicate that identification of CVD risk factors is profoundly recommended to prevent mortality and morbidity related to cardiovascular events during the Hajj period.

1 INTRODUCTION

Almost 2 million Muslims from more than 183 countries conduct Hajj to Mecca, Saudi Arabia, every year (Shafi et al. 2016). Hajj is the major religious ritual for every Muslim who is physically, psychologically, and financial stable, at least once in their lifetime. This event carries significant public-health challenges – not only infectious-disease transmission but also exacerbation of noninfectious disease or noncommunicable diseases. The disease pattern among Hajj pilgrims attending medical facilities of Pakistani was respiratory disease, musculoskeletal disorder, gastrointestinal disease, ear nose throat disease, skin disease, cardiovascular disease and others, with frequencies of 29%, 18%, 15%, 8%, 6% and 2%, respectively (Raja et al. 2017). Although the prevalence of cardiovascular disease was not high, it confers a high mortality rate, as reported by another study (Aljoudi 2013).

Cardiovascular disease (CVD) including coronary heart disease (CHD), stroke, and heart failure, is the leading cause of morbidity and mortality during Hajj (Al Shimemeri 2012, Al Masud et al. 2016). The mortality rate of CVD with hypertension during Hajj was reported as high as 45.8%, meanwhile cerebrovascular disease mortality rate was 3.4% (Arabi & Alhamid 2006). Another study also supports that noncommunicable diseases, mainly cardiovascular events, are the major causes of morbidity and mortality among young and elderly pilgrims with pre-existing illnesses (Memish et al. 2014). The CVD has been reported as the cause of 64% of admissions to intensive care units (Madani et al. 2007). The outcome of the CVD among Hajj pilgrims admitted in All-Nurr Hospital, Makkah, was discharged with stable condition (84%), discharged against medical advice (9%), unable to perform Hajj (4.5%), and died as high as 2.5% (Serafi 2010). The median age of French pilgrims was 61 years old, while pilgrims aged from 65 to 74 years old had a high number of age-related chronic illnesses such as diabetes mellitus (21%) and hypertension (21%) (Gautret 2009).

The Hajj ritual consists of several activities which require significant amounts of optimal physical exertion. Walking around the *Kaaba*, a cube-shaped building in Mecca considered as the most sacred site, is followed by walking between two hills (Safa and Marwa) for about 3.15 km is not only the activity of the Hajj. Another exhausting activity is the 14.5 km journey to Arafat desert which is followed by the day when the Hajj pilgrims should spend a night at Muzdalifah to throw gravel at Mena (Mulyana & Gunawan 2010). Hence, people joining the Hajj pilgrimage should be in a good health and able to afford to take the journey. Screening for several chronic illnesses before joining the Hajj pilgrimage is obligatory and must be conducted one or two months prior to the journey. Heart failure, serious arrhythmia, uncontrolled hypertension, unstable angina, recent cardio and cerebrovascular diseases, and other illnesses which make one unfit to fly are also considered to make one unfit to undertake the Hajj pilgrimage (Chamsi-Pasha 2014).

Indonesia contributed as many as 10% of the Hajj pilgrims before 2008, with the mortality rate for the Indonesian pilgrims ranging from 200 to 380 deaths per 100,000 pilgrims during 10-week Hajj periods (Pane 2013). According to this study, in 2008, among 206,831 Indonesian pilgrims, there were 446 deaths ("equivalent to 1,968 deaths per 100,000 pilgrims"). The highest of the death factors were related to CVD (66%) followed by respiratory disease (28%). The high mortality rate among Indonesian pilgrims occurs despite the predeparture health screening. Therefore, more extensive identification of cardio-cerebrovascular risk factors among pre-Hajj pilgrims are profoundly recommended. The aim of this study is to identify major risk factors of cardiovascular events among Hajj pilgrims from Denpasar Bali in 2018.

2 METHODS

2.1 Data collection

A cross-sectional study was conducted with a total of 99 participants. Participants were screened and recruited in Western Denpasar Community Health Centers, where the Hajj preparations take place. Pilgrims older than 18 years of age were included after signing the informed consent. Demographic data include age, gender, and date of birth. History of cigarette smoking and alcohol consumption were obtained from the participants. Result of physical examination, laboratory testing, electrocardiography, and radiography were also collected. Body mass index (BMI) was calculated as weight in kilograms divided by the square of height in meters using Microsoft Office Excel 2010.

Past medical history or preexisting illness information were obtained from the participants; included history of cardiac diseases, diabetes mellitus, hypertension, dyslipidemia, chronic kidney disease, chronic liver disease, chronic respiratory disease, chronic diarrhea, and stroke. Five ml fasting blood samples were taken to be tested for complete blood count, kidney function test, liver function test, blood sugar level, lipid profile, and uric acid level. Electrocardiography was done by a professional nurse. The participants older than 40 years of age also underwent a chest x-ray in the radiologic center.

2.2 Statistical analysis

Pearson's chi-square or Fisher's exact tests were used to compare univariate categorical data. All statistical analyses were performed using SPSS 23.0. The results were expressed in terms of number and percentage.

CVD risk was defined as the finding of at least one of the variables: cigarette smoking, alcohol consumption, obesity, hypertension, diabetes mellitus, dyslipidemia, hyperuricemia, or ECG or x-ray abnormalities).

2.3 Ethical considerations

The study, the collection of clinical and epidemiological data, was submitted for ethical approval to the Research Ethic Committee of Universitas Udayana, Denpasar. Enrolment of the study participants was conditional on appropriate consent.

3 RESULTS

From one public health center in Western Denpasar, 99 Hajj pilgrims were enrolled to this study. Most of the Hajj pilgrims were 41–50 years old; more than half of them were female (Table 1).

Table 2 showed that the main chronic diseases history in Hajj Pilgrims from Bali were hypertension (26.3%), diabetes mellitus (15.2%), and chronic lung disease (7.1%). Ten percent of the Hajj had a history of cigarette smoking (Table 2). The most common current illness with a cardiovascular risk factor in Hajj Pilgrim from Denpasar was dyslipidemia (77.8%), followed by obesity (62.7%) and hypertension (37.4%).

Table 1. Demographic characteristics of participants.

Variable	Description	N (%)
Age	18–30 years old	3 (3.1)
	31–40 years old	10 (10.1)
	41–50 years old	34 (34.4)
	51–60 years old	25 (25.3)
	>60 years old	8 (8.1)
Gender	Male	42 (42.4)
	Female	57 (57.6)

Table 2. Cardiovascular risk factors of Hajj pilgrims.

Variable	Description	N (%)
Past medical history	Hypertension	26 (26.3)
	Diabetes mellitus	15 (15.2)
	Chronic lung disease	7 (7.1)
	Cardiac disease	2 (2.1)
	Dyslipidemia	1 (1.0)
	Stroke	1 (1.0)
	Chronic kidney disease	0 (0.0)
	Chronic liver disease	0 (0.0)
Social history	Cigarette smoking	10 (10.1)
	Alcohol consumption	1 (1)
Current illness	Dyslipidemia	77 (77.8)
	Obesity	62 (62.7)
	Hypertension	37 (37.4)
	X-ray abnormalities	36 (36.4)
	ECG abnormalities	32 (32.3)
	Hyperglycemia	19 (19.2)
	Hyperuricemia	15 (15.2)

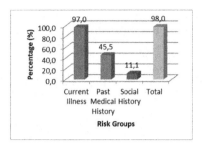

Figure 1. Percentage of Hajj pilgrims with cardiovascular risk factors.

Figure 1 showed the percentage of Hajj pilgrims who had at least one risk factor based on their current illness, past medical history, and social history. The total number of Hajj Pilgrims that were at risk of having cardiovascular event was 97 persons (98%).

4 DISCUSSION

Over the past few years, cardiovascular diseases have emerged as an important cause both of intensive care unit (ICU) admission and of mortality during Hajj. More than 60% of patient admissions to ICU were caused by myocardial infarction and failure of left ventricle (Al Shimemeri 2012). The most common current illnesses of Denpasar pilgrims in 2018 with a CVD risk factor were dyslipidemia and obesity, with more than 50% prevalence each.

The main items in the past medical history of the pilgrims that increased the risk of having cardiovascular event were hypertension and diabetes mellitus with the percentage of 26.3% and 15.2%, respectively. This result was in accordance with a study involving 379 Malaysian pilgrims in 2013, which stated that the most common health problems before pilgrimage were hypertension (48.3%) and diabetes (25.2%). That study also look at health problems during pilgrimage among Malaysian pilgrims. They discovered that the main problem was respiratory diseases (77.5%), followed by fever (15.0%) (Radhiah et al. 2017).

The limitations of our study were the total number of samples, which was relatively small for a survey and we did not obtain data about diseases experienced by the pilgrims during pilgrimage. The next study should involve more health care centers and participants to discover more reliable results. Also it is important to gather data after pilgrimage to explore the incidence of cardiovascular events during pilgrimage, also morbidity and mortality rates. The proportion of Hajj pilgrims having cardiovascular risk factors who actually suffer from cardiovascular diseases during pilgrimage can also be calculated.

5 CONCLUSION

The finding of significant numbers of Hajj pilgrims with CVD risk indicate that identification of CVD risk factors is profoundly recommended to prevent mortality and morbidity related to cardiovascular events during the Hajj period. Health education and precautionary measures can be shared with the Hajj pilgrims with cardiovascular risk factors to prevent the risks becoming an illness during their pilgrimage.

REFERENCES

Al Masud, S. M. R., Bakar, A. A., & Yussof, S. 2016. Determining the types of diseases and emergency issues in Pilgrims during Hajj: A literature review. *Stat Inf*, 5(7).

Al Shimemeri, A. 2012. Cardiovascular disease in Hajj pilgrims. *Journal of the Saudi Heart Association*, 24(2): 123–127.

Aljoudi, A. S. 2013. A university of the Hajj? *The Lancet*, 382(9906): 1689.

Arabi, Y. M., & Alhamid, S. M. 2006. Emergency room to the intensive care unit in Hajj. *Saudi Med J*, 27(7): 937–941.

Chamsi-Pasha, H., Ahmed, W. H., & Al-Shaibi, K. F. 2014. The cardiac patient during Ramadan and Hajj. *Journal of the Saudi Heart Association* 26(4): 212–215.

Gautret, P., Soula, G., Delmont, J., Parola, P., & Brouqui, P. 2009. Common health hazards in French pilgrims during the Hajj of 2007: A prospective cohort study. *Journal of Travel Medicine* 16(6): 377–381.

Madani, T. A., Ghabrah, T. M., Albarrak, A. M., Alhazmi, M. A., Alazraqi, T. A., Althaqafi, A. O., & Ishaq, A. H. 2007. Causes of admission to intensive care units in the Hajj period of the Islamic year 1424 (2004). *Annals of Saudi medicine*, 27(2): 101–105.

Memish, Z. A., Zumla, A., Alhakeem, R. F., Assiri, A., Turkestani, A., Al Harby, K. D., ... & McCloskey, B. 2014. Hajj: infectious disease surveillance and control. *The Lancet*, 383(9934): 2073–2082.

Mulyana, W. W., & Gunawan, T. S. 2010. Hajj crowd simulation based on intelligent agent. In *International Conference on Computer and Communication Engineering (ICCCE'10)* (pp. 1–4). IEEE.

Pane, M., Imari, S., Alwi, Q., Kandun, I. N., Cook, A. R., & Samaan, G. 2013. Causes of mortality for Indonesian Hajj Pilgrims: Comparison between routine death certificate and verbal autopsy findings. *PLoS One*, 8(8).

Radhiah, N., Rosminah, M., Suhaimi, A. W., & Omar, O. 2017. The fundamental of Hajj demand for health care services within congestion in Makkah. *Malaysian Journal of Public Health Medicine*, 17(1): 84–93.

Raja, W., Aziz, A., Hassan, T. B., Jalil, M., Niazi, I. U., Jawaid, N., & Tariq, M. 2017. Disease patterns among Hajj pilgrims attending medical facilities of Pakistan Hajj medical mission 2016 (1437 HIJRI). *Pakistan Armed Forces Medical Journal*, 67(5): 825–831.

Serafi, A. S. 2010. Pattern of cardiovascular diseases in pilgrims admitted in Al-Noor hospital Makkah during Hajj 1429H. *Pakistan Journal of Physiology*, 6(1): 14–17.

Shafi, S., Dar, O., Khan, M., Khan, M., Azhar, E. I., McCloskey, B., ... & Petersen, E. 2016. The annual Hajj pilgrimage—minimizing the risk of ill health in pilgrims from Europe and opportunity for driving the best prevention and health promotion guidelines. *International journal of infectious diseases*, 47: 79–82.

Public health and occupational health

Medical Technology and Environmental Health – Abdullah, Widiaty & Abdullah (eds)
© *2020 Taylor & Francis Group, London, ISBN 978-0-367-86053-0*

Anticancer effect of 1,2-epoxy-3(3-(3,4-dimethoxyphenyl)-4H-1-benzopiran-4on) Propane (EPI) and combination with Doxorubicin on HTB183 lung cell cancer culture

A.F. Sobandi, R.B. Soeherman & L. Yuniarti
Universitas Islam Bandung, Bandung, Jawa Barat, Indonesia

F.A.F. Mansoer
Rumah Sakit Al Ihsan Bandung, Bandung, Jawa Barat, Indonesia

ABSTRACT: Lung cancer is a cancer with the highest incidence in the world. Conventional chemotherapy for lung cancer has side effects, high resistance, and toxicity, so it is necessary to search for sensitive and effective chemotherapy drugs from natural ingredients that are widely available in Indonesia. This study aims to examine the anticancer effect of compound 1,2-epoxy-3(3-(3,4-dimethoxyphenyl)-4H-1-benzopiran-4on) propane (EPI), which is synthesized from clove leaf oil, and its combination effects with doxorubicin on HTB183 lung cancer cells. This research is a pure in vitro experimental study of HTB183 lung cancer culture cells. Toxicity tests were carried out using the method of tetrazolium 3-(4,5-dimethylthiazol-2-il) 2,5-diphenyltetrazolium bromide (MTT) to calculate cell viability. IC50 value was obtained by analysis using probit regression calculation using SPSS software. The synergism of this compound with doxorubicin was determined based on the value of the Combination Index (IK) using a combination test with series 1/2IC50, 3/8IC50, 1/4IC50, and 1/8IC50, and the data was analyzed using Compusyn 1.0 software. The results showed IC50 EPI compounds is 1.41(±0.86) μg/mL, and doxorubicin is 2.21(±0.98) μg/mL. The IK value of EPI-doxorubicin has an average value of 0.3–0.7 which means it has a synergistic effect, and the lowest value is obtained from a combination of EPI 0.63 μg/mL and doxorubicin 0.26 μg/mL with CI = 0.066. In conclusion EPI compounds have anticancer effects potent to HTB183 lung cancer cells and synergistic with doxorubicin.

1 INTRODUCTION

Lung cancer is the most common malignancy among men and women in this century, with a total of 960,000 new cases and 850,000 deaths each year in men, and 390,000 cases and 330,000 deaths in women. The survival rate of lung cancer is very small, ranging from 5–10% at five years of survival (Boyle & Levin 2008).

Doxorubicin is a chemotherapy drug of anthracycline antibiotics with antineoplastic activity. However, doxorubicin has side effects on the body, such as toxicity to the heart and resistance, that make its use limited (Thorn et al. 2011).

One of the active compounds isoflavone, which has a chemical structure similar to genistein, namely 1,2-epoxy-3(3-(3,4-dimethoxyphenyl)-4H-1-benzopiran-4on) propane (EPI). EPI compounds are isoflavones that are synthesized from clove leaf oil. Some studies have shown a synergistic effect on the combination of EPI with several other anticancer drugs such as doxorubicin and cisplatin against several cancers such as HeLa cervical cancer cells (Yuniarti et al. 2018).

The purpose of this study was to examine the anticancer effect of 1,2-epoxy-3 (3-(3,4-dimethoxyphenyl)-4H-1-benzopiran-4on) propane (EPI) compounds and their combined effect with doxorubicin on lung cancer cell cultures HTB183.

2 METHOD

This study is a pure in vitro experimental study of HTB183 lung cancer culture. The research activity was carried out at the Parasitology Laboratory of the Faculty of Medicine, Gadjah Mada University, in August 2018. The anticancer test was carried out using the tetrazolium 3-(4,5-dimethyltiazol-2-il) 2,5-diphenyltetrazolium bromide (MTT) method and cell viability was measured using ELISA reader with wavelength λ = 550-600 nm (595 nm). The test was carried out at EPI concentration series 250 µg/mL; 125 µg/mL; 62.5 µg/mL, 31.25 µg/mL, 15.625 µg/mL, 7.8125 µg/mL, and 3.906 µg/mL. Doxorubicin tests were carried out in series of 100 µg/mL, 50 µg/mL, 25 µg/mL 12.5 µg/mL 6.25 µg/mL, 3.125 µg/mL, and 1.56 µg/mL. Concentrations that can inhibit 50% of cells (IC50) are determined using probit regression calculations using SPSS software. Then the determination of synergism is measured using a combination test with the concentrations series 1/2 IC50, 3/8 IC50, 1/4 IC50, and 1/8 IC50 both on doxorubicin and EPI. Cell viability was determined using an ELISA reader with a wavelength of λ = 550–600 nm (595 nm). IK values were analyzed using Compusyn software with the toxicity criteria based on Baharum et al., namely 20 µg/mL, potent; > 20–100 µg/mL, moderate; > 100–1000 µg/mL, weak; and> 1000 µg/mL, inactive (Baharum et al. 2014), and synergism using Chou criteria, namely (CI = 1), synergistic (C <1), and antagonist (C> 1) (Chou 2010).

3 RESULTS AND DISCUSSION

Lung cancer includes all malignant diseases in the lungs (Komite Penaggulangan Kanker Nasional 2017). Lung cancer is the most common cause of cancer deaths in men and women, and was responsible, for example, for the death of 1.3 million people worldwide in 2011 (Ravichandiran & Nirmala 2011). To increase the efficacy and reduce the side effects of therapy, it is necessary to develop drugs from natural ingredients.

This study was a cytotoxic test of 1,2-epoxy-3(3- (3,4-dimethoxyphenyl)-4h-1-benzopiran-4on) propane (EPI) compound and its combination with doxorubicin on HTB183 lung cancer cells. The cytotoxic test results of 1,2-epoxy-3(3-(3,4-dimethoxyphenyl)-4h-1-benzopiran-4on) propane (EPI) compounds against HTB183 lung cancer cell culture using the MTT method resulted in IC50 values of 1.41 (± 0.86) µg/mL against HTB183 lung cancer, which shows potent anticancer effects (Baharum et al. 2014).

EPI (1,2-epoxy-3(3-(3,4-dimethoxyphenyl)-4H-1-benzopiran-4on) propane) is a compound synthesized from clove leaf oil with a chemical structure similar to genistein. Genistein belongs to the class of flavonoids that selectively bind to estrogen receptors with different affinities, and modulate estrogen receptor signals (Polkowski & Mazurek 2000). A study shows that genistein induces p53 independently and is associated with induction of p21WAF1, thereby inducing apoptosis in large-cell lung cancer H460 (Lian et al. 2009).

Several in vitro and in vivo studies show the protective effect of genistein from pulmonary carcinogenesis when this compound is used alone or with other compounds (Mahmood et al. 2011, Zhu et al. 2012). Genistein shows anticancer effects on small-cell lung cancer (SCLC) H446; molecules that induce cell-cycle arrest and apoptosis cycles, deregulation of Forkhead box M1 (FoxM1) proteins and their target genes (for example, the 25B cell division cycle (Cdc25B), cyclin B1, and surviving). In A549, genistein lung cancer cells (5–10 mM) increase apoptosis induced by trichostatin A (TSA) and increase expression of TNF receptor 1 TNF receptors (TNFR-1), which mediate the extrinsic apoptotic pathway (Shiau et al. 2010, Wu et al. 2012). If EPI has a structure similar to genistein, EPI allows it to have genistein-like activity against cancer cells. EPI has an anticancer effect on HeLa cervical cancer cells with a mechanism similar to genistein and has anticancer properties through induction of intrinsic pathways and extrinsic apoptosis in tumor cells that vary through inhibition of NF-Kappa β, activation of p53 signaling, activation of p53, and regulation of the formation of Bcl-2 and ROS (Yuniarti et al. 2018).

The results of this study also showed IC50 doxorubicin of 2.21 (± 0.98) µg/mL against HTB183 lung cancer cells. This shows that doxorubicin has a potent anticancer effect

(IC50≤20μg/mL) (Baharum et al. 2014). IC50 bar graph of EPI and doxorubicin compound can be seen in Figure 1.

Combined cytotoxic curves of EPI and doxorubicin compound can be seen in Figure 2. Isobologram normalization curve of EPI and doxorubicin compound can be seen in Figure 3.

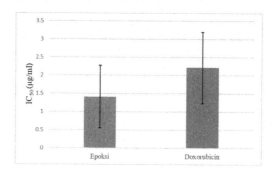

Figure 1. IC_{50} bar graph of EPI and doxorubicin compound.

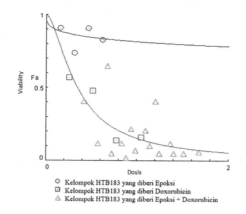

Figure 2.   Combined cytotoxic curves of EPI and doxorubicin compound.

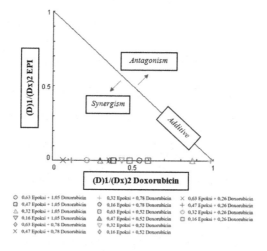

Figure 3.   Isobologram normalization curve of EPI and doxorubicin compound.

329

Doxorubicin is a broad-spectrum anticancer drug that can be used in lung cancer therapy (Johnson-Arbor & Dubey 2017). There are two mechanisms of the doxorubicin process working against cancer cells; namely, doxorubicin can enter the cell nucleus and damage topoisomerase-II, then cause damage to the DNA and cell death. In addition, there are other mechanisms; in particular, doxorubicin is oxidized to semiquinone, an unstable metabolite, which is converted back into doxorubicin in the process of releasing reactive oxygen species. Reactive oxygen species can cause lipid peroxidation and membrane damage, DNA damage, and oxidative stress, and can trigger apoptotic pathways for cell death. Doxorubicin triggers cell death through activation of p53. The role of p53 apoptosis might be in its ability to disrupt the balance between antiapoptotic proteins (such as Bcl-XL, Bcl-2, and Mcl-1) and pro-apoptotic proteins (such as Bax and Bak) (Rathos et al. 2013).

In this study, the combination test results showed that the combination of EPI with doxorubicin had a synergistic effect against HTB183 lung cancer cells with IK values of 0.3–0.7 (Chou 2010).

It has been proven that with combination therapy, the efficacy of the drug can be increased and side effects can be minimized (Chou 2010). Research shows that doxorubicin has a synergistic effect with several other compounds, which thus increases the efficacy and decreases side effects, as in Park et al.'s research on the combination of doxorubicin therapy with gamitrinib on several types of cancer cells, namely cervical, ovarian. and prostate cancer (Domanski 2014). Studies on the effect of the combination of doxorubicin with genistein were carried out by Yun et al. on MFC-7 breast cancer cells and showed there is a synergistic effect by the mechanism whuereby genistein is thought to increase doxorubicin accumulation in cells. A study conducted by Yuniarti et al. (2018) on the combination of EPI with doxorubicin on HeLa cancer cells showed a very strong synergistic effect. This combination can increase the expression of p53, TIMP-3, and MiR-34a proteins in HeLa cells (Yuniarti et al. 2018). Thus, EPI compounds with structures that resemble genistein have anticancer effects against HTB183 lung cancer cells. And EPI can increase the anticancer activity of doxorubicin, which means it has a synergistic effect against HTB183.

4 CONCLUSIONS

The conclusion of this study is EPI compound (1,2-epoxy-3(3-(3,4-dimethoxyphenyl)-4H-1-benzopiran-4on) propane) has potent anticancer properties against HTB183 lung cancer cell culture and a combination of EPI compounds and doxorubicin have a synergistic effect against HTB183 lung cancer cells.

REFERENCES

Baharum, Z., Akim, A. M., Taufiq-Yap, Y. H., Hamid, R. A., & Kasran, R. 2014. In vitro antioxidant and antiproliferative activities of methanolic plant part esxtracts of Theobroma cacao. *Molecules*, 19(11): 18317–18331.

Boyle, P., & Levin, B. 2008. *World Cancer Report 2008*. IARC Press, International Agency for Research on Cancer.

Chou, T. C. 2010. Drug combination studies and their synergy quantification using the chou-talalay method. *Cancer Research*, 70(2): 440–446.

Domanski, R. C. N. 2014. *The Changing Map of Europe, the Trajectory Berlin-Poznań-Warszawa: A Tribute to Antoni Kukliński*. Ofycina Wydawnicza Rewasz, 1–9.

Johnson-Arbor, K., & Dubey, R. 2017. *Doxorubicin. StatPearls*. StatPearls Publishing.

Komite Penaggulangan Kanker Nasional 2017. *Pedoman nasional pelayanan kedokteran kanker paru (Hasil dan diskusi. Edisi pertama)*. Jakarta: Kementerian Kesehatan Republik Indonesia, 1–75.

Lian, F., Li, Y., Bhuiyan, M., & Sarkar, F. H. 2009. P53-independent apoptosis induced by genistein in lung cancer cells. *Nutrition and Cancer*, 33(2): 125–131.

Mahmood, J., Jelveh, S., Calveley, V., Zaidi, A., Doctrow, S. R., & Hill, R. P. 2011. Mitigation of radiation-induced lung injury by genistein and EUK-207. *International Journal of Radiation Biology*, 87(8): 889–901.

Polkowski, K., & Mazurek, A. P. 2000. Biological properties of genistein. A review of in vitro and in vivo data. *Acta Poloniae Pliarmaceutica—Drug Research*, 57(2): 135–155.

Rathos, M. J., Khanwalkar, H., Joshi, K., Manohar, S. M., & Joshi, K. S. 2013. Potentiation of in vitro and in vivo antitumor efficacy of doxorubicin by cyclin-dependent kinase inhibitor P276-00 in human non-small cell lung cancer cells. *BMC Cancer*, 13(1): 1.

Ravichandiran, V., Ahamed, H.J. & Nirmala, S. 2011. Natural flavonoids and lung cancer. *Pharmacie Globale (IJCP)*, 6(02): 1–9.

Shiau, R. J., Chen, K. Y., Wen, Y. Der, Chuang, C. H., & Yeh, S. L. 2010. Genistein and β-carotene enhance the growth-inhibitory effect of trichostatin A in A549 cells. *European Journal of Nutrition*, 49(1): 19–25.

Thorn, C. F., Oshiro, C., Marsh, S., Hernandez-Boussard, T., McLeod, H., Klein, T. E., & Altman, R. B. 2011. Doxorubicin pathways: Pharmacodynamics and adverse effects. *Pharmacogenetics and Genomics*, 21(7): 440.

Wu, T. C., Yang, Y. C., Huang, P. R., Wen, Y.Der, & Yeh, S. L. 2012. Genistein enhances the effect of trichostatin A on inhibition of A549 cell growth by increasing expression of TNF receptor-1. *Toxicology and Applied Pharmacology*, 262(3): 247–254.

Yuniarti, L., Mustofa, M., Aryandono, T., & Mubarika, S. 2018. Synergistic action of 1, 2-Epoxy-3 (3-(3, 4-dimethoxyphenyl)-4H-1-benzopiyran-4-on) propane with doxorubicin and cisplatin through increasing of p53, TIMP-3, and MicroRNA-34a in cervical cancer cell line (HeLa). *Asian Pacific Journal of Cancer Prevention: APJCP*, 19(10): 2955.

Zhu, H., Cheng, H., Ren, Y., Liu, Z. G., Zhang, Y. F., & De Luo, B. 2012. Synergistic inhibitory effects by the combination of gefitinib and genistein on NSCLC with acquired drug-resistance in vitro and in vivo. *Molecular Biology Reports*, 39(4): 4971–4979.

Larvae and pupae in Bandung city: Can they be the indicators of Dengue outbreak?

T. Respati
Universitas Islam Bandung, Bandung, Jawa Barat, Indonesia

A. Raksanagara
Universitas Padjadjaran, Bandung, Jawa Barat, Indonesia

ABSTRACT: Mosquito-breeding sites are closely related to macro and micro environmental factors, especially the lack of availability of water, which makes people store water in containers. This study aims to analyze the factors that affect the presence of larvae and pupae. This is a survey using stratified random sampling method in 2,035 houses and inspected 5,984 containers. The analysis was generalized estimating equation (GEE). *Aedes aegypti* were found in 90% of containers, *Aedes albopictus* in 3.5 %, other *Aedes* species in 5% and *Culex* in 4.2%. The *Stegomyia* index showed high infestation of mosquito in all study areas. Volume influences the presence of larvae in containers whereas the presence of pupae as a proxy of adult mosquitoes is related to the volume of containers, plant-covered containers in total and in part, and containers that are entirely and partially closed. It is concluded that containers containing larvae and pupae are large-volume containers that are not closed or partially closed with water sources from rainwater and well water. Bandung city has a high risk of DHF outbreaks.

1 INTRODUCTION

Mosquito-breeding sites are closely related to macro and micro environmental factors that are determined by human activities individually or collectively (Luz et al. 2011, Arunachalam et al. 2012). Social factors include eradication and prevention programs; provision of public facilities such as sanitation and clean water; and community participation including prevention activities, community knowledge, and behavior regarding dengue (Faridah et al. 2017, Respati et al. 2018).

The increased population and the need for a place to live cause the growth of settlements. The lack of availability of basic sanitation facilities in the form of waste management, clean water facilities available at all times, and wastewater disposal facilities are among factors that rarely receive attention (Tana et al. 2012). The poor areas in many countries have the minimum of clean water facilities, garbage disposal, and basic sanitation facilities (Respati et al. 2018). Water supply is usually limited to certain times so that people collect water in containers. These containers are vulnerable to contamination by several biological agents, including mosquito larvae (Balasubramanian et al. 2015). Several studies have shown that management of the utilization of water containers in households is an essential determinant in the dynamic population of mosquitoes (Toledo et al. 2007, Erlanger et al. 2008, Arunachalam et al. 2012). This article aims to analyze the factors that affect the presence of larvae and pupae.

2 METHOD

This research is a survey with a stratified random sampling method in 16 villages in Bandung city, Indonesia. The research is part of a study entitled "Eco-health System Dynamics Model to Reduce Mosquito Breeding Places as an Effort to Cut Transmission of Dengue Haemor-rhagic Fever."

Entomological surveys were carried out using modified data collection instruments from the World Health Organization on behalf of the Special Program for Research and Training in Tropical Diseases (Tana et al. 2012). Each container in the house was grouped according to place, type of material, diameter, capacity, closed status (closed/semi-closed/open), color (light/dark), exposure to sunlight, as well as prevention measures such as the use of Abate, or the activity of raising fish.

Only containers containing water were inspected. The larva survey method used is visual. The survey method is as follows: each container containing water is observed in a state of calm water; if no larvae were found, wait for 30 seconds to one minute, then repeat the process. If larvae were found, larvae and pupae were taken and then put into clear identified plastic bags. All plastic bags containing larvae and pupae were sent and stored in the Laboratory of the Parasitology Department of the Faculty of Medicine, Padjadjaran University, to breed. The results calculated larval free numbers (ABJ), housing index (HI), Breteau index (BI), and container index (CI).

Statistical analysis used was correlation analysis and generalized estimating equations (GEE) using SPSS Ver. 17. This study has received a research permit from the National Unity and Protection Society Agency (BKBPMD) of the Government of West Java Province no. 070/3799/BKBPM.

3 RESULTS

The survey was carried out in 2,035 houses from March to July 2015 at the end of every week (Saturday and Sunday) to ensure that the occupants of the house were in their homes. The number of containers found and examined was 5,984 pieces. Below is the result of an entomology survey. Mosquito larvae obtained from the research area were bred in the Parasitology Lab, Faculty of Medicine, Padjadjaran University with the following results:

From Table 1, it appears that the predominant mosquito species found in breeding sites are *Aedes aegypti*, followed by *Aedes albopictus*, other *Aedes* species, and *Culex*. The data show that the overall entomology measure is above the standard given by the Ministry of Health with average HI 23.6, BI 35.5, CI 11.9, and ABJ 76.4. Some areas show very high numbers such with HI above 30%, which means that 30% of houses are infested with mosquito larvae. One area has a BI of 86.7% which shows that a large number of houses in the region have at least one container with mosquito larvae. Table 2 and Table 3 showed factors contribute to the presence of larvae and pupae.

4 DISCUSSION

Rapid population growth causes increasingly widespread use of land as a settlement in the city of Bandung. More horizontal settlements of housing complexes or clusters cause land functions to change. Land use as settlement has close links to DHF because it is related to the habitat of the *Aedes* spp. as a dengue vector. Dense settlements, inadequate basic sanitation

Table 1. Mosquito species in study area.

Species	Number	%
Aedes aegypti	260	90.3
Aedes albopictus	11	3.8
Aedes sp	5	1.7
Culex sp	12	4.2
Total	288	100

Table 2. Factors contributes to the presence of larvae.

Effect	Univariable			Model		
	Est. Par.	RSE	P Value	Est. Par.	RSE	P Value
Containers category						
A	0.1004	0.2086	0.6304			
B	0.2476	0.2247	0.2704			
C	Reference	-	-			
Log volume	0.3041	0.0409	< 0.0001	0.2966	0.0421	< 0.0001
Container materials						
Cement	Reference	-	-			
Ceramic	0.0676	0.1358	0.6188			
Plastic	-0.9315	0.1209	< 0.0001			
Rubber	0.6236	0.5299	0.2393			
Fiber	0.0191	0.2703	0.9436			
Others	-0.7195	0.2090	0.0006			
Color						
Black	Reference	-	-			
Blue	0.0061	0.0831	0.9412			
Red	-0.1007	0.1526	0.5093			
Green	-0.3341	0.1420	0.0186			
Pink	0.5599	0.2013	0.0054			
Yellow	0.5602	0.3602	0.1199			
Others	-0.0548	0.0939	0.5597			
Water sources						
State water	Reference	-	-	Reference	-	-
Others	0.2822	0.1392	0.0426	0.4613	0.2211	0.0369
Rainwater	1.1229	0.2274	< 0.0001	0.9353	0.3673	0.0109
Wells	0.4814	0.1106	< 0.0001	0.4906	0.1256	< 0.0001
Covered by trees						
Total	0.7920	0.2553	0.0019	0.6585	0.4090	0.1074
Half	0.5256	0.1899	0.0056	0.1236	0.2599	0.6345
No	Reference	-	-	Reference	-	-
Placed outside						
Yes	0.3847	0.1310	0.0033	0.0240	0.1835	0.8960
No	Reference	-	-	Reference	-	-
Chemical treatment						
Yes	0.5455	0.2254	0.0155	0.2115	0.3298	0.5214
No	Reference	-	-	Reference	-	-
With lids						
Total	0.3550	0.1196	0.0030	0.5000	0.1671	0.0028
Half	0.6991	0.1879	0.0002	0.8272	0.2405	0.0006
No	Reference	-	-	Reference	-	-

Notes: RSE = robust standard error. Category A = container for everyday use; B = container filled with water (e.g. flower-pot); C = specific use (e.g. fishpond)

facilities, along with poorly maintained environmental hygiene increase the risk of dengue virus transmission (Respati et al. 2017).

Rain affects larvae in two ways, namely causing a decrease in temperature and increasing the relative humidity of the air. *Aedes* spp. survive at low temperatures, but their metabolism will decrease if the temperature drops below the critical temperature. Temperature higher than 35°C will slow down the physiological process. With an average optimum temperature of mosquito growth of 25–27°C, then the City of Bandung is an ideal area for mosquito growth (Tana et al. 2012).

Table 3. Factors contributes to the presence of pupae.

Effect	Univariable			Model		
	Est. Par.	RSE	P Value	Est. Par.	RSE	P Value
Containers category						
A	0.3354	0.3045	0.2706			
B	0.2461	0.3129	0.4317			
C	Reference	-	-			
Log volume	0.4185	0.0526	< 0.0001	0.4309	0.0557	< 0.0001
Container materials						
Cement	Reference	-	-			
Ceramic	0.2982	0.1931	0.1225			
Plastic	-0.7236	0.1791	< 0.0001			
Rubber	1.1049	0.5642	0.0502			
Fiber	-0.0176	0.4378	0.9679			
Others	-0.5183	0.3028	0.0869			
Color						
Black	Reference	-	-			
Blue	0.2379	0.1228	0.0527			
Red	-0.5126	0.2890	0.0761			
Green	-1.2703	0.2568	< 0.0001			
Pink	0.0561	0.3814	0.8832			
Yellow	-0.2339	0.6608	0.7234			
Others	-0.3924	0.1560	0.0119			
Water sources						
State water	Reference -	-	-	Reference -	-	-
Others	0.6016	0.2150	0.0051	0.8078	0.4123	0.0501
Rainwater	0.6093	0.2935	0.0379	-0.1534	0.5555	0.7825
Wells	-0.5048	0.1655	0.0023	-0.7881	0.1835	< 0.0001
Covered by trees						
Total	1.0859	0.3493	0.0019	1.3809	0.4796	0.0040
Half	0.7722	0.2447	0.0016	0.8249	0.3203	0.0100
No	Reference	-	-	Reference	-	-
Placed outside						
Yes	0.2857	0.1720	0.0968	0.1193	0.2483	0.6310
No	Reference	-	-	Reference	-	-
Chemical treatment						
Yes	0.2819	0.2761	0.3074			
No	Reference	-	-			
With lids						
Total	0.4236	0.1692	0.0123	0.6680	0.2565	0.0092
Half	1.0009	0.2942	0.0007	0.9123	0.4174	0.0288
No	Reference	-	-	Reference	-	-

Notes: RSE = robust standard error. Category A = container for everyday used; B = container filled with water (e.g. flower pot); C = specific use (e.g. fish pond)

The results of the entomology survey show a low rate of larval-free containers (ABJ), while container index (CI), house index (HI), and Breteau index (BI) are very high. This means that the number completely exceeds the standardized number determined to measure the transmission rate. The size of HI for all research areas is 23.6%, and this figure is far above the number considered as the threshold limit of dengue transmission of 5%. However, this limit is still debatable. In some countries, the 5% limit is being targeted to start dengue control programs such as in Quito, Ecuador (Chan et al. 2011). On the other hand, in Singapore, even though the HI level has reached <1%, and in Cuba with a HI index between 0.05%–0.91%, dengue outbreaks still occur. Brazil, on the other hand, shows that outbreaks do not occur when HI reaches less than 1% (Ooi et al. 2006). WHO recommends the use of HI and BI to determine high-risk areas to establish prevention programs. If the prevention program is not carried out, the possibility of an outbreak will be substantial (Tana et al. 2012, Respati et al. 2016)

The rate of *Aedes aegypti* infestation in Indonesia is measured using the ABJ value that shows the percentage of houses free of mosquito larvae. The standard given by the Ministry of Health for ABJ is 80% for non-endemic areas and ≥ 95% for endemic areas (Kementrian Kesehatan Republik Indonesia 2008). The city of Bandung is a dengue-endemic area that the figure to be achieved is ≥ 95%. The results of the study showed that none of the regions achieved this figure. These results are not following data from the Bandung City Health Office, which state that the entire area of Bandung City achieved ABJ > 95%, which means that it is already reaching national standards. Some of the reasons for the difference in numbers include the first method of routine inspection carried out by health cadres to allow the homeowner to clean the breeding place because the inspection schedule was provided beforehand, while the research data did not provide an opportunity for residents to clean the breeding place before the inspection. Considering the size of its *Stegomyia* index, which was entirely above the above standards, it should be noted that the city of Bandung is very prone to dengue outbreaks. Seeing the differences in research results and report results in ABJ figures, the use of ABJ as an indicator of the success of PSN in preventing dengue transmission needs evaluation.

Breeding places outside the house are affected by terraced houses, perhaps because residents of terraced houses pay less attention to breeding places on the upper floors of the house compared to residents of one-story houses because they are not directly visible to residents. Field observations show that the second floor of a house tends to be used as a storage area for unused items such as flower pots, as well as drying and washing. This finding can be one of the concerns when delivering community education and dengue programs; in general, to pay more attention to the condition of high-rise buildings.

5 CONCLUSION

Containers containing larvae and pupae are containers with large volumes that are not closed or partially closed with water sources from rainwater and well water. Types of productive containers as mosquito breeding places are large volume containers, partially covered either by plants or closed containers. Based on entomological measures, the city of Bandung has a high risk of dengue outbreaks.

REFERENCES

Arunachalam, N., Tyagi, B. K., Samuel, M., Krishnamoorthi, R., Manavalan, R., Tewari, S. C., ... & Petzold, M. (2012). Community-based control of *Aedes aegypti* by adoption of eco-health methods in Chennai City, India. *Pathogens and Global Health*, 106(8): 488–496.

Balasubramanian, R., Anukumar, B., & Nikhil, T. L. 2015. *Stegomyia* indices of Aedes mosquito infestation and container productivity in Alappuzha district Kerala. *Int J Mosquito Res*, 2(2): 14–8.

Chan, E. H., Sahai, V., Conrad, C., & Brownstein, J. S. 2011. Using web search query data to monitor dengue epidemics: A new model for neglected tropical disease surveillance. *PLoS Neglected Tropical Diseases*, 5(5).

Erlanger, T. E., Keiser, J., & Utzinger, J. 2008. Effect of dengue vector control interventions on entomological parameters in developing countries: A systematic review and meta-analysis. *Medical and Veterinary Entomology*, 22(3): 203–221.

Faridah, L., Respati, T., Sudigdoadi, S., & Sukandar, H. 2017. Gambaran Partisipasi Masyarakat terhadap Pengendalian Vektor Melalui Kajian Tempat Perkembangbiakan Aedes aegypti di Kota Bandung. *Majalah Kedokteran Bandung*, 49(1): 43–47.

Kementrian Kesehatan Republik Indonesia. 2008. *Indonesia: dengue trend from 1985 to 2006 (Profil Kesehatan Indonesia 2012)*. [Online]. http://www.pppl.depkes.go.id/wc77a-0fe2faba2.htm.

Luz, P. M., Vanni, T., Medlock, J., Paltiel, A. D., & Galvani, A. P. 2011. Dengue vector control strategies in an urban setting: An economic modelling assessment. *The Lancet*, 377(9778): 1673–1680.

Ooi, E., Goh, K., & Gubler, D. J. 2006. Dengue prevention and 35 years of vector control in Singapore. *Emerging Infectious Diseases*, 12(6): 887–893.

Respati, T., Feriandi, Y., Ndoen, E., Raksanegara, A., Djuhaeni, H., Sofyan, A., & Dale, P. (2018). A qualitative ecohealth model of dengue fever (DF) in Bandung, Indonesia. *Int J Trop Dis*, 1 (008): 1–12.

Respati, T., Nurhayati, E., Feriandi, Y., Yulianto, F., & Sakinah, K. 2016. Pemanfaatan Kalender 4M Sebagai Alat Bantu Meningkatkan Peran Serta Masyarakat dalam Pemberantasan dan Pencegahan Demam Berdarah. *Global Medical and Health Commmunication*, 4(2): 121–128.

Respati, T., Raksanagara, A., Djuhaeni, H., & Sofyan, A. 2017. Spatial distribution of dengue hemorrhagic fever (DHF) in urban setting of Bandung City. *Global Medical & Health Communication, (GMHC)* 5(3): 212.

Tana, S., Abeyewickreme, W., Arunachalam, N., Espino, F., Kittayapong, P., Wai, K., … & Sommerfeld, J. 2012. Eco-bio-social research on dengue in Asia: General principles and a case study from Indonesia. In *Ecohealth Research in Practice. Innovative Applications of an Ecosystem Approach to Health* (pp. 173,184). New York: Springer.

Toledo, M. E., Vanlerberghe, V., Baly, A., Ceballos, E., Valdes, L., Searret, M., … & Van der Stuyft, P. (2007). Towards active community participation in dengue vector control: Results from action research in Santiago de Cuba, Cuba. *Transactions of the Royal Society of Tropical Medicine and Hygiene*, 101(1): 56–63.

Medical Technology and Environmental Health – Abdullah, Widiaty & Abdullah (eds)
© 2020 Taylor & Francis Group, London, ISBN 978-0-367-86053-0

Can traditional psychoeducation reduce the stigma against depression among teenagers?

E. Nurhayati, T. Respati, F.A. Yulianto, B. Budiman & Y. Feriandi
Universitas Islam Bandung, Bandung, Jawa Barat, Indonesia

ABSTRACT: Negative stigma against depression, especially among teenagers, may lead to inappropriate professional treatment, thus the symptoms get worse and eventually lead to suicidal tendencies. Psychoeducation is one of many ways used by healthcare professionals to reduce stigma against depression especially among teenagers. This study was accomplished to measure the effectiveness of psychoeducation to reduce stigma against depression by comparing the level of stigma against depression pre- and post-psychoeducation. This was analytic, cross-sectional research. The study focused on 37 teenagers who were given psychoeducation by psychology professionals. The tool that was used to measure stigma was the Depression Stigma Scale (DSS). It has two nine-item subscales assumed to measure personal stigma and perceived stigma. The DSS was given before and after the psychoeducation. Results obtained from unpaired t-tests showed two-tailed p value 0.90 (95% CI –1.63 to 1.43) for personal stigma and 0.35 (95% CI –1.09 to 3.05) for perceived stigma. This study showed that psychoeducation cannot reduce the level of stigma against depression among teenagers. We need more modern mental health promotion with longer duration to gain a more positive perception.

1 INTRODUCTION

Depression is one of the mental health disorders that can affect teenagers. Based on data released by the World Health Organization (WHO), the number of people with depression in the world today is around 154 million people, of which 20% are teenagers (United Nations 2014). National Basic Health Research (Riskesdas) 2013 showed that the number of people with depression and anxiety in Indonesia in 2013 was around 14 million people, of whom 6% were in the 15 years and over age group (Kemenkes 2013).

The problem of depression is not only limited to the prevalence, but also the delay or lack of management. These problems are caused by negative stigma attached to depression. Society's negative stigma related to depression affects patients who then to not get help from health professionals, and often the symptoms worsened and can even end in suicide (Zartaloudi & Madianos 2010, Del Casale et al. 2013, Corrigan et al. 2014, Klarić & Lovrić 2017).

Stigma related to depression among teenagers can be overcome with adequate mental health knowledge. Del Casale quotes Barney as saying that adequate knowledge about mental health can reduce misunderstandings and worries about mental health disorders. Their research was conducted by involving 44 students, aged 16–18 years, who were given psychoeducation. The instrument used to measure the level of stigma was the Haghighat's Standardized Stigmatization Questionnaire (Del Casale et al. 2013).

Teenagers are targeted to be the subject of promotion of mental health education because the population tends to increase. Teenagers nowadays are also considered to have advantages over the previous generation because they have better overall health, they are more educated, they understand and embody urban values, they have broader access to knowledge, and they are more vulnerable to mental disorders (United Nations 2014).

Psychoeducation is one of many ways to provide an understanding of mental health and reduce the negative stigma of depression and other mental disorders. A similar study was conducted by Kassam in which he and his team used psychoeducation to 408 medical students using the Knowledge Quiz and the Mental Illness: Clinicians Attitudes (MICA) Scale. The results of his research showed that the psychoeducation method is effective enough to motivate better understanding (Wassink 2014).

This study's objective was to measure the effectiveness of psychoeducation as a tool to reduce the level of stigma against depression among teenagers by comparing the level of stigma against depression before and after psychoeducation.

2 METHODS

This study was a part of a research study to develop a board game to reduce negative stigma against depression among teenagers. The original research was done by comparing two groups of teenagers. One group played the board game that was developed by the research team. The other group received traditional psychoeducation and became the respondents of this study.

This study was conducted on 37 first-year students of the Faculty of Psychology, Bandung Islamic University. Respondents were taken by convenience sampling through recruitment using Google form.

The study was conducted in Faculty of Medicine Building, Universitas Islam Bandung (UNISBA), Indonesia, on December 13, 2018. The instrument used to measure the level of stigma against depression was the Depression Stigma Scale (DSS) (Subramaniam et al. 2017).

The DSS questionnaire consisted of two subscales: personal and perceived scale. Each of the scales consisted of nie questions with a Likert scale answer model from 1 (strongly disagree) to 5 (strongly agree) (Subramaniam et al. 2017).

Data was taken by giving questionnaires to respondents before and after the psychoeducation. Psychoeducation was given for 1 hour with a traditional interactive lecture method. The lecture described mental disorders that commonly happened in teenagers and depression. The resource person who gave the material was a psychologist who was an expert in developmental psychology.

Data were calculated using Microsoft Excel to assess the results of the Likert scale. The results were then analysed with the unpair T test method using Stata.

3 RESULTS

The total number of respondents involved in this study was 37 people, consisting of 36 females and 1 male with average age 18.1 years old. The results of the study using the DSS questionnaire are shown in Table 1.

Table 1 shows that perceived stigma was higher that personal stigma in both pre- and posttest. Statistical test using a two-tailed unpair T test showed that there was no significant difference on the level of either perceived or personal stigma.

Table 1. Item endorsement on depression stigma scale.

Item	Personal Stigma		Perceived Stigma	
	Pre Test	Post Test	Pre Test	Post Test
Average	15.2	15.3	21.95	20.97
Standard Deviation	3.41	3.18	4.46	4.49
Median	15	16	23	21
P value	0.9 (95% CI -1.63 to 1.43)		0.35 (95% CI -1.09 to 3.05)	

4 DISCUSSION

Stigma against depression is one of the factors that can influence mental health care. The results showed that the mean value for perceived stigma was higher compared to personal stigma both at pre- and posttest. This proved that there is tendency for negative stigma against people with depression. This negative stigma can lead to discrimination. People with mental disorders can get even worse or patients can become involved in criminal acts such as drug abuse or alcohol addiction and even end up committing suicide (Kakuma et al. 2010, United Nations 2014).

Various ways have been developed to reduce the level of stigma against depression, especially in teenagers. One of the ways is psychoeducation. The results showed that the traditional psychoeducation method could not reduce the level of stigma, both personal and perceived. These results contradict the results of Del Casale's study that examined the reduction of stigma levels against depression in 24 teenagers aged 18 years. The results of his research showed that campaigns providing adequate information about depression and mental health can reduce the level of stigma (Del Casale et al. 2013).

This difference is due to the differences in the duration of psychoeducation provided. Del Casale provided psychoeducation in four sessions where each session lasted 1.5 hours, while this study only provided one psychoeducation session with a duration of 1 hour. Another factor is the type of psychoeducation given, not only by interactive lecture presentations but also showing films that fit the theme. The type of questionnaire used was also different because Del Casale used the Haghighat's Standardized Stigmatization Questionnaire (SSQ). All of these factors can cause differences in results (Del Casale et al. 2013).

The results of this study also differ from Pinfold et al. who conducted stigma-level research in high school students in the United Kingdom. Total of 472 respondents were given pre-test questionnaires, then an intervention, and post-test questionnaires. The psychoeducation provided by Pinfold was carried out in stages consisting of two phases and then followed up one week after the intervention. Pinfold even brought in former mental-health sufferers and opened discussion sessions for respondents. Pinfold's research results showed that brief psychoeducation can bring positive change (Pinfold et al. 2003).

The comparison of the results of research conducted with the results of Del Casale's and Pinhold's research shows that there are many factors that can influence the success of psychoeducation in reducing the level of stigma. These factors include the duration, the material, and how the psychoeducation is provided. Psychoeducation interventions must be made attractive and adapted to the age of the respondents so respondent have to adjust to the preferred technology.

5 CONCLUSION

The traditional psychoeducation method cannot reduce the level of stigma against depression among teenagers. More interesting methods and longer duration of psychoeducation are urgently needed to increase more positive perception.

ACKNOWLEDGEMENT

We acknowledge the Research Unit of Faculty Medicine, Bandung Universitas Islam Bandung (UNISBA), for sponsoring this research. Psychoeducation can be done with the support of the Faculty of Psychology, UNISBA.

REFERENCES

Corrigan, P. W., Druss, B. G., & Perlick, D. A. 2014. The impact of mental illness stigma on seeking and participating in mental health care. *Psychological Science in the Public Interest*, 15(2): 37–70.

Del Casale, A., Manfredi, G., Kotzalidis, G. D., Serata, D., Rapinesi, C., Caccia, F., & Girardi, P. 2013. Awareness and education on mental disorders in teenagers reduce stigma for mental illness: A preliminary study. *Journal of Psychopathology*, 19(3): 208–212.

Kakuma, R., Kleintjes, S., Lund, C., Drew, N., Green, A., & Flisher, A. J. 2010. Mental health stigma: What is being done to raise awareness and reduce stigma in South Africa? *African Journal of Psychiatry*, 13(2).

Kemenkes, R. I. 2013. *Riset Kesehatan Dasar (RISKESDAS) 2013, Laporan Nasional 2013*. Jakarta: Kementerian Kesehatan Republik Indonesia, 1–384.

Klarić, M., & Lovrić, S. 2017. Methods to fight mental illness stigma. *Psychiatr Danub*, 29(Suppl 5): 910–917.

Pinfold, V., Toulmin, H., Thornicroft, G., Huxley, P., Farmer, P., & Graham, T. 2003. Reducing psychiatric stigma and discrimination: Evaluation of educational interventions in UK secondary schools. *The British Journal of Psychiatry*, 182(4): 342–346.

Subramaniam, M., Abdin, E., Picco, L., Pang, S., Shafie, S., Vaingankar, J. A., ... & Chong, S. A. 2017. Stigma towards people with mental disorders and its components–a perspective from multi-ethnic Singapore. *Epidemiology and Psychiatric Sciences*, 26(4): 371–382.

Wassink, A. K. 2014. *Using a Mental Health Board Game Intervention to Reduce Mental Illness Stigma Among Nursing Students*. Michigan: Grand Valley State University, 1–89.

United Nations 2014. *Mental Health Matters*. New York: United Nations.

Zartaloudi, A., & Madianos, M. 2010. Stigma related to help-seeking from a mental health professional. *Health Science Journal*, 4(2): 77.

How employment periods and working posture lead to musculoskeletal disorders

Y. Susanti, F.A. Dewi, M.A. Djojosugito & S.N. Irasanti
Universitas Islam Bandung, Bandung, Indonesia

S.A. Adianto
Cibabat Hospital, Bandung, Indonesia
Universitas Islam Bandung, Bandung, Indonesia

ABSTRACT: Work-related musculoskeletal disorders (WRMSDs) are a frequent health problem in the industrial sector and associated with work patterns like body position, repetitive movement, and duration of work. The objective of this research was to explore how employment periods and working posture can lead to musculoskeletal disorder. This research used a cross-sectional study design in which the data were collected through observational interviews with 50 tailors. Data regarding working posture were collected using rapid upper limb assessment (RULA), and the musculoskeletal problems were assessed with Nordic body maps (NBM). The results indicated that as many as 40% of respondents had employment periods of more than 10 years, and 88% of respondents had risk categories in working posture. Based on a chi-square test, there was no relationship between employment periods and musculoskeletal disorder nor working posture and musculo-skeletal disorder. Even though the risk category in working posture was high, there was no correlation with musculoskeletal complaint due to body adaptation to daily working process.

1 INTRODUCTION

Work-related musculoskeletal disorders (WRMSDs) have emerged as a major health problem among workers. Common risk factors of WRMSDs are awkward working postures, repetitive movements, and vibration (Lietz et al. 2018).

Musculoskeletal diseases are defined as a group of diseases and complaints that affect different structures of the musculoskeletal system. These include nerves, tendons, muscles, joints, ligaments, bones, blood vessels, and supporting structures such as intervertebral discs (Aminian et al. 2012, Aljanakh et al. 2015).

WRMSDs mainly affect the lower back, neck, and upper and lower extremities. Lower back and neck are the most affected areas among ready-made-garment workers (Hossain et al. 2018). The neck, followed by the lower back, shoulder, and the upper back are affected most often in dental professionals (Lietz et al. 2018).

Many studies have linked musculoskeletal problems with time exposure. The current study identified that the length of employment/work experience is an important determinant of back pains and lower extremity disorders. Working posture considerably contributed to the risk of developing back and lower extremity pains.

Rapid upper limb assessment (RULA) was developed to investigate the exposure of individual workers to risk factors associated with work-related upper-limb disorders. Musculoskeletal disorders (MSDs) can be assessed using Nordic body maps (NBM), one of the most popular survey tools for detecting severity and duration of MSDs.

This research's objective was to analyze the impact of length of employment and working posture that lead to musculoskeletal disorder.

2 METHODS

This research used a cross-sectional study design carried out by using observational and interview methods. We observed and tested 50 tailors in a fertilizer-bag department with a consecutive random sampling technique. Data regarding working posture were collected using RULA.

The RULA ergonomic assessment methods are a single page worksheet used to evaluate body posture, force, and repetition. The examiner filled in the scores for each body region in section A for the arm and wrist analysis, and section B for neck, trunk, and leg analysis. After the data for each region was collected, tables on the form were used to compile the risk factors and generate a single score that represents the risk level (Golchha et al. 2014).

There were four category levels that indicate MSD risk (no risk = score 1–2, low risk = score 3–4, medium risk = score 5–6, high risk = score >6; Ergonomics plus).

The NBM is a screening instrument to assess musculoskeletal problems from 28 locations in the body. From each location, we filled grade of complaints (A = no complaint, B = mild pain, C = moderate pain, D = severe pain).

The data was collected in PT Pupuk Kujang from June to August 2018 and we analyzed the data using SPSS with chi-square test.

3 RESULTS

The subjects in this study were 50 tailors from the fertilizer-bag department and all respondents were male. Table 1 shows characteristics of the subjects in this research.

RULA score interpretation results can give information about working posture risk level to MSD as shown in Figure 1.

Table 1. Tailor characteristics.

Characteristic	n	%
Age (years)		
20–30	14	28
31–50	19	38
> 50	20	40
Total	50	100
Employment periods (years)		
< 5	11	22
6–10	19	38
> 10	20	40
Total	50	100

Figure 1. Working posture risk level (RULA).

The NBM gives information about severity of pain levels and regions in musculoskeletal problem that are shown in Table 2 and Figures 1 and 2.

RULA score results in Figure 1 show the subjects have moderate (58%) and high-risk posture (30%) that can lead to musculoskeletal disorder.

Table 2 shows severity levels in musculoskeletal problems: most subjects have mild pain level in upper musculoskeletal and no pain category in lower musculoskeletal.

Musculoskeletal complaints predominantly occur in upper musculoskeletal region (Figures 2–3): locations of the problem are left and right shoulder, waist, lower neck, and right calf.

Table 2. Musculoskeletal problem.

Pain level	n	%
Upper musculoskeletal		
no pain	18	36
mild	32	64
Total	50	100
Lower musculoskeletal		
no pain	26	52
mild	13	26
moderate	9	18
severe	2	4
Total	50	100

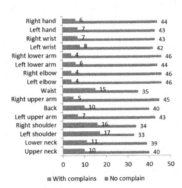

Figure 2. Upper musculoskeletal complaint.

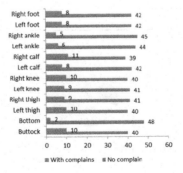

Figure 3. Lower musculoskeletal complaint.

Table 3. The association between employment periods and working posture with MSDs.

| Variable | Musculoskeletal disorder | | | | | | P-value |
| | Yes | | No | | Total | | |
	n	%	n	%	n	%	
Employment periods							
≤ 5 years	8	72.7	3	27 3	11	100	0.163
6–10 years	11	57 9	8	42 1	19	100	
> 10 years	17	85 0	3	15 0	20	100	
Working posture							
Risk	32	72 7	12	27 3	44	100	0 545
No risk	4	66 7	2	33 3	6	100	

Table 4. The association between age and lower musculoskeletal disorders.

| Variable | Lower MSDs | | | | | P-value |
	Severe	Moderate	Mild	No pain	Total	
Ages (years)						
20–30	0	1	6	7	14	0 035
31–50	2	8	7	13	30	
> 50	0	0	0	6	6	

Based on the results presented in Table 3, there was no correlation between employment periods with musculoskeletal disorder (P= 0.163), and there was also no correlation between working posture and musculoskeletal disorder (P= 0.545).

Table 4 shows there were correlation between age and lower musculoskeletal disorder: the most affected age category was 31–50 years old.

4 DISCUSSION

The results of this research showed that 30% of the subjects had high risk and 58% had moderate risk posture. Unsuitable working posture had a considerable effect on musculoskeletal disorder. These findings were similar with those found among ready-made-garment (RMG) workers in Bangladesh with medium risk score, and the risk factor resulted in musculoskeletal complaints (Rafie et al. 2015, Hossain et al. 2018, Lietz et al. 2018).

Different factors such as hereditary, stress, unsuitable working posture, and lack of regular exercise can affect incidence of musculoskeletal disorders. Work position was the only factor that predicts the occurrence of musculoskeletal disorders in dentists (Ambarwati 2017).

Following musculoskeletal complaints in body regions, this study found that shoulder, waist, and neck were the most affected regions. Previous research showed neck, lower back, shoulder, and upper back were the most affected body regions (Lietz et al. 2018, Mohammadipour et al. 2018).

Back and lower extremity musculoskeletal problems were associated with age, alcohol use, safety training, work postures, and length of employment in Ethiopian barbers (Mekonnen 2019).

In this research, we found that there was no correlation between employment periods with MSDs, nor with working posture to MSDs. Workers who have been employed for longer periods of time had fewer occupational injuries than recently employed workers because there was adaptation to ergonomic risk factors in workplaces. Acceptable working posture during simulation may reduce development of work-related musculoskeletal disorders (Gandavadi et al. 2007, Hossain et al. 2018).

Findings of our study showed that there was statistically significant relationship between age and musculoskeletal disorder. These findings were similar to a study among RMG workers in Bangladesh (Hossain et al. 2018).

5 CONCLUSION

This study revealed fertilizer-bag tailors have moderate to high risk working posture, and the most affected body regions with musculoskeletal complaints are shoulder, waist, and neck. There is no correlation between working posture and employment periods to musculoskeletal disorder. The other factor that affects musculoskeletal disorder in tailors is age.

REFERENCES

Aljanakh, M., Shaikh, S., Siddiqui, A. A., Al-Mansour, M., & Hassan, S. S. 2015. Prevalence of musculoskeletal disorders among dentists in the Ha'il Region of Saudi Arabia. *Annals of Saudi Medicine*, 35(6): 456–461.

Ambarwati, T., Wicaksena, B., Sopianah, Y., & Miko, H. 2018. Posture work to complaint musculoskeletal disorders at the dentist. *Journal of International Dental and Medical Research*, 11(1): 57–61.

Aminian, O., Alemohammad, Z. B., & Hosseini, M. H. 2015. Neck and upper extremity symptoms among male dentist and pharmacist. *Work*, 51(4): 863–868.

Ergonomic plus 2018. *A step-by-step guide rapid upper limb assessment (RULA)*. [Online]. http://ergoplus.com/wp-content/uploads/RULA-A-Step-by-Step-Guide1.

Gandavadi, A., Ramsay, J. R. E., & Burke, F. J. T. 2007. Assessment of dental student posture in two seating conditions using RULA methodology... A pilot study. *British Dental Journal*, 203(10): 601.

Golchha, V., Sharma, P., Wadhwa, J., Yadav, D., & Paul, R. 2014. Ergonomic risk factors and their association with musculoskeletal disorders among Indian dentist: A preliminary study using rapid upper limb assessment. *Indian Journal of Dental Research*, 25(6): 767.

Hossain, M. D., Aftab, A., Al Imam, M. H., Mahmud, I., Chowdhury, I. A., Kabir, R. I., & Sarker, M. 2018. Prevalence of work related musculoskeletal disorders (WMSDs) and ergonomic risk assessment among readymade garment workers of Bangladesh: A cross sectional study. *PloS one*, 13(7).

Lietz, J., Kozak, A., & Nienhaus, A. 2018. Prevalence and occupational risk factors of musculoskeletal diseases and pain among dental professionals in Western countries: A systematic literature review and meta-analysis. *PloS one*, 13(12).

Mekonnen, T. H. 2019. The magnitude and factors associated with work-related back and lower extremity musculoskeletal disorders among barbers in Gondar town, northwest Ethiopia, 2017: A cross-sectional study. *PloS one*, 14(7): e0220035.

Mohammadipour, F., Pourranjbar, M., Naderi, S., & Rafie, F. 2018. Work-related musculoskeletal disorders in Iranian office workers: Prevalence and risk factors. *Journal of Medicine and Life*, 11(4): 328.

Rafie, F., Zamani Jam, A., Shahravan, A., Raoof, M., & Eskandarizadeh, A. 2015. Prevalence of upper extremity musculoskeletal disorders in dentists: Symptoms and risk factors. *Journal of Environmental and Public Health*, 2015.

Medical Technology and Environmental Health – Abdullah, Widiaty & Abdullah (eds)
© 2020 Taylor & Francis Group, London, ISBN 978-0-367-86053-0

Achievement evaluation in minimum service standards of health services: Mental illness patient service using shackles – expectation and reality

A. Widodo
Universitas Muhammadiyah Surakarta, Sukoharjo, Indonesia

ABSTRACT: Minimum service standards (MMS) in the health sector are regency/city government references for providing health services appropriate for every citizen. Mental illness patients have the right to minimum service according to the standard provided by doctors and nurses in community health centers. Method of this research is qualitative, by conducting in-depth interviews with participants including health-field stakeholders, caregivers, nurses, and health cadres in a community health center of Sukoharjo Regency and Klaten Regency. Results of this study are: 1) the head of Sukoharjo Public Health Office (PHO) and the head of the Klaten PHO, the head of community health center, caregivers, and health workers were committed to a good policy of fostering community mental health but were still limited to general policies; 2) MSS of mental health had not been implemented in either the Klaten PHO nor in Sukoharjo PHO; 3) The application of health promotion involving families and the community to prevent recurrence and reentrance of post-mental illness patients has been carried out in accordance with the basic principles of the community mental health program. MMS of mental health had not been administered based on community expectations. The community expected that mental health services would be carried out at the nearest health center. MMS of mental health needs to be optimized by involving nurses, other health workers, and the community.

1 INTRODUCTION

Mental health covers emotional, psychological, and social health conditions, which can be seen from fulfilling interpersonal relationships, behavior, effective coping, positive self-concept, and emotional stability (Videbeck 2006). According to Law Number 18 of 2014 concerning Mental Health, Chapter 1 Article 1 states that mental health is a state of a particular individual who is able to develop physically, mentally, spiritually, and socially, so he is aware of his ability, can overcome pressure, work productively, and be able to contribute to the community (Kementeri Hukum dan Hak Asasi Manusia 2014).

From being condemned to the stocks, physical restraint of individuals with mental illness has been happening in Indonesia and is still happening today. Violations of human rights against patients with mental illness occur in Indonesia as Indonesia's law has not been sufficient to provide adequate protection for mental illness patients. Human rights commissions, community organizations, health professions, and health services must play significant roles in assisting mental illness patients who often experience relapse and are condemned to the shackles (Irmansyah et al. 2009).

Many households with members experiencing severe mental illness who are shackled indicate that the Mental Health Law Number 18 of 2014 has not been optimally implemented. This is in accordance with the findings of Nurjannah et al. (2015), which identified that laws regarding human rights were not being implemented in policies and procedures.

Patient family participation needs to be given mental health promotion as an important area of nursing. Several studies have shown that mental health education in families has a positive impact in reducing the recurrence rate of mental disorders (Sudiyanto 1998, Widodo 2004). The community mental health team, if it is fixed, proves to be more advantageous than hospital treatment (Malone et al. 2010). Several studies have shown that involving the community has been proven to be more effective in treating people with mental illness (Black & Rose 2002, Stacciarini et al. 2011).

Government health policies are not fully concerned about mental illness in low and middle-income countries (Patel 2007). The emergence of programs in several regions concerning shackles-free treatment in Indonesia shows the Indonesian government's concern about mental health problems in the community. The 2019 shackles-free Indonesia program will remain a vacuum without the help of families and community. Family and community are expected to participate (social participation and involvement) in prevention, therapy, and rehabilitation, and welcome patients into family and community environments (Bengtsson-tops 2004). Through appropriate therapy and methods, schizophrenia can be controllable and manageable, although more likely noncurable. Therapy for schizophrenic mental illness not only lies in psychopharmaceutical drug therapy and other types of therapy but also in the role of family and community (Hawari 2012). Cheng et al. (2014) conclude that the prevalent problems of quality of life among mental illness patients are mostly due to vulnerable environmental factors and are significantly related to family support and relationships with close neighbors.

Quality of life is an independent factor associated with therapeutic outcomes in mental illness (Aloba et al. 2013). Many factors affect quality of life, such as finance, health, security, environmental conditions, and others. Financial security is not always directly proportional to quality of life. Many other factors affect quality of life, such as the level of mental health. Mental illness significantly affects many aspects of life, especially quality of life (Kian et al. 2014).

Mental health services in Indonesia are available at primary, secondary, and tertiary levels. At the primary level, health service is administered by community health centers, at the secondary level it is conducted at public hospitals, and at the tertiary level it is carried out at psychiatric hospitals. In 2002, the community health center program priority was physical health. Although the community health center included 18 programs, the mental health program was always the last priority. According to the regulation of the minister of health Number 43 of 2016 concerning minimum service standards (MSS) in the health sector, mental health services are included in 10 of the 12 programs that began in 2016.

Health facility research data in 2011 reported that the number of community health centers in Indonesia with mental health service programs was 64% (out of 8981 community health centers). However, only 21.47% of community health centers have actively implemented mental health services (Idaiani & Riyadi 2018). The treatment approach to mental illness patients in the community health center has been using a referral system – the community health center is the primary service center that solely examines mental illness patients who then are referred to psychiatric hospitals. Mental health services cannot only be provided by psychiatric hospitals, but also by community health centers that are obligated to provide mental health services for patients in surrounding areas. Mental health integration in primary health services is defined as the existence of mental health services at the primary level or community health center (Idaiani 2016).

In a previous study, researchers found that community health centers in the Sukoharjo and Klaten regency had not implemented mental health service programs. Nurses, doctors, and other health workers revealed that the problem was at the level of public health offices in Sukoharjo and Klaten with no mental health specialists. Furthermore, they did not have trained health workers to diagnose patients with mental illness and some did not have storage facilities in accordance with standard operational procedure (SOP) regarding the storage of psychotropic drugs. When visiting the community health center for treatment, both new and old patients would be referred to the closest public hospital or psychiatric hospital. This problem was the condition most criticized by patients with mental illness and their families. They expected that mental health services to be carried out at community health centers as primary services (Widodo et al. 2019). The experience of the community health center in the Yogyakarta region with trained health workers in the management of mental illness patient services encouraged researchers to seek the implementation of minimum service standards of mental illness in Sukoharjo and Klaten regency.

2 METHODS

This research used descriptive-qualitative research method that covers the following; 1) commitment of stakeholders in the Sukoharjo and Klaten public health office, head of the community health center, and health workers regarding the implementation of the shackle-free program; 2) community perspective (community leaders, religious leaders, and health cadres) regarding shackle-free programs and community needs regarding shackle-free programs; 3) the opinion of post-shackle patients' families regarding the need and necessity of shackle-free program to prevent recurrence and improve the quality of life of post-shackle patients.

Data collection was conducted by collecting secondary data from related institutions such as public health offices and community health centers. In addition, data were also collected by conducting in-depth interviews with participants. The data obtained were recorded audio, and the recordings were transcribed. The data from the transcript results were coded according to the qualitative coding guidelines, after which categories and content analysis were carried out. Qualitative analysis was completed employing content analysis of data obtained through participation observation, focus group discussion (FGD), and in-depth interviews presented in the form of excerpts (Moleong 2001).

3 RESULTS AND DISCUSSION

The results of the qualitative study carried out are as follows:

3.1 Commitment and policy towards community mental health efforts and the 2017 shackle-free Indonesia program

Stakeholders in the Sukoharjo and Klaten health sectors had good levels of commitment and policies regrading community mental health efforts, but were still limited to general policies. Government policies regarding the 2017 shackle-free Indonesia program had not specifically become a program that seriously and continuously followed up as a continuation of the 2014 shackle-free Java program launched by the governor of Central Java in 2012. The organizational structure of the regional committee began in 2018, and started the establishment of the noncontagious diseases section and mental health. This is in accordance with the informant from Sukoharjo public health office as follows:

"In the structure of last year, mental health became a part of basic services, for now, starting in October 2017 there is a new organizational structure of committee, that non-contagious diseases and mental health have its own division in the public health office and the person in charge at community health center" (informant 1, location 1).

Implementation of efforts to prevent and overcome mental health problems in the community, should be done via cross-sectoral coordination and cooperation, according to the health officer at one of the community health center as follows:

"The preventive, promotive, curative, and rehabilitative efforts on community mental health problems should be done in coordination across sectors. If it is preventive, it promotes cooperation with social services, village officials, and council for regency. If curative can be routinely controlled to the health center, if the health center is unable, the patient can be referred. Rehabilitation can be done by returning the patient to the community, to the authorities in the village, so that the community can welcome, can learn skills" (informant 2, location 1).

According to the noncontagious diseases division and the Sukoharjo public health office, community mental health coordination, especially concerning the shackle problem, already exists but has not been optimal in practice, as stated by the informant:

"Coordination of mental health issues is already underway, with Social Service. There is a social welfare staff at the sub-district level with municipal police and police officers cooperated on the 2012 stock-free Central Java program. Now, if mental illness patients are condemned to the stocks, the cost of transportation to take patients to the psychiatric hospital is a problem because it is not guaranteed by the Social Security Administrative Body" (informant 3 location 1).

Not all of the written policies have impacted the families and post-shackle patients the researchers met. From 18 post-shackle patients the researchers met, only 10 had been visited and given mental health socialization by health workers from the community health center and the psychiatric hospital. Six post-shackle patients stated that they were only visited and recorded as mental illness patients, the rest were never visited. Some answers regarding the existence of community mental health program from the public health center or psychiatric hospital that patients and the community receive are as follows:

"In my opinion the implementation of the community mental health program from the public health office was only once held in providing health counseling door-to-door" (informant 4, location 1).

"So far, what I received from the community health office is that there was no community mental health implementation program, but from the psychiatric hospital there has been someone who came to houses and gave health promotion and that kind of was rare" (informant 5, location 1).

"Usually just recording and giving a small socialization regarding the problem of mental illness patients" (informant 6, location 1).

3.2 Sources of funds in the implementation of community mental health promotion efforts

In the more than 74 years since the independence of Indonesia, the funds allocated for public health development have never exceeded 3–3.5%, although WHO sets a minimum of 5% of the state budget. Funding for health and education sector in developed countries is still on top of priority even if it reaches 30–40% of the state budget. Fund allocation for mental health in each city is widely used for the activities of the community mental health team. According to the Ministry of Health/Kementrian Kesehatan RI (2009), community mental health team funding can be obtained through the state budget, regional budget, and other funding sources (from the community, sponsors, and others). This is in accordance with the Head of non-contagious diseases division and Sukoharjo public health office board as following:

"Funding for mental health promotion activities is part of the demands to achieve minimum service standards in the health sector, which must be achieved by community health center. We, in the non-contagious disease division and Sukoharjo public health office community realize it through community mental health team activities" (informant 3 location 1).

Regulation of Ministry of Health Number 43 of 2016 concerning MSS in the health sector has stated that mental health services for severely mentally ill patients are provided by nurses and doctors in community health centers in their working area (Kementrian Kesehatan RI 2009), but not all community health centers in Sukoharjo Regency have been able to serve because there were no drugs available for mental illness patients, as stated by the informants below:

"If I take my child to go to the psychiatric hospital, then we have to take transportation, and rent motorcycle, the distance is far enough, so that it may spend no less than Rp. 50,000" (informant 7, location 1).

"I once brought my younger sibling who was sick to check up at the community health center, but because there was no medicine available at community health center for mental patients, he was finally referred and we were forced to have to re-control every month at the psychiatric hospital" (informant 8, location 1).

"Because medicines for mental illness patients include psychotropic drugs, it requires strict storage, and there is no psychiatrist specialist who is authorized to prescribe, so if mental patients come to the health center, we will refer them to the psychiatric hospital" (informant 3, location 1).

Only part of the community health center in Klaten regency has been able to provide outpatient treatment and re-control of mental illness patients, as stated by the informant from location 2 below:

"There are already a number of community health center that can serve mental patients for outpatient treatment, including providing mental health drugs which are the implementation of referral activities, that is Manisrenggo community health center, Cawas 1, and Jogonalan 2" (informant 1, location 2).

4 CONCLUSION

– Heads of the Sukoharjo and Klaten public health office, the head of the community health center, caregivers, and health workers were committed to good community mental health policy efforts but were still limited to general policies.
– Minimum service standards of mental health were still not implemented at Klaten and Sukoharjo community health center.
– The implementation of health promotion involving families and the community for the prevention of recurrence and re-loading of post-passive ODG patients had been carried out in accordance with the basic principles of the community mental health program.

ACKNOWLEDGEMENT

The writer acknowledge the contribution of Indonesia Ministry of Research and Technology in funding the main author's educational expenses, the rector of Universitas Muhammadiyah Surakarta for supporting the education of the main author, and as well as Sukoharjo and Klaten district governments for providing permission for this study in their areas.

REFERENCES

Aloba, O., Fatoye, O., Mapayi, B., & Akinsulore, S. 2013. A review of quality of life studies in Nigerian patients with psychiatric disorders. *African Journal of Psychiatry*, 16(5): 333–337.

Bengtsson-tops, A. 2004. Mastery in patients with schizophrenia living in the community: Relationship to sociodemographic and clinical characteristics, needs for care and support, and social network. *Journal of Psychiatric and Mental Health Nursing*, 11(3): 298–304.

Black, B. L., & Rose, S. M. 2002. *Advocacy and empowerment: Mental health care in the community.* Routledge.

Cheng, Y., Li, X., Lou, C., Sonenstein, F. L., Kalamar, A., Jejeebhoy, S., ... & Ojengbede, O. 2014. The association between social support and mental health among vulnerable adolescents in five cities: Findings from the study of the well-being of adolescents in vulnerable environments. *Journal of Adolescent Health*, 55(6): S31–S38.

Hawari, D. 2012. *Skizofrenia (pendekatan holistik bio-psiko-sosial-spiritual).* Jakarta: Badan Penerbit FKUI.

Idaiani, S., 2016. Penyakit-Penyakit di Bidang Psikiatri yang Harus Dituntaskan di Puskesmas. *Jurnal Kebijakan Kesehatan Indonesia: JKKI*, 5(4): 168–175.

Idaiani, S., & Riyadi, E. I. 2018. Sistem Kesehatan Jiwa di Indonesia: Tantangan untuk Memenuhi Kebutuhan. *Jurnal Penelitian dan Pengembangan Pelayanan Kesehatan*, 70–80.

Irmansyah, I., Prasetyo, Y., & Minas, H. 2009. Human rights of persons with mental illness in Indonesia: more than legislation is needed. *International Journal of Mental Health Systems*, 3(1): 14.

Kementeri Hukum dan Hak Asasi Manusia 2018. *Undang-Undang Kesehatan Republik Indonsia, No. 18 Tahun 2014 tentang Kesehatan Jiwa 2014.* Tidak dipublikasikan, tersimpan dalam Lembaran Negara RI Tahun 2014, nomor 185.

Kementrian Kesehatan RI 2009. *Keputusan Menteri Kesehatan R.I., Nomor 406/Menkes/SK/VI/2009 tentang Pedoman Pelayanan Kesehatan Jiwa Komunitas.* Tidak dipublikasikan, tersimpan dalam Lembaran Negara RI.

Kian, S. F., Bahar, M. M., & Habib, A. B. (2014). An approach to observing effective demographic factors in the quality of life for schizophrenic patients. *Life Science Journal*, 11(1s).

Malone, D., Marriott, S. V., Newton-Howes, G., Simmonds, S., & Tyrer, P. 2007. Community mental health teams (CMHTs) for people with severe mental illnesses and disordered personality. *Cochrane Database of Systematic Reviews* (3).

Moleong, L. 2001. *Metode Penelitian Kualitatif.* Bandung: Remaja Rosdakarya.

Nurjannah, I., Mills, J., Park, T., & Usher, K. 2015. Human rights of the mentally ill in Indonesia. *International Nursing Review*, 62(2): 153–161.

Patel, V. 2007. Mental health in low-and middle-income countries. *British Medical Bulletin*, 81(1): 81–96.

Stacciarini, J. M. R., Shattell, M. M., Coady, M., & Wiens, B. 2011. Review: Community-based participatory research approach to address mental health in minority populations. *Community Mental Health Journal*, 47(5): 489–497.

Sudiyanto, A., 1998. *Pendidikan Kesehatan Jiwa Keluarga untuk Mencegah Kekambuhan Penderita Gangguan Afektif Berat.* Yogyakarta: Universitas Gadjah Mada.

Videbeck, S.L. 2006. *Psychiatric Mental Health Nursing (3rd edition).* Philadelpia: Lippincott Williams & Wilkins.

Widodo, A. 2004. Perbandingan Efektivitas Antara Metode Ceramah Modul, Dengan Ceramah Video Film dalam Memberikan Pendidikan Kesehatan Jiwa Bagi Keluarga Penderita Skizofrenia. *Varidika - FKIP University Muhammadiyah of Surakarta* ISSN 0852.

Widodo, A., Prabandari, Y. S., Sudiyanto, A., & Rahmat, I. 2019. Increasing the quality of life of post shackling patients through multilevel health promotion of shackling prevention. *Medical Journa, l* 8(2): 542–549.

Medical Technology and Environmental Health – Abdullah, Widiaty & Abdullah (eds)
© 2020 Taylor & Francis Group, London, ISBN 978-0-367-86053-0

Do fast food consumption and physical activities associate with blood pressure of senior high school students in South Tangerang, Indonesia?

S. Sugiatmi & M. Fauziah
Universitas Muhammadiyah Jakarta, Jakarta, Indonesia

ABSTRACT: Prevalence of hypertension in adolescents is increasing. There are only a few studies of hypertension prevalence and its association with fast food consumption and physical activity in Indonesian adolescents. This study determines blood pressure status and its association with fast food consumption and physical activity in adolescents. This cross-sectional study was conducted during January and August 2018 in Senior High School in South Tangerang, Indonesia. We selected 220 students by proportional stratified random sampling. Blood pressure measurements were categorized as normal, prehypertension, or hypertension using the 2004 Fourth Report blood pressure screening recommendations. Fast food consumption was collected using the Food Frequency Questionnaire and divided into often and rare. Physical activity was categorized using the Baecke questionnaire into active and not active. Associations between fast food consumption and physical activity with blood pressure were determined using a chi-square test. Prevalence of prehypertension was 11.5% and hypertension was 12.0%. Fast food consumption and physical activity were not associated with pre-hyper-tension/hypertension (p value > 0.05). Hypertension and prehypertension were highly prevalent in the studied adolescents. There were no associations between fast food consumption and physical activity with adolescent blood pressure.

1 INTRODUCTION

Hypertension affects high morbidity and mortality and is a threat to public health in developed and developing countries. Hypertension plays an important role in the development of ischemic heart disease, kidney and heart failure (Danasekaran et al. 2016).

Globally, the prevalence of hypertension in people aged 20 years and older is increasing. In 2000, 26.4% of the adult population in the world suffered from hypertension (26.6% of men and 26.1% of women). By 2025, the prevalence of hypertension is predicted to increase to 29.2% (29.0% in men and 29.5% in women) (Kearney et al. 2005). Prevalence of high blood pressure in adolescents globally was 11.2%, 13% for boys and 9.6% for girls (de Moraes 2014). Study among high school students in a middle Anatolian province of Turkey, found that prevalence of high blood pressure was 4.4% (Nur et al. 2008), while in Egyptian adolescents, the prevalence rates of prehypertension and hypertension were 5.7% and 4.0%, respectively (Abolfotouh et al. 2011), and in Enugu, South East Nigeria, the prevalence was 13.5% and 74.5% prehypertension and hypertension, respectively (Ujunwa et al. 2013).

In Indonesia, the prevalence of hypertension in people aged >18 years was 25.8% while prevalence in adolescents age 15–17 years was 5.3% (men 6.0% and women 4.7%), rural (5.6%) higher than urban (5.1%) areas in 2013 (Kemenkes 2013). Prevalence of hypertension in adolescents in Depok City, West Java Province, was 42.4% (Angesti & Sartika 2018), while in Pangkalpinang, Bangka Belitung Province, it was 14.17% and 8.33% in men and women, respectively (Yusrizal et al. 2016).

Many factors contribute in developing hypertension in adolescents. Alcohol consumption, physical activity level, dietary pattern, and genes are some of the factors related to hypertension (Mü Ller-Riemenschneider et al. 2010, Rosner et al. 2013, Ewald & Haldeman 2016, Falkner 2018). Hypertension in adolescents will continue into adulthood (Redwine & Falkner 2012).

There is a few information about prevalence of hypertension among adolescents in Banten Province of Indonesia. Considering that hypertension in adolescence is a risk factor for hypertension in adulthood, which will increase mortality due to cardiovascular disease, this study aims to identify the prevalence of hypertension and its association with fast food consumption patterns and physical activity in adolescents in Senior High School in South Tangerang, Indonesia.

2 METHODS

2.1 Population and sample

This cross-sectional study was conducted during January and August 2018 in Senior High School in South Tangerang, Indonesia. We selected 220 students by proportional stratified random sampling.

2.2 Blood pressure

Students' blood pressures were measured using a digital monitoring blood pressure kit. Blood pressure was measured three times with 5-minute intervals. Systolic blood pressure and diastolic blood pressure were defined as the average of three measurements. Blood pressure measurements were categorized as normal, prehypertension, or hypertension using the 2004 Fourth Report blood pressure screening recommendations (Falkner et al. 2004). In further analysis, blood pressure was categorized into two: normal and prehypertension/hypertension.

2.3 Fast food consumption

Fast food consumption was collected using the Food Frequency Questionnaire and divided into often (consumption of fast food ≥ 3 times per week) and rare (consumption of fast food < 3 times per week). Fast food includes fried chicken, chicken nuggets, hamburgers, sandwiches, hot dogs, spaghetti, and French fries.

2.4 Physical activity

Physical activity of the students was all their activity for a day. It was measured using Baecke et al.'s (1982) questionnaire. The activities were divided into three: (1) work activity, (2) sports activity, and (3) leisure activity. Physical activity levels were divided into active and not active.

2.5 Statistical analysis

Mean ± SD were used to describe blood pressure, while fast food consumption and physical activity were described using proportion. A chi-square test was used to determine associations of fast food consumption with blood pressure and physical activity with blood pressure. Association was considered significant when p value < 0.05.

2.6 Ethical clearance

This study has been approved by the ethical committee of Medicine and Health Faculty of Universitas Muhammadiyah Jakarta with ethical clearance number 074/PE/KE/FKK-UMJ /IV/2018.

3 RESULTS

Table 1 shows the characteristics of the study population. Prevalence of prehypertension 11.5% while hypertension 12%. Proportion of students who often consumed fast food was 56.4% while 20.2% of students were considered not active.

Table 2 shows the association between fast food consumption and physical activity with blood pressure. There was no significant association between fast food consumption and blood pressure (p > 0.05) or between physical activity and blood pressure (p > 0.05).

4 DISCUSSION

Prevalence of prehypertension and hypertension in this research were 11.5% and 12%, respectively. A study in Depok city found prevalence of hypertension in adolescents aged 17 years was 42.4% (Angesti & Sartika 2018), while a study in Pangkalpinang Bangka Belitung Province determined prevalence of hypertension in overnourished adolescents was 14.17% and 8.33% in men and women, respectively (Yusrizal et al. 2016). Indonesian basic health research found that prevalence of hypertension in adolescents age 15–17 years was 5.3% (men 6.0% and women 4.7%), rural (5.6%) higher than urban (5.1%) areas in 2013 (Kemenkes 2013). It

Table 1. Characteristics of students.

Variable	Value[a]
Age (yr)	16.09 ± 0.79
Systolic Blood Pressure (mmHg)	112.27 ± 13.40
Diastolic Blood Pressure (mmHg)	69.63 ± 9.35
Blood Pressure Classification	
Normal	166 (76.5)
Prehypertension	25 (11.5)
Hypertension	26 (12.0)
Fast Food Consumption Classification	
Often	124 (56.4)
Rare	96 (43.6)
Physical Activity Classification	
Active	162 (79.8)
Not Active	41 (20.2)

[a]Values are mean ± SD or No. (%)

Table 2. Association between fast food consumption and physical activity with pre/hypertension in senior high students.

| | Prehypertension/Hypertension | | | | Total | | |
| | Pre/Hypertension | | Normal | | | | |
Variable	n	%	n	%	n	%	Pvalue
Fast Food Consumption							
Often	113	10.5	111	89.5	124	100	1.000
Rare	0	10.4	86	89.6	96	100	
Physical Activity							
Not Active	4	9.8	37	90.2	41	100	1.000
Active	8	11.1	144	88.9	162	100	

can be shown that prevalence of hypertension in adolescents in this research is higher than the national prevalence.

In this study, 54.6% adolescents consumed fast food more than 3 times per week. Fast food is food that contains little or no nutritional value but contains other food additives that are not healthy when consumed regularly, and it has an impact on developing hypertension (Ashakiran & Deepthi 2012). However, there is no significant association between fast food consumption and hypertension in this study. Results of the present study are in accordance with results of studies conducted by Khan et al. 2010, Payab et al. 2014, and Zhao et al. 2017, which found no significant relationship between fast food consumption and hypertension in adolescents.

There was no association between physical activity and hypertension in adolescents in this study. This is probably because many adolescents are active in moving but are limited to routine physical activities such as studying, watching television, and doing homework. Sports were participated in only as necessary without meeting health standards, without paying attention to the type of exercise, time, frequency, and intensity. This study is in accordance with the research results found by Durrani & Fatima 2015 and Yusrizal et al. 2016.

5 CONCLUSION

Prevalence of prehypertension was 11.5% and hypertension was 12.0%. Fast food consumption and physical activity were not associated with prehypertension/hypertension (p value > 0.05).

ACKNOWLEDGEMENT

We would like to thank the research and community service institute of the Universitas Muhammadiyah Jakarta for financial support and to the principal and students of SMA Negeri 6 South Tangerang for their participation in this research.

REFERENCES

Abolfotouh, M. A., Sallam, S. A., Mohammed, M. S., Loutfy, A. A., & Hasab, A. A. 2011. Prevalence of elevated blood pressure and association with obesity in Egyptian school adolescents. *International Journal of Hypertension*, 2011(i).

Angesti, A. N., & Sartika, R. A. D. 2018. Riwayat Hipertensi Keluarga Sebagai Faktor Dominan Hipertensi Pada Remaja Kelas XI SMA Sejahtera 1 Depok Tahun 2017. *Buletin Penelitian Kesehatan*, 46(1).

Ashakiran, D. R., & Deepthi, R. 2012. Fast foods and their impact on health. *Journal of Krishna Institute of Medical Sciences University*, 1(2): 7–15.

Baecke, J. A., Burema, J., & Frijters, J. E. 1982. A short questionnaire for the measurement of habitual physical activity in epidemiological studies. *The American Journal of Clinical Nutrition*, 36(5): 936–942.

Danasekaran, R., Mani, G., & Annadurai, K. 2016. Adolescent hypertension: A challenge for the future. *Bangladesh Journal of Medical Science*, 15(1): 5–9.

de Moraes, A. C. F., Lacerda, M. B., Moreno, L. A., Horta, B. L., & Carvalho, H. B. 2014. Prevalence of high blood pressure in 122,053 adolescents. *Medicine*, 93(27): e232.

Durrani, A. M., & Fatima, W. 2015. Effect of physical activity on blood pressure distribution among school children. *Advances in Public Health*, 2015.

Ewald, D. R., & Haldeman, L. A. 2016. Risk Factors in Adolescent Hypertension. *Global Pediatric Health*, 3: 1–26.

Falkner, B. 2018. The childhood role in development of primary hypertension. *American Journal of Hypertension*, 31(7): 762–769.

Falkner, B., Daniels, S. R., Flynn, J. T., Gidding, S., Green, L. A., Ingelfinger, J. R., ... & Rocchini, A. P. 2004. The fourth report on the diagnosis, evaluation, and treatment of high blood pressure in children and adolescents. *Pediatrics*, 114(2 III): 555–576.

Kearney, P. M., Whelton, M., Reynolds, K., Muntner, P., Whelton, P. K., & He, J. 2005. Global burden of hypertension: Analysis of worldwide data. *The Lancet*, 365: 7.

Kemenkes, R. I. 2013. *Riset Kesehatan Dasar (RISKESDAS) 2013*. Laporan Nasional 2013, 1–384.

Khan, M. I., Lala, M. K., Patil, R., Mathur, H. N., & Chauhan, N. T. (2010). A study of the risk factors and the prevalence of hypertension in the adolescent school boys of Ahmedabad City. *J Clin Diagn Res* 4: 3348–54.

Mü Ller-Riemenschneider, F., Nocon, M., & Willich, S. N. 2010. Prevalence of modifiable cardiovascular risk factors in German adolescents. *European Journal of Cardiovascular Prevention and Rehabilitation*, 17(17): 204–210.

Nur, N., Çetinkaya, S., Yilmaz, A., Ayvaz, A., Bulut, M. O., & Sümer, H. 2008. Prevalence of hypertension among high school students in a middle Anatolian province of Turkey. *J Health Popul Nutr*, 26(1): 88–94.

Payab, M., Kelishadi, R., Qorbani, M., Motlagh, M. E., Ranjbar, S. H., Ardalan, G., ... & Heshmat, R. (2014). Association of junk food consumption with high blood pressure and obesity in Iranian children and adolescents: the CASPIAN-IV Study. *Journal de pediatria*, 91(2): 196–205.

Redwine, K. M., & Falkner, B. 2012. Progression of prehypertension to hypertension in adolescents. *Current Hypertension Reports*, 14(6): 619–625.

Rosner, B., Cook, N. R., Daniels, S., & Falkner, B. 2013. Childhood blood pressure trends and risk factors for high blood pressure: The NHANES experience, 1988–2008. *Hypertension*, 62(2): 247–254.

Ujunwa, F. A., Ikefuna, A. N., Nwokocha, A. R., & Chinawa, J. M. 2013. Hypertension and prehypertension among adolescents in secondary schools in Enugu, South East Nigeria. *Italian Journal of Pediatrics*, 39(1): 70.

Yusrizal, M., Indarto, D., & Akhyar, M. 2016. Risk of hypertension in adolescents with over nutritional status in Pangkalpinang, Indonesia. *Journal of Epidemiology and Public Health*, 1(1): 27–36.

Zhao, Y., Wang, L., Xue, H., Wang, H., & Wang, Y. 2017. Fast food consumption and its associations with obesity and hypertension among children: Results from the baseline data of the Childhood Obesity Study in China mega-cities. *BMC Public Health*, 17(1): 933.

Is there a correlation between patients' knowledge and attitudes about tuberculosis?

N. Romadhona, T. Respati, Y. Triyani & W. Purbaningsih
Universitas Islam Bandung, Bandung, Jawa Barat, Indonesia

ABSTRACT: Various efforts, ranging from prevention to treatment, have been made by the government to reduce the incidence of tuberculosis (TB), which is still high in Indonesia, but the target has still not been reached. Good levels of knowledge and the positive attitude of TB patients are expected to reduce TB transmission. The purpose of this study was to analyze the correlation between knowledge and attitudes of patients regarding TB. This research method is the Spearman correlation test; respondents were 60 tuberculosis patients coming to the clinic at one of the hospitals in Bandung Regency, Indonesia. The study was conducted in March 2018. The results of this study found characteristics of the respondents: most were women, the most common age was in the range of 18–35 years, with higher levels of education, and many were employed. Correlation test obtained r-value of 0.771, with p 0.014, which means that there is a strong significant correlation with a positive direction. We conclude, the higher the level of patients' knowledge is, the better their attitude regarding TB gets. Attitude is one of the factors that determine a person's behavior. In determining attitudes, knowledge of a particular subject/object is needed. A person's knowledge can be influenced by several factors, including the level of education.

1 INTRODUCTION

Tuberculosis (TB), caused by *Mycobacterium tuberculosis*, is one of the top 10 infectious diseases that cause death worldwide (Viney et al. 2014). TB is a disease that can be treated and cured. Between 2000 and 2017, an estimated 54 million lives were saved through TB diagnosis and treatment (WHO 2018).

The results of research on people who have TB in the South Pacific Islands shows that the knowledge and attitudes of these communities determine behavior in seeking healthcare, adherence to TB treatment, and TB transmission (Viney et al. 2014). Research on TB patients in Nigeria shows that the level of patient knowledge of TB is unsatisfactory and is significantly related to the delay in patients seeking care (Biya et al. 2014). Research on TB patients in rural areas in Southwest Ethiopia states that the level of patient knowledge about tuberculosis is still low. Researchers found that treatment-seeking behavior and stigma were not appropriate in these patients (Abebe et al. 2010). The results of research on TB patients in India showed 73% feel a stigma and 98% have a discriminatory attitude towards TB patients. Stigma and discrimination towards TB patients remains high in the Indian population. Only 17% of respondents have adequate knowledge about TB, with lower rates observed among women, those in rural areas, and respondents from lower income groups (Sagili et al. 2016).

The initiation of treatment and treatment compliance of TB patients among South African respondents is more influenced by perception than knowledge of TB. The level of knowledge of TB patients in this community is quite good, but respondents' perceptions (stigma) affect detection and case finding. Future interventions must be directed at improving attitudes and perceptions to reduce stigma so that case finding and treatment can begin early (Cramm et al. 2010).

This is reflected in the national TB Control Strategy consisting of strengthening TB program leadership; increasing access to quality TB services; controlling TB risk factors; increasing TB partnerships; strengthening TB program management, and increasing community independence in TB control (Dinas Kesehatan Jawa Barat 2017). The community is expected to have good levels of knowledge about TB, which creates a positive attitude, so that TB treatment behavior can be lived well.

In the studies above, no one has directly examined the correlation between knowledge and patient attitudes about TB. Based on this, the purpose of this study was to determine the correlation between knowledge and patient attitudes about TB.

2 METHODS

This research was conducted in a clinic at a hospital in Bandung Regency, Indonesia, in March 2018. This is an observational analytic study with cross-sectional design. The inclusion criteria in this study was tuberculosis patients who came to the clinic. The number of samples is 60 people. The sample size in this study uses the correlation test formula. The independent variable in this study is the knowledge level of tuberculosis patients, while the dependent variable is the attitude of tuberculosis patients. The instrument of this study was a questionnaire that had been tested for validity and reliability. The questionnaire consisted of respondent characteristics, knowledge, and attitudes of patients regarding tuberculosis. Respondent characteristics consisted of gender, age, level of education, and work status. Patient knowledge consists of 14 questions covering causes, transmission, symptoms, and treatment of tuberculosis. The correct answer on the knowledge variable is given a value of 1 and an incorrect answer is given a value of 0, then the score is added up, so the data included is numeric. Patient attitude consists of 10 statements covering transmission, symptoms, and treatment. The statement about attitude is accompanied by five possible closed answers of the Likert scale model (Likert's summated ratings/LSR) – strongly agree, agree, neutral, disagree, and strongly disagree – which are given a score of 5 to 1, then added up (numerical data). The analysis of this study uses the Spearman correlation test to determine the correlation coefficient and is processed with tools such as **IBM SPSS** version 23.

3 RESULTS

Based on Table 1, respondents were mostly female (32 people; 58.3%), age range 18–35 (50%), higher education level (43 people; 71.67%), and 31 were working people (48.33%).

Based on the correlation test Table 2, a p-value < 0.05 means that the correlation is significant. Correlation value (r) 0.771 means that it has strong correlational strength, and a positive correlational direction.

4 DISCUSSION

Correlation test results between knowledge and the attitude of tuberculosis patients showed a strong correlation with a positive direction. This means that the higher the level of a patient's knowledge about tuberculosis, the more positive the patient's attitude.

Knowledge, according to Peter Senge, is the ability to take effective action (Young & Milton 2011). Another definition states that knowledge is familiarity, awareness, or understanding of someone or something, such as facts, information, or skills descriptions, obtained through experience or education, by understanding, discovering, and learning. Knowledge is obtained from various sources. First, sensory perception is obtained through sight, hearing, smell, taste, or touch where humans feel stimulation from outside and inside the body. Second, intuition is a direct perception of facts and truth, which is independent of any reasoning process. Third, authority is based on testimony by individuals deemed authorities in their

Table 1. Characteristics of respondents by gender, age, education level, and working status.

Variable	n	%
Gender		
Man	28	46.7
Woman	32	53.3
Age		
18–35	30	50
36–50	21	35
> 50	9	15
Education Level		
Low	17	28.33
High	43	71.67
Work Status		
Working	31	51.67
Not working	29	48.33
Total	60	100.00

Table 2. Correlation test knowledge with patient attitudes about tuberculosis.

Variable	r-Value	p	Comment
Knowledge and attitude	0.771	0.014	There is a positive correlation with a strong relationship

particular field and on verified facts from exceptional and extraordinary fields of various sciences. Finally, the reasons for gaining intellectual knowledge, either by direct understanding of the first principle or by arguments to form conclusions (Shafie 2004).

Knowledge and attitude are behaviors that cannot be observed by others, while behavior in the form of actions is something that can be observed by others. Attitude is a person's belief about an object or situation, accompanied by a certain feeling, and it provides a basis for that person to respond or behave in a particular way. Manifestations cannot be directly seen. Attitude is not an action, but is a predisposition to an action or behavior (Wijaya 2012).

One of the factors that influence knowledge and attitude is education. Educational level affects the ability to receive and understand information. Respondents with higher education are more receptive to an idea than respondents with less education. In this study, there were many respondents in the higher education category, but in this study there was no study of the relationship of education with the level of patient knowledge (Nurkomarasari et al. 2014).

Another factor influencing knowledge and attitude is age. In this study, the largest age range was 18–35 years. In most cases, tuberculosis patients get cured in their productive age. Patients of childbearing age are highly mobility so they may not regularly participate in TB-drug education (Nurkomarasari et al. 2014).

The majority of TB patients in this study were women. This is consistent with studies in Sudan, India, and Ethiopia, which report that the level of lack of knowledge about TB is highest in women. But the level of awareness to seek treatment is better in women than men (Agho et al. 2014).

The number of TB patients based on work status was almost the same, only slightly higher among respondents who worked. The results of other studies mention unemployment is associated with poor patient knowledge and attitudes towards TB (Agho et al. 2014).

The results of research on TB patients in developing countries stated that there are several factors that cause patient noncompliance with TB treatment behavior. These factors are socioeconomic factors such as gender, age, lack of food, transportation costs, social support, employment; individual behavioral factors such as stigma, feeling better after a few weeks of treatment, tobacco and alcohol use; patient knowledge about TB disease and its treatment; poor communication between patients and health workers; distance from the treatment center; and the side effects of TB treatment (Habteyes Hailu et al. 2015). Other research in Nigeria's Plateau State states that TB treatment interruptions are associated with great distances from care, including transportation costs; poor knowledge about the length of TB treatment; smoking; and the attitude of health workers who are not friendly to patients (Ibrahim et al. 2014).

Behavior is influenced by attitudinal factor, as stated in Sinaga's research that shows that with a positive attitude or support for pulmonary TB treatment, it is possible for sufferers to be twice as likely to succeed in treating pulmonary TB (Nurkomarasari et al. 2014). Another study of TB patients in Peru states that low levels of knowledge about TB can predict the risk of TB recurrence doubling (Westerlund et al. 2015). Behavior influenced by knowledge factors is in line with the results of research on TB patients in Nigeria, which states that patient delay in seeking TB care is associated with unsatisfactory levels of knowledge about TB (Biya et al. 2014).

This study is not in line with the results of research in Nigeria that shows that most respondents know about TB and believe that TB is a contagious disease that can be cured. This finding is consistent with that of research in India and in Ethiopia, where the majority of participants in the survey have heard about TB and believe that the disease can be cured. However, a large number of respondents will hide the fact that family members suffer from TB. This means that the majority of respondents showed an unfavorable attitude (stigmatization) toward TB patients that caused delays in seeking treatment. There are some respondents who believe that TB can spread through sexual contact, food, or mosquito bites, and by touching TB patients (Agho et al. 2014). Research in Nigeria is also in line with the results of research in India that stigma and discrimination against TB patients remains high among the general population in India. This attitude does not depend on knowledge about TB. It is possible that current knowledge disseminated about TB, especially from a medical point of view, might not adequately address the factors that lead to stigma and discrimination against TB patients. Therefore, there is an urgent need to review the messages and strategies currently used to disseminate knowledge about TB among the general population and revise it appropriately. Knowledge that is disseminated is not only about the medical aspects, but must include the psychosocial and economic aspects so it can be expected to eliminate the stigma and discrimination of TB patients (Sagili et al. 2016).

5 CONCLUSION

The conclusion of this study is that knowledge correlates strongly with patients' attitudes regarding tuberculosis. As knowledge increases, patients' attitude about TB is more positive.

REFERENCES

Abebe, G., Deribew, A., Apers, L., Woldemichael, K., Shiffa, J., Tesfaye, M., ... & Aseffa, A. 2010. Knowledge, health seeking behavior and perceived stigma towards tuberculosis among tuberculosis suspects in a rural community in southwest Ethiopia. *PloS one*, 5(10): e13339.

Agho, K. E., Hall, J., & Ewald, B. 2014. Determinants of the knowledge of and attitude towards tuberculosis in Nigeria. *Journal of Health, Population, and Nutrition*, 32(3): 520.

Biya, O., Gidado, S., Abraham, A., Waziri, N., Nguku, P., Nsubuga, P., ... & Sabitu, K. 2014. Knowledge, care-seeking behavior, and factors associated with patient delay among newly-diagnosed pulmonary tuberculosis patients, Federal Capital Territory, Nigeria, 2010. *The Pan African Medical Journal*, 18(Suppl 1): 6.

Cramm, J. M., Finkenflügel, H. J., Møller, V., & Nieboer, A. P. 2010. TB treatment initiation and adherence in a South African community influenced more by perceptions than by knowledge of tuberculosis. *BMC Public Health*, 10(1): 72.

Dinas Kesehatan Jawa Barat 2017. *Profil Kesehatan.* [Online] http://www.diskes.jabarprov.go.id/index.php/arsip/categories/MTE4/profile-kesehatan.

Habteyes Hailu, T. O. L. A., Azar, T. O. L., & Davoud Shojaeizadeh, G. G. 2015. Tuberculosis treatment non-adherence and lost to follow up among TB patients with or without HIV in developing countries: A systematic review. *Iranian Journal of Public Health,* 44(1): 1.

Ibrahim, L. M., Hadejia, I. S., Nguku, P., Dankoli, R., Waziri, N. E., Akhimien, M. O., ... & Nsubuga, P. 2014. Factors associated with interruption of treatment among pulmonary tuberculosis patients in Plateau State, Nigeria. 2011. *Pan African Medical Journal,* 17(1): 1–8.

Nurkomarasari, N., Respati, T., & Budiman, B. 2014. Karakteristik Penderita Drop Out Pengobatan Tuberkulosis Paru di Garut. *Global Medical & Health Communication* 2(1): 21–26.

Sagili, K. D., Satyanarayana, S., & Chadha, S. S. 2016. Is knowledge regarding tuberculosis associated with stigmatising and discriminating attitudes of general population towards tuberculosis patients? Findings from a community based survey in 30 districts of India. *PLoS One,* 11(2): e0147274.

Shafie, A. B. 2004. *The Educational Philosophy of al-Shaykh Muhammad'Abduh.* International Islamic University Malaysia.

Viney, K. A., Johnson, P., Tagaro, M., Fanai, S., Linh, N. N., Kelly, P., ... & Sleigh, A. 2014. Tuberculosis patients' knowledge and beliefs about tuberculosis: a mixed methods study from the Pacific Island nation of Vanuatu. *BMC Public Health,* 14(1): 467.

Westerlund, E. E., Tovar, M. A., Lönnermark, E., Montoya, R., & Evans, C. A. 2015. Tuberculosis-related knowledge is associated with patient outcomes in shantytown residents; results from a cohort study, Peru. *Journal of Infection,* 71(3): 347–357.

WHO 2018. *10 Facts on Tuberculosis.* [Online]. https://www.who.int/features/factfiles/tuberculosis/en/.

Wijaya, I. 2012. *Hubungan Pengetahuan, Sikap dan Motivasi Kader Kesehatan dengan Aktivitasnya dalam Pengendalian Kasus Tuberkulosis di Kabupaten Buleleng* (Doctoral dissertation). Surakarta: Sebelas Maret University.

Young, T., & Milton, N. 2011. Principles of knowledge management. *Knowledge Management for Sales and Marketing,* 1–23.

Santri health cadre as innovation toward healthy religious boarding schools (pesantren)

Y. Triyani, W. Purbaningsih, T. Respati & I. Safrudin
Universitas Islam Bandung, Bandung, Indonesia

ABSTRACT: The incidence of scabies and pediculosis are related to poor hygiene and unhealthy living behavior, caused by various factors including the absence of a teaching material module for clean and healthy living behaviors (PHBS) (Healthy Santri pocket books), and health facilities in pesantren. The purpose of this program is providing solutions to health problems in the pesantren. The approach method included phase I: the making of the PHBS teaching materials module used in phase II, which was the formation of a tutorial group of students, and phase III: santri (student) training practices as health cadres. The results of this PKM produce students as health cadres who are expected to be pioneers and role models of PHBS in pesantren with the PHBS module guidelines made by the PKM team. It is hoped this will make the pesantren healthy, not only free from scabies and pediculosis.

1 INTRODUCTION

Scabies is a skin disease caused by *Sarcoptes scabiei* ectoparasites and pediculosis caused by *Pediculosis humanus*, which is usually found in crowded environments such as boarding schools (pesantren) (Yingklang et al. 2018, Nanda et al. 2016).

The incidence of scabies in Islamic boarding schools in West Java generally, including the city of Bandung, is still high. Based on the results of scabies selection in 20 pesantren by the 2018 FK Unisba service team, it was found to vary by 10–88% (Triyani et al. 2019). Factors that cause the incidence of scabies are still challenging to eliminate, both internal (host/santri), and external (host and environment), which are interrelated. Host factors include poverty, lack of knowledge, and good clean and healthy living behavior from both students and teaching staff. Environmental factors related to scabies infestation are lack of clean water facilities and slum environments, low temperatures, humid climate, and dense bedrooms. All of those facilitate the transmission of scabies from infected students to other healthy students because of the lack of specific teaching materials that can be used as practical aids (Heukelbach & Feldmeier 2006). From the survey, levels of student knowledge about PHBS in daily life are low. The students only get teaching material about cleanliness from the lessons of Bab Thaharah, which is delivered orally by the instructor from the book of Fiqh.

In one of the pesantren in Bandung, Indonesia, which had a high incidence of scabies, they found that they did not have PHBS teaching materials for their students.

This article aims to describe the program that has been carried out in an Islamic boarding school to reduce the incidence of scabies through several activities.

2 METHODS

This research used a participatory action research (PAR) approach, chosen to ensure that the implementation of the program, including the prevention and control of disease, was successful with community participation. The study was conducted at one of the pesantren in the city of Bandung in 2017 and a survey in 2019. The research was carried out in phases:

phase I – the making of the PHBS teaching materials module used in phase II, the formation of a tutorial group of students; and phase III: santri training practices as health cadres. The PHBS teaching materials module was accommodated by a pre-test before training and a post-test after training. The two people with the best results from each tutorial group were to be trained as health cadres in phase III.

3   RESULTS AND DISCUSSION

3.1   *Characteristics of santri in MH Islamic boarding school in Bandung*

The number of students (santri) in Manarul Huda Islamic boarding school in Bandung, West Java, in 2017 was 120 people, consisting of 80 men and 40 women, ranging in age from 9 to 20 years. They come from various cities in West Java, and tuition fees are free because this pesantren is committed to providing Islamic religious teaching and education to disadvantaged people. The results of scabies and pediculosis netting survey data obtained from 86 students, examined with the results of scabies and pediculosis selection can be seen in Table 1, Table 2 and Table 3.

Table 1.   Santri characteristics at 2017.

Diagnosis	Male		Female		Total	
	n	%	n	%	n	%
Scabies	45	52	5	6	50	58
Pediculosis	0	0	1	1	1	1
Scabies and pediculosis	3	3	31	36	34	40
Healthy	1	1	0	0	1	1
Total	49	57	37	43	86	100

* Survey 2017

Table 2.   Santri characteristics at 2018.

Diagnosis	Male		Female		Total	
	n	%	n	%	n	%
Scabies	18	25	2	3	20	28
Pediculosis	0		0			
Scabies and pediculosis	2	3	15	21	17	24
Healthy	27	38	7	10	34	48
Total	47	66	24	34	71	100

* Survey 2018

Table 3.   Santri characteristics at 2019.

Diagnosis	Male		Female		Total	
	n	%	n	%	n	%
Scabies	13	21	0	0	13	21
Pediculosis	0		7	11	7	11
Scabies and pediculosis	0	0	0	0	0	0
Healthy	25	40	17	27	42	68
Total	38	61	24	39	62	100

* Survey 2019

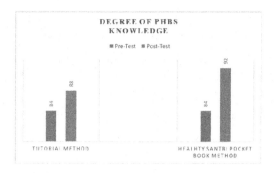

Figure 1. Degree of PHBS knowledge.

3.2 Degree of PHBS knowledge

Degree of PHBS knowledge was assessed by comparing the pre- and post-test results. Pre-test was done before students were given guidance in the form of tutorials and before handbooks were given as teaching materials as guidelines for students to practice daily PHBS. The post-test was conducted after 3 weeks of tutorials and 1 week after being given the handbook. The results obtained can be seen in Figure 1.

The incidence of scabies and pediculosis is influenced by many factors including cold climate, poverty, lack of affordable health facilities, crowded environment, lack of clean water, and the myth that scabies is a blessing, especially in religious boarding school (Yusof et al. 2015, Imartha 2016). In addition to these factors, it turns out that the existence of teaching materials and coaching play an important role in practicing clean and healthy life behavior. This can be seen by the decrease of the incidence of scabies and pediculosis in guided pesantren in 2019 (Tresnasari et al. 2019). Coaching to educate santri a health cadre is beneficial since santri act as the main subject not just an object in this study. This is a breakthrough to help eradicate scabies and pediculosis in pesantren (Engelman & Steer 2018, Triyani et al. 2019). The existence of santri handbooks as guidelines for clean and healthy living behavior is also important. The guidance illustrated knowledge about scabies from causes, modes of transmission, and ways of prevention to its treatment. The handbooks are very helpful to assist students so that they have increased knowledge about healthy living behavior. This handbook also becomes the santri health cadre references (Martino et al. 2018).

4 CONCLUSION

The PAR method that treated santri as a subject to disseminate the knowledge and insight into the behavior of clean and healthy living is the right approach. The provision of teaching materials in the form of a healthy santri handbook can also help reduce the incidence of scabies and increase santris' knowledge about clean and healthy living behaviour.

ACKNOWLEDGEMENT

This work was supported by the grant from Kemenristekdikti (The Ministry of Research, Technology, and Higher Education), the Republic of Indonesia.

REFERENCES

Engelman, D., & Steer, A. C. 2018. Diagnosis, treatment, and control of scabies: Can we do better? *The Lancet Infectious Diseases*, 18(8): 822–823.

Heukelbach, J., & Feldmeier, H. 2006. Scabies. *The Lancet*, 367(9524): 1767–1774.

Imartha, A. G. 2016. Faktor-Faktor Yang Berhubungan Dengan Kejadian Skabies Di Pondok Pesantren Jabal an-Nur Al-Islami Kecamatan Teluk Betung Barat Kota Bandar Lampung. *Jurnal Fakultas Kedokteran Unila*, 7: 1–8.

Martino, Y. A., Sulistiowati, E., & Purnomo, Y. 2018. Model Pemberdayaan Santri Ponpes Al-Hidayah Batu Alang Sebagai Kader Kesehatan Berbasis Terapi Herbal. *JIPEMAS: Jurnal Inovasi Hasil Pengabdian Masyarakat*, 1(2): 86.

Nanda, F. D., Murti, B., & Dharmawan, R. 2016. Path analysis on factors associated with the risk of scabies among students at Darussalam Islamic Boarding School, Blokagung, Banyuwangi, Indonesia. *Journal of Epidemiology and Public Health*, 1(1): 18–26.

Tresnasari, C., Respati, T., Maulida, M., Triyani, Y., Tejasari, M., Kharisma, Y., & Ismawati, I. 2019. Understanding scabies in religious boarding school (pesantren). *Social and Humaniora Research Symposium (SoRes 2018)*, 307(SoRes 2018): 520–522.

Triyani, Y., Yuniarti, L., Tejasari, M., Purbaningsih, W., Ismawati, I., & Respati, T. 2019. A journey to a better community service in religious boarding school pesantren. *Social and Humaniora Research Symposium (SoRes 2018)*, 307(SoRes 2018): 497–499.

Yingklang, M., Sengthong, C., Haonon, O., Dangtakot, R., Pinlaor, P., Sota, C., & Pinlaor, S. 2018. Effect of a health education program on reduction of pediculosis in school girls at Amphoe Muang, Khon Kaen Province, Thailand. *PLoS ONE*, 13(6): 1–15.

Yusof, M., Fitri, S., & Damopolii, Y. 2015. A study on knowledge, attitude and practice in preventing transmission of scabies in pesantren Darul Fatwa, Jatinangor. *Althea Medical Journal*, 2(1): 131–137.

Hospital and nursing management

Medical Technology and Environmental Health – Abdullah, Widiaty & Abdullah (eds)
© 2020 Taylor & Francis Group, London, ISBN 978-0-367-86053-0

Understanding the referring system for knee osteoarthritis patients from primary health care's physician: A pilot study at Al-Ihsan hospital, Indonesia

S. Waspodo
Muhammadiyah Hospital, Bandung, Indonesia
Universitas Islam Bandung, Bandung, Indonesia

A. Rachmi
Al Ihsan Hospital, Bandung, Indonesia
Universitas Islam Bandung, Bandung, Indonesia

ABSTRACT: Osteoarthritis (OA) is a chronic disease that needs nonpharmacological therapy or a rehabilitation program for management. This study aims to know which specialist actually refers knee osteoarthritis patients to physiatrists in Al Ihsan Hospital. From January to June 2019, data were collected from all OA cases. This study uses a quantitative research method with observational study design from patients' medical records. The results show that 81% of patients with knee osteoarthritis were referred to a physiatrist from a eurologist in Al Ihsan hospital. This study concludes that there is still a lack of understanding of the referring system for knee osteoarthritis patients by the primary health care physician. Further information and standards are required to help primary health care physicians refer to the right specialist for OA cases.

1 INTRODUCTION

Musculoskeletal complaints are a common clinical symptom of patients in daily clinical practice (Indonesian Rheumatology Association 2014). One musculoskeletal problem that is quite often found is osteoarthritis (OA), a degenerative joint disease characterized by damage-prone joints and subchondral bones and which causes pain in the joints (Indonesian Rheumatology Association 2014). Osteoarthritis is a health problem that is often encountered in daily practice. OA involves failure of the structure and function of synovial joints and is characterized by degradation of articular cartilage (Cifu & Lew 2017). Treatment of osteoarthritis cannot depend on medical treatment only. Treatment of OA also requires education and lifestyle modification, medical rehabilitation management, or even surgery (Cifu & Lew 2017). Health workers require a good understanding of the condition for management of osteoarthritis to be thorough and for the patient to get the right therapeutic choice for pain so the patient's quality of life can be better. Based on the competency standards of Indonesian general practitioners, the level of ability they possessed is to be able to determine the most appropriate referral for patient treatment of osteoarthritis.

Research funded by *King Fund* about the quality of general practitioner diagnosis and referral shows that musculoskeletal problems often lack clarity as to whether patients need to be referred to orthopedics, rheumatology, or physiotherapy (Foot et al. 2010). Referring osteoarthritis patients is strongly influenced by disease severity as well as the perspectives of the patients about the referral (Musila et al. 2011). The need for medical rehabilitation as one of the treatments for nonpharmacological management of osteoarthritis is stated in Decree of Ministry of Health Hk.02.02/Menkes/514/2015 (Ministry of Health Indonesia 2015). In Indonesia, 85% of patients who come to the hospital are BPJS participants. BPJS rules state patients who will be referred from PPK 1 (general practitioners) to medical rehabilitation services must be referred from another specialist. This study wanted to find out whether patients diagnosed with OA of the knee were mostly sent to specialists in physical medicine and rehabilitation and from which specialists in Al Ihsan Hospital.

2 METHODS

The data were collected from central referral West Java's hospital, which is RSUD Al Ihsan. Osteoarthritis patients were calculated from January to June 2019 with total sampling as sampling technique. This is a cross-sectional descriptive study where we only calculate the percentage and number of OA cases that went to a physiatrist in the Physical Medicine and Rehabilitation Department for further therapy.

3 RESULTS AND DISCUSSION

There were 2,559 OA patients who came to the oupatient department in Al Ihsan Hospital. The percentage of referral was 52% with total cases of OA as high as 1,342 patients. Al Ihsan Hospital was a secondary referral from orthopedics (total case 170 cases), neurology (total case 2,074), and internal medicine departments (total case 315). Table 1 shows that nerve specialists diagnose knee OA more than internal medicine and orthopedics so that many patients sent to a physiatrist come from neurologists compared to internal medicine doctors or other specialists.

Osteoarthritis is experienced by one third of the population over the age of 65 years and constitutes one of the five main causes of disability in the elderly population in the United States. In Indonesia, osteoarthritis is the case of rheumatic disease most often encountered. This disease can affect both sexes, although it often occurs in women; and is generally found in the elderly population. With an increasingly aging population in various countries in the world, the number of patients suffering from osteoarthritis continuously increased. OA is mostly felt in the knee, pelvis, spine, and ankle. In Indonesia, the prevalence of knee OA found radiologically in people aged 40–60 years reached 15.5% in men and 12.7% in women.

Osteoarthritis causes chronic pain and disability and can affect the quality of life of patients. Bearing in mind the burden, the large epidemiology, and that the chronic pain it causes can decrease the quality of life, this rheumatic disease requires attention. Osteoarthritis (OA) constitutes the most frequent form of arthritis found in the community and is chronic, which has a big impact on public health (IRA, 2014).

A referring system from general practitioners to specialists is necessary when cases cannot be managed by the general practitioners themselves. Good referral to specialists can be defined in terms of necessity, timelines, destination, and process, even though there is no common agreed-upon criteria (Foot et al. 2010). Destination of referring patients is the main criteria discussed here, as shown in Table 1: the majority of referrals came from neurologists. This is a new finding as arthritis cases are usually referred to internists or rheumatologists as the main nonsurgical references or orthopedics for surgical stherapies (Musila et al. 2011). The decisions GPs make about referral are crucial both for the patients' progress through the

Table 1. Outpatient osteoarthritis patients January–June 2019.

Outpatient department patient first referral	Al Ihsan Hospital	
	Number of cases	%
Internal medicine	315	12.3
Orthopedics	170	6.7
Neurologist	2,074	81.0
Total	2,559	100

system, and for their effect on the costs incurred by the system (Davies et al. 2011). Referral was common with knee or shoulder musculoskeletal condition but not correlated with clinical outcomes in Boston (Solomon et al. 2001).

Getting the right specialty destination for referral showed itself to be more challenging in medical specialties than surgical (Jenkins 1993). This is particularly an issue for musculoskeletal problem if the referral is misdirected during hospital-based treatment due to a lack of clarity over which specialist needs to be refer to (Speed & Crisp 2005). Speed's research shows that 73% of patients with musculoskeletal problems who are referred to orthopedics actually should be referred to a rheumatologist or surgeon. Referral of musculoskeletal problems to physical therapy can be from surgical orthopedics or primary health care in United States (Freburger et al. 2003). The decisions for orthopedists and primary physicians show that it depends on insurance status and specific characteristics of physicians during the consultation to refer. That system is different in Indonesia, as public insurance requires patients who come for treatment in a physical rehabilitation department not be directly referred from a primary physician (Ministry of Health Indonesia 2015).

Musila's research on recommendations for referral for osteoarthritis states that patients with OA of the knee should only be guided by symptom severity and patients' referral preferences (Musila et al. 2011). Consensus on the appropriateness of referral for patients is lacking. Although the Indonesian Association of Rheumatologists has made an IRA Recommendation for Diagnosis and Management of osteoarthritis, in general, there are still physicians who do not know about these recommendations. A better understanding of nonoperative and operative treatment options for patients with OA should prove to be beneficial to the general practitioner.

4 CONCLUSION

There is still a lack of understanding of the correct process of referral for knee disease in Al Ihsan Regional Hospital. Future studies can be done to find out more about general practitioners' knowledge for knee OA referral system.

REFERENCES

Cifu, D. X., & Lew, H. L. 2017. *Braddom's Rehabilitation Care: A Clinical Handbook E-Book*. Elsevier Health Sciences.
Davies, P., Pool, R., & Smelt, G. 2011. What do we actually know about the referral process? *Br J Gen Pract*, 61(593),752–753.
Foot, C., Naylor, C., & Imison, C. 2010. The quality of GP diagnosis and referral, *HAL Archive ouverte en Sciences de l'Homme et de la Société*.
Freburger, J. K., Holmes, G. M., & Carey, T. S. 2003. Physician referrals to physical therapy for the treatment of musculoskeletal conditions. *Archives of Physical Medicine and Rehabilitation*, 84(12): 1839–1849.
Indonesian Rheumatology Association. 2014. *Diagnosis and Management of Osteoartritis/Diagnosis dan Penatalaksanaan Osteoartritis*, Perhimpunan Reumatologi Indonesia: Viii–32.
Jenkins, R. M. 1993. Quality of general practitioner referrals to outpatient departments: Assessment by specialists and a general practitioner. *Br J Gen Pract*, 43(368): 111–113.
Ministry of Health Indonesia. 2015. *Decree of the Minister of Health of the Republic of Indonesia Number Hk.02.02/Menkes/514/2015 About Clinical Practice Guidelines for Doctors in the Service Facilities Health First Level*.
Musila, N., Underwood, M., McCaskie, A. W., Black, N., Clarke, A., & van der Meulen, J. H. 2011. Referral recommendations for osteoarthritis of the knee incorporating patients' preferences. *Family Practice*, 28(1): 68–74.
Solomon, D. H., Bates, D. W., Schaffer, J. L., Horsky, J., Burdick, E., & Katz, J. N. 2001. Referrals for musculoskeletal disorders: patterns, predictors, and outcomes. *The Journal of Rheumatology*, 28(9): 2090–2095.
Speed, C. A., & Crisp, A. J. 2005. Referrals to hospital-based rheumatology and orthopaedic services: Seeking direction. *Rheumatology*, 44(4): 469–471.

Medical Technology and Environmental Health – Abdullah, Widiaty & Abdullah (eds)
© 2020 Taylor & Francis Group, London, ISBN 978-0-367-86053-0

Why do patients want to upgrade the service level in hospitals?

S.N. Irasanti, Y. Susanti & Y.D. Suryani
Universitas Islam Bandung, Bandung, Indonesia

ABSTRACT: The era of National Health Insurance (BPJS Kesehatan) has seen rising demand arising for health services increasing decisions to upgrade service levels. The aim of the study is to explore the factors that influence patients opting to upgrade their service level in hospital. This study was an observational study with cross-sectional design in which data were collected through a questionnaire with 172 respondents from one of the public hospitals in Indonesia. The results revealed seven determinants of hospital service level upgrades: availability of hospital rooms with better service levels, complete facilitation of the new hospital room with better services levels, affordable price, additional insurance, quality of health services, comfortable places, and private health room services. The reason patients most often gave for upgrading their service level was the complete facilitation of the new hospital room with better service level (90.7%), followed by affordable prices, comfortable places, private health room services, quality of health services, the availability of hospital room with better service levels, and additional insurance. Rate adjustment and standardization of hospital services, monitoring class availability in hospital, as well as the National Health Insurance premium adjustment are required in order to implement better social security.

1 INTRODUCTION

Competition was expected to make care more responsive to patients and, among other things, improve efficiency (including cost decreases), quality and (in the UK) equity of healthcare (Dixon et al. 2010, Vrangbaek et al. 2012).

Choice has become more important in healthcare, as patients can compare hospitals and healthcare providers much more easily on the internet. The goal of emphasizing patient choice was to protect and promote the position of patients in healthcare (Victoor et al. 2012). Secondly, engaging patients in their own healthcare is also seen as the best way to ensure sustainability of health systems, to promote quality improvement, and to shorten waiting times (Magnussen at al. 2009, Victoor et al. 2012, Vrangbaek et al. 2012).

When patients decide to upgrade the service level, they already have an expectation due to the cost. When the expectations do not meet the reality, dissatisfaction is most likely to appear.

In contrast, evidence about the effects of engaging patients in treatment choices has accumulated over some time, but the findings have been largely ignored. Although only a small minority of people want to switch providers, patient surveys show a large unmet demand for greater involvement in treatment decisions that has persisted over the past eight years (Coulter 2010).

Requiring patients to pay a portion of their medical bill out of pocket, also known as cost sharing, sharply reduces their use of healthcare resources. The use of this strategy by health insurance plans to lower expenditures is controversial: proponents argue that healthcare consumers will appropriately ration their use of medical services; critics fear that this financial disincentive will lead patients to use less care than may be necessary and will result in worse health outcomes. Until now, there have been a few studies about the patients' reasons to upgrade the service level in hospital (Wong et al. 2001).

To be able to evaluate whether promoting patient choice of healthcare providers has its desired effects, it must be clear exactly which effects are desired. A variety of patient characteristics determines whether patients make choices, are willing and able to choose, and how they choose. Patients take into account a variety of structural, process, and outcome characteristics of providers, differing in the relative importance they attach to these characteristics.

Comparative information seems to have a relatively limited influence on the choices made by many patients, and patients base their decisions on a variety of provider characteristics instead of solely on outcome characteristics. For patients who are relatively well, they can choose or increase their comfort by upgrading the service level in the hospital.

Our national medical insurance is *Program Jaminan Kesehatan Nasional - Badan Penyelenggara Jaminan Sosial (JKN-BPJS)*. There is Health ministry regulation No.1 about implementation guidelines concerning the improvement of the class of hospital service. Patients should pay the difference tariffs between the hospital rate and the service rate by BPJS (cost sharing) (Palupi et al. 2016).

There are complaints about hospital service fees. The complaints are mostly about service level upgrade costs, which are unpredictable and depend on hospital policies. Based on the research, 20% of patients upgrade their service level in the hospital per month (Rita & Afconneri 2009).

The aim of the study is to explore the factors that influence the patients' decision of choosing to upgrade their service level in Hospital.

2 METHODS

This study was an cross-sectional observational study in which the data were collected through questionnaires with 172 respondents from one of the public hospitals, namely Al Ihsan Hospital in West Java, Bandung, Indonesia. This research identified a list of 15 items of questionnaires from 5 dimension of patients' quality and expectation of healthcare services, some of which are responsiveness component, the assurance component, empathy component and component reliability. The research was conducted by interviewing the respondents using a questionnaire at the time of discharge from the hospital or during the outpatient process.

3 RESULTS

The results revealed seven determinants of hospital service level upgrade. The biggest factors that influence patients to upgrade service level were the complete facilitation of the new hospital room with better service levels (90.7%), followed by affordable prices, comfortable places, private health room services, quality of health services, the availability of hospital room with better service level, and additional assurance (Figure 1).

The age range of the respondents between 36–50 years old was most commom, at 44%, followed by 21–35 yo, > 50 yo, and < 20 yo (Figure 2).

The majority of respondents had a bachelor degree, at 34.9%, followed by senior high school, associate degree, junior high school, master degree, doctoral degree, and elementary school (Figure 3).

Most respondents were employed in the private sector; other respondent occupations were civil service employee, housewife, entrepreneur, unemployed, labourer, primary health employee, students, nurse, manager, and contract temporary teacher (Figure 4).

There was a significant relationship between several respondent characteristics and patient reason to upgrade service level. There was a significant relationship between age and complete facilitation ($p = 0.047$); there was significant relationship between education and complete facilitation ($p = 0.03$); there was significant relationship between education and privacy ($p = 0.039$); there was significant relationship between occupation and quality of health services ($p = 0.004$); and there was significant relationship between occupation and comfortable places ($p = 0.05$) (Table 1).

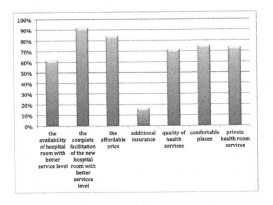

Figure 1.   Factors that influence patients to upgrade the service level.

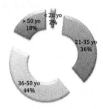

Figure 2.   Patient age.

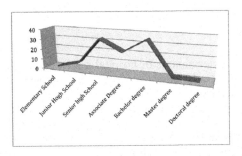

Figure 3.   Patient education.

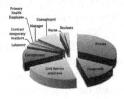

Figure 4.   Patient occupation.

Table 1. Chi-square test (p value (sig)) between respondent characteristics and patient reason to upgrade services level.

	The availability of hospital room	Complete facilitation	Affordable prices	Additional Insurance	Quality of health services	Comfortable places	Privacy
Age	0.870	0.047	0.502	0.608	0.893	0.795	0.834
Sex	0.509	0.170	0.391	0.410	0.196	0.281	0.663
Education	0.417	0.03	0.091	0.369	0.366	0.093	0.039
Occupation	0.892	0.560	0.507	0.278	0.004	0.05	0.133

4 DISCUSSION

Cost sharing was associated with lower rates of patients seeking care for serious symptoms, but only at the highest cost-sharing level (Wong et al. 2001). Beyond the quality of medical care, patient experience is hugely impacted by the quality of the healthcare physical environment, the management of facilities, and the maintenance of assets. According to a report by the American Hospital Association (AHA) and the American Society for Healthcare Engineering (ASHE), patient experience comprises three important and related factors: people, process, and place.

This indicates that caring staff, patient-centered operations, and well-designed facilities are the keys to providing a best-in-class patient experience.

This research found that the factor with the greatest influence on patients deciding to upgrade the service level was the complete facilitation of the new hospital room with better service levels. Hospitals in Indonesia have different facilities in hospital rooms based on service class to serve and gain the better revenue from the patients. This factor is reasonable because in Indonesia income is very variable; however, it is in contradiction with the available number of hospital rooms with desired and necessary facilities.

The significant relationship between age and complete facilitation can reveal that older patients want to choose complete facilitation for their health service. This research also reveals that higher the education level of the patients, the better and more complete facilitation they wanted to have; particularly related to their privacy in their health services. This differs from a study by Martin et al. (2001) stating that patient age, gender, and education level were not relevant to perceptions that healthcare professionals promote patient involvement in their own care. However, this result is similar to that of a study by Haris & Dalton (2014), which showed that university students with higher education levels expect a high degree of confidentiality.

In our finding, it has also been revealed that patient occupation was related to their tendency to choose better quality and have more comfortable places of health services. Sharon Wilson (1997) conducted research about comforting nurse–patient relationships and find that comfort is personal and contextual; therefore, practice and quality improvement decisions must be developed in the context of patients' underlying healthcare condition, culture, and care setting. Wensley et al. (2017) agree that assessment of patient comfort needs should consider patient interactions with all healthcare staff because influences on comfort were not specific to any one discipline, or clinical staff.

5 CONCLUSION

Patients' choices to upgrade service levels are determined by a complex interplay between a variety of patient characteristics. Age, education, and occupation seem to have significant correlation with patient choices to upgrade service levels, although there were some statements from other researchers that there is no such thing as the typical patient: different patients make different choices in different situations. Patients often attach greater importance to their own previous healthcare experiences or to GP recommendations than to comparative information. Additionally, patients base their decisions not only on outcome indicators but also on a variety of provider characteristics. It can thus be argued that the choice process is much more complex than is often assumed. It might be because of a number of gaps in current patient knowledge. Rate adjustment and standardization of hospital services, monitoring class availability in hospitals, as well as the National Health Insurance premium adjustments are required in order to implement better social security.

ACKNOWLEDGEMENT

We would like to thank the authors of all the studies we reviewed. Finally, we would like to thank the Research Unit, Faculty of Medicine, Unisba, which provided funding for this review.

REFERENCES

Coulter, A. 2010. Do patients want a choice and does it work? *Bmj*, 341.

Dixon, A., Robertson, R., & Bal, R. 2010. The experience of implementing choice at point of referral: a comparison of the Netherlands and England. *Health Economics, Policy and Law*, 5(3): 295–317.

Harris, G. E., & Dalton, S. 2014. University student expectations of confidentiality when disclosing information to their professors. *Higher Education Studies*, 4(1): 43–50. ISSN 1925-4741 E-ISSN 1925-475X. http://dx.doi.org/10.5539/hes.v4n1p43.

Magnussen, J., Vrangbaek, K., & Saltman, R. 2009. *Nordic Health Care Systems: Recent Reforms and Current Policy Challenges: Recent Reforms and Current Policy Challenges*. McGraw-Hill Education (UK).

Martin, L. R., Robin Di Matteo, M., & Lepper, H. S. 2001. Facilitation of patient involvement in care: development and validation of a scale. *Behavioral Medicine*, 27(3): 111–120.

Palupi, J. K. N., Wardhani, V., & Andarini, S. 2016. Determinan Pilihan Naik Kelas Perawatan Rumah Sakit dari Kelas I ke Kelas VIP. *Jurnal Kebijakan Kesehatan Indonesia: JKKI*, 5(4): 176–183.

Rita, N., & Afconneri, Y. 2019. Faktor-Faktor yang Mempengaruhi Keputusan Pasien Memilih Jasa Pelayanan Kesehatan. *Jurnal Endurance: Kajian Ilmiah Problema Kesehatan*, 4(1): 132–140.

Sharon Wilson, R. N. 1997. The comforting interaction: Developing a model of nurse–patient relationship. *Research and Theory for Nursing Practice*, 11(4): 321.

Victoor, A., Delnoij, D. M., Friele, R. D., & Rademakers, J. J. 2012. Determinants of patient choice of healthcare providers: a scoping review. *BMC Health Services Research*, 12(1): 272.

Victoor, A., Friele, R. D., Delnoij, D. M., & Rademakers, J. J. 2012. Free choice of healthcare providers in the Netherlands is both a goal in itself and a precondition: Modelling the policy assumptions underlying the promotion of patient choice through documentary analysis and interviews. *BMC health services research*, 12(1): 441.

Vrangbaek, K., Robertson, R., Winblad, U., Van de Bovenkamp, H., & Dixon, A. 2012. Choice policies in Northern European health systems. *Health Economics, Policy and Law*, 7(1): 47–71.

Wensley, C., Botti, M., McKillop, A., & Merry, A. F. 2017. A framework of comfort for practice: An integrative review identifying the multiple influences on patients' experience of comfort in healthcare settings. *International Journal for Quality in Health Care*, 29(2): 151–162.

Wong, M. D., Andersen, R., Sherbourne, C. D., Hays, R. D., & Shapiro, M. F. 2001. Effects of cost sharing on care seeking and health status: results from the Medical Outcomes Study. *American Journal of Public Health*, 91(11):1889–1894.

Medical Technology and Environmental Health – Abdullah, Widiaty & Abdullah (eds)
© 2020 Taylor & Francis Group, London, ISBN 978-0-367-86053-0

Author Index

Printed in the United States
by Baker & Taylor Publisher Services